P9-DTS-941

U.S. HEALTH
IN
INTERNATIONAL PERSPECTIVE

Shorter Lives, Poorer Health

Panel on Understanding Cross-National Health Differences
Among High-Income Countries

Steven H. Woolf and Laudan Aron, *Editors*

Committee on Population
Division of Behavioral and Social Sciences and Education

Board on Population Health and Public Health Practice
Institute of Medicine

NATIONAL RESEARCH COUNCIL *AND*
INSTITUTE OF MEDICINE
OF THE NATIONAL ACADEMIES

THE NATIONAL ACADEMIES PRESS
Washington, D.C.
www.nap.edu

b13701988   9

RA
114008233   445
.U17
2013

832425056

THE NATIONAL ACADEMIES PRESS   500 Fifth Street, NW   Washington, DC 20001

NOTICE: The project that is the subject of this report was approved by the Governing Board of the National Research Council, whose members are drawn from the councils of the National Academy of Sciences, the National Academy of Engineering, and the Institute of Medicine. The members of the panel responsible for the report were chosen for their special competences and with regard for appropriate balance.

This study was supported by the John E. Fogarty International Center, the National Center for Complementary and Alternative Medicine, the National Institute on Aging, and the Office of Behavioral and Social Sciences Research, all within the National Institutes of Health, and the Office of Women's Health within the U.S. Department of Health and Human Services through Contract No. N01-OD-4-2139 Task Orders # 237 and 271 and Contract No. HHSN26300011 between the National Academy of Sciences and the U.S. Department of Health and Human Services. Any opinions, findings, conclusions, or recommendations expressed in this publication are those of the author(s) and do not necessarily reflect the views of the organizations or agencies that provided support for the project.

International Standard Book Number-13:   978-0-309-26414-3
International Standard Book Number-10:   0-309-26414-6

Library of Congress Cataloging-in-Publication data are available from the Library of Congress.

Additional copies of this report are available from the National Academies Press, 500 Fifth Street, NW, Keck 360, Washington, DC 20001; (800) 624-6242 or (202) 334-3313; http://www.nap.edu.

Copyright 2013 by the National Academy of Sciences. All rights reserved.

Printed in the United States of America

Suggested citation: National Research Council and Institute of Medicine. (2013). *U.S. Health in International Perspective: Shorter Lives, Poorer Health.* Panel on Understanding Cross-National Health Differences Among High-Income Countries, Steven H. Woolf and Laudan Aron, Eds. Committee on Population, Division of Behavioral and Social Sciences and Education, and Board on Population Health and Public Health Practice, Institute of Medicine. Washington, DC: The National Academies Press.

THE NATIONAL ACADEMIES
Advisers to the Nation on Science, Engineering, and Medicine

The **National Academy of Sciences** is a private, nonprofit, self-perpetuating society of distinguished scholars engaged in scientific and engineering research, dedicated to the furtherance of science and technology and to their use for the general welfare. Upon the authority of the charter granted to it by the Congress in 1863, the Academy has a mandate that requires it to advise the federal government on scientific and technical matters. Dr. Ralph J. Cicerone is president of the National Academy of Sciences.

The **National Academy of Engineering** was established in 1964, under the charter of the National Academy of Sciences, as a parallel organization of outstanding engineers. It is autonomous in its administration and in the selection of its members, sharing with the National Academy of Sciences the responsibility for advising the federal government. The National Academy of Engineering also sponsors engineering programs aimed at meeting national needs, encourages education and research, and recognizes the superior achievements of engineers. Dr. Charles M. Vest is president of the National Academy of Engineering.

The **Institute of Medicine** was established in 1970 by the National Academy of Sciences to secure the services of eminent members of appropriate professions in the examination of policy matters pertaining to the health of the public. The Institute acts under the responsibility given to the National Academy of Sciences by its congressional charter to be an adviser to the federal government and, upon its own initiative, to identify issues of medical care, research, and education. Dr. Harvey V. Fineberg is president of the Institute of Medicine.

The **National Research Council** was organized by the National Academy of Sciences in 1916 to associate the broad community of science and technology with the Academy's purposes of furthering knowledge and advising the federal government. Functioning in accordance with general policies determined by the Academy, the Council has become the principal operating agency of both the National Academy of Sciences and the National Academy of Engineering in providing services to the government, the public, and the scientific and engineering communities. The Council is administered jointly by both Academies and the Institute of Medicine. Dr. Ralph J. Cicerone and Dr. Charles M. Vest are chair and vice chair, respectively, of the National Research Council.

www.national-academies.org

PANEL ON UNDERSTANDING CROSS-NATIONAL HEALTH DIFFERENCES AMONG HIGH-INCOME COUNTRIES

STEVEN H. WOOLF (*Chair of Panel*), Department of Family Medicine, Virginia Commonwealth University

PAULA A. BRAVEMAN, School of Medicine, University of California, San Francisco

KAARE CHRISTENSEN, Institute of Public Health, University of Southern Denmark

EILEEN M. CRIMMINS, Davis School of Gerontology, University of Southern California

ANA V. DIEZ ROUX, School of Public Health, University of Michigan

DEAN T. JAMISON, Department of Global Health, University of Washington

JOHAN P. MACKENBACH, Department of Public Health, Erasmus University, Rotterdam, The Netherlands

DAVID V. McQUEEN, Global Consultant, Atlanta, GA

ALBERTO PALLONI, Department of Sociology, University of Wisconsin–Madison

SAMUEL H. PRESTON, Department of Sociology, University of Pennsylvania

LAUDAN ARON, *Study Director*
DANIELLE JOHNSON, *Senior Program Assistant*

v

COMMITTEE ON POPULATION
2012

LINDA J. WAITE (*Chair*), Department of Sociology, University of Chicago

CHRISTINE BACHRACH, School of Behavioral and Social Sciences, University of Maryland

JERE BEHRMAN, Department of Economics, University of Pennsylvania

PETER J. DONALDSON, Population Council, New York, NY

KATHLEEN HARRIS, Carolina Population Center, University of North Carolina at Chapel Hill

MARK HAYWARD, Population Research Center, University of Texas, Austin

CHARLES HIRSCHMAN, Department of Sociology, University of Washington

WOLFGANG LUTZ, World Population Program, International Institute for Applied Systems Analysis, Laxenburg, Austria

ROBERT MARE, Department of Sociology, University of California, Los Angeles

SARA McLANAHAN, Center for Research on Child Wellbeing, Princeton University

BARBARA B. TORREY, Independent Consultant, Washington, DC

MAXINE WEINSTEIN, Center for Population and Health, Georgetown University

DAVID WEIR, Survey Research Center, Institute for Social Research, University of Michigan

JOHN R. WILMOTH, Department of Demography, University of California, Berkeley

BARNEY COHEN, *Director* (until August 2012)
THOMAS PLEWES, *Director* (after August 2012)

BOARD ON POPULATION HEALTH
AND PUBLIC HEALTH PRACTICE
2012

ELLEN WRIGHT CLAYTON (*Chair*), Center for Biomedical Ethics and Society, Vanderbilt University
MARGARITA ALEGRÍA, Cambridge Health Alliance, Somerville, MA
SUSAN M. ALLAN, Northwest Center for Public Health Practice, University of Washington
GEORGES C. BENJAMIN, American Public Health Association, Washington, DC
BOBBIE A. BERKOWITZ, School of Nursing, Columbia University
DAVID R. CHALLONER, Vice President for Health Affairs, Emeritus, University of Florida
R. ALTA CHARO, University of Wisconsin Law School
JOSE JULIO ESCARCE, Department of General Internal Medicine and Health Services Research, Department of Medicine, University of California, Los Angeles
ALVIN D. JACKSON, Ohio Department of Health, Fremont, OH
MATTHEW W. KREUTER, George Warren Brown School of Social Work, Washington University in Saint Louis
HOWARD MARKEL, University of Michigan Medical School
MARGARET E. O'KANE, National Committee for Quality Assurance, Washington, DC
SUSAN L. SANTOS, School of Public Health, University of Medicine and Dentistry of New Jersey
MARTIN JOSE SEPÚLVEDA, Integrated Health Services, International Business Machines Corporation, Somers, NY
SAMUEL SO, School of Medicine, Stanford University
ANTONIA M. VILLARRUEL, School of Nursing, University of Michigan
PAUL J. WALLACE, The Lewin Group, Falls Church, VA

ROSE MARIE MARTINEZ, *Director*

Foreword

The United States spends much more money on health care than any other country. Yet Americans die sooner and experience more illness than residents in many other countries. While the length of life has improved in the United States, other countries have gained life years even faster, and our relative standing in the world has fallen over the past half century.

What accounts for the paradoxical combination in the United States of relatively great wealth and high spending on health care with relatively poor health status and lower life expectancy? That is the question posed to the panel that produced this report, *U.S. Health in International Perspective: Shorter Lives, Poorer Health*. The group included experts in medicine, epidemiology, and demography and other fields in the social sciences. They scrutinized the relevant data and studies to discern the nature and scope of the U.S. disadvantage, to explore potential explanations, and to point the way toward improving the nation's health performance.

The report identifies a number of misconceptions about the causes of the nation's relatively poor performance. The problem is not simply a matter of a large uninsured population or even of social and economic disadvantage. It cannot be explained away by the racial and ethnic diversity of the U.S. population. The report shows that even relatively well-off Americans who do not smoke and are not overweight may experience inferior health in comparison with their counterparts in other wealthy countries. The U.S. health disadvantage is expressed in higher rates of chronic disease and mortality among adults and in higher rates of untimely death and injuries among adolescents and small children. The American health-wealth paradox is a pervasive disadvantage that affects everyone, and it has not been improving.

The report describes multiple, plausible explanations for the U.S. health disadvantage, from deficiencies in the health system to high rates of unhealthy behaviors and from adverse social conditions to unhealthy environments. The panel painstakingly reviews the quality and limitations of evidence about all of the factors that may contribute to poor U.S. health outcomes. In this, and in earlier work the panel cites, many remediable shortcomings have been identified. Thus, the report advances an agenda for both research and action.

The report was made possible by the dedicated work of the panel and staff who conducted this study and by the generous support of the Office of Behavioral and Social Sciences Research and other units of the National Institutes of Health. The National Research Council and the Institute of Medicine are very much indebted to all who contributed.

The nation's current health trajectory is lower in success and higher in cost than it should be. The cost of inaction is high. We hope this report deepens understanding and resolve to put America on an economically sustainable path to better health.

Harvey V. Fineberg
President, Institute of Medicine

Robert M. Hauser
Executive Director, Division of
Behavioral and Social Sciences and
Education, National Research Council

Preface

In 2011 the Office of Behavioral and Social Sciences Research (OBSSR) of the National Institutes of Health (NIH) asked the National Research Council (NRC) and the Institute of Medicine (IOM) to undertake a study on understanding cross-national health differences among high-income countries. The NRC's Committee on Population and the IOM's Board on Population Health and Public Health Practice established our panel for this task.

The impetus for this project came from a recently released NRC report that documented that life expectancy at age 50 had been increasing at a slower pace in the United States than in other high-income countries. The charge to our panel was to probe further and to determine whether the same worrying pattern existed among younger Americans, to explore potential causes, and to recommend future research priorities.

As readers who know this issue can appreciate, this is a daunting and complex charge. The questions put to the panel involve many fields, including medicine and public health, demography, social science, political science, economics, behavioral science, and epidemiology. They require the examination of data from many countries, drawn from disparate sources. The panel was given 18 months for the task, enough time to pull back the curtain on this issue but not to conduct a systematic review of every contributory factor and every relevant study or database. This report serves only to open the inquiry, with the invitation to others to probe deeper and with the disclaimer that the evidence cited here can only skim the surface of highly complex issues.

The report that follows could not have been produced without the help of many dedicated individuals. We begin by thanking the report's sponsor, OBSSR, and also the National Institute on Aging (NIA), which contributed financing for our work and was the primary sponsor of the prior NRC report that led to this study. We are especially grateful for guidance and contributions from Robert M. Kaplan, director, and Deborah H. Olster, deputy director of OBSSR, and Richard M. Suzman, director of the Division of Behavioral and Social Research at NIA. Ronald Abeles and Ravi Sawhney, both formerly with NIH, were also instrumental in conceiving of this project and seeing it get off the ground.

In fulfilling its charge, the panel also relied heavily on presentations and background papers and analyses from many of the world's leading experts on the social and health sciences that relate to cross-national health disparities. Specifically, the panel benefited greatly from presentations by Michele Cecchini, OECD; Neal Halfon, University of California, Los Angeles; Ronald Kessler, Harvard University; Sir Michael Marmot, University College London; Ellen Nolte, RAND Europe; Robert Phillips, Robert Graham Center; Cathy Schoen, Commonwealth Fund; and David Stuckler, Cambridge University. Also critical to the panel's deliberations and thinking were presentations and commissioned background papers from Clare Bambra, University of Durham; Jason Beckfield, Harvard University; and Russell Viner, University College London.

Several postdoctoral and graduate students worked intensively with a number of panel members to produce unique and compelling data analyses that appear throughout this report. We thank these contributors: Jessica Ho, University of Pennsylvania, who collaborated with Samuel Preston on developing much of the evidence presented in Chapter 1; Stéphane Verguet, University of Washington, who collaborated with Dean Jamison on a "years-behind" analysis presented in Chapter 1; James Yonker, University of Wisconsin, who collaborated with Alberto Palloni on an extensive analysis of health indicators across the life course presented in Chapter 2; and Aïda Solé Auró, University of Southern California, who collaborated with Eileen Crimmins on evaluating the health of adults at age 50.

Several other individuals at the home institutions of panel members contributed to their analyses for this report. In particular, the panel thanks Jung Ki Kim at the University of Southern California for assisting Eileen Crimmins; Malavika Subramanyam at the University of Michigan for assisting Ana Diez Roux with her review of environment factors for Chapter 7; and Karen Simpkins at the University of California, San Francisco, for assisting Paula Braveman with tables and figures for Chapter 6.

We also thank the authors of two background papers the panel commissioned: Russell Viner, University College London, for an assessment of cross-national differences in adolescent health and the importance of adolescence in shaping life-long health outcomes; and Clare Bambra, Durham University, and Jason Beckfield, Harvard University, for an analysis of how cross-national differences in political systems, governance structures, and public policy making might influence health at the national level.[1]

During the course of this project, the panel also benefited from targeted consultations with national experts to help make sense of data uncovered in this review. In particular, the panel thanks Sheldon H. Danziger, University of Michigan; Thomas Getzen, International Health Economics Association; and Timothy M. Smeeding, Institute for Research on Poverty, University of Wisconsin–Madison, for their advice on interpreting poverty statistics and Clemencia Cosentino de Cohen for her advice on interpreting data on educational attainment. We also thank J. Michael McGinnis, senior scholar at the IOM, for the valuable advice he offered this panel and for serving as a discussant at a crucial panel meeting.

This report would not have been possible without the support of NRC staff. I first thank Laudan Aron, our study director, who toiled over every page of this document. The panel is also indebted to Barney Cohen, former director of the NRC's Committee on Population; Thomas Plewes, who succeeded him and shepherded the report to its release; and Rose Marie Martinez, senior director of IOM's Board on Population Health and Public Health Practice, who provided oversight and support of this project at every level. The panel also thanks Wendy Jacobson and Robert Pool for assistance with background research and writing; Danielle Johnson for administrative and logistical support and formatting of references, figures, and tables; Alina Baciu, Amy Geller, and Keiko Ono, for assembling the bibliography; Amy Geller, Hope Hare, and Rose Marie Martinez for assistance with graphics; Kirsten Sampson Snyder for guiding the report through review; Eugenia Grohman for editing; Yvonne Wise for managing the production process; and Sara Frueh, Patricia Morison, Lauren Rugani, Christine Stencel, and Steve Turnham for help with communications.

This report has been reviewed in draft form by individuals chosen for their diverse perspectives and technical expertise, in accordance with procedures approved by the National Academies' Report Review Committee. The purpose of this independent review is to provide candid and critical comments that will assist the institution in making its published report as sound as possible and to ensure that the report meets institutional standards for objectivity, evidence, and responsiveness to the study charge. The review comments and draft manuscript remain confidential to protect the integrity of the deliberative process.

We thank the following individuals for their review of this report: James Banks, Department of Economics, Institute for Fiscal Studies, Uni-

---

[1] All background papers and analyses are available directly from the authors.

versity College London; Daniel G. Blazer, Duke University Medical Center; James S. House, Survey Research Center, University of Michigan Institute for Social Research; David A. Kindig, School of Medicine, University of Wisconsin–Madison; Cato T. Laurencin, University of Connecticut Health Center; David Melzer, Department of Epidemiology and Public Health, Exeter University; Carlos Mendes de Leon, University of Michigan; Angelo O'Rand, School of Social Sciences, Duke University; Mauricio Avendano Pabon, Center for Population and Development Studies, Harvard University; David Vlahov, School of Nursing, University of California, San Francisco; and John R. Wilmoth, Department of Demography, University of California, Berkeley. Dana Glei of Georgetown University also provided a focused mid-project technical review of the commissioned data analysis conducted by Jessica Ho and Samuel Preston for Chapter 1.

Although the reviewers listed above provided many constructive comments and suggestions, they were not asked to endorse the conclusions or recommendations, nor did they see the final draft of the report before its release. Robert Wallace, College of Public Health, University of Iowa, and Patricia Danzon, Health Care Management Department, The Wharton School, University of Pennsylvania, oversaw the review of this report. Appointed by the NRC and the IOM, they were responsible for ensuring that this report underwent an independent examination in accordance with institutional procedures and that all review comments were carefully considered. Responsibility for the final content of this report rests entirely with the authoring panel and the institution.

Finally, I would like to thank my fellow panel members for their wisdom, collegiality, and energy in producing this important report. Every member was immensely helpful, but I wish to specifically acknowledge Samuel Preston, Alberto Palloni, Paula Braveman, and Ana Diez Roux for their first drafts of Chapters 1, 2, 6, and 7, respectively. This report is truly an ensemble effort. I hope that readers will notice the interdisciplinary collaboration reflected in the pages of this document. The panel members, all highly regarded experts in their fields, contributed wonderful insights and the literatures of their disciplines to give our discussions and data analysis the holistic perspective this topic deserves. I am indebted to these colleagues, who despite many demanding responsibilities, gave generously of themselves and operated under a very demanding timeline. I am sure I speak for the panel and staff in collectively thanking our spouses and families for the disruption in lives this undertaking required.

Our panel was unprepared for the gravity of the findings we uncovered. We hope that others will take notice. Our charge was to give advice to the scientific community, and this report fulfills that charge by outlining ways that the NIH, other research agencies, and investigators can collect new data and advance understanding of the causes of cross-national health disparities. But the gravity of our findings also deserves attention outside the scientific community. A broader audience—most importantly the American public—should know what this report says. Concerted action is required on many levels of society if the nation is to change the conditions described here and to give the people of the United States—particularly the nation's children—the superior health and life expectancy that exist elsewhere in the world.

Steven H. Woolf, *Chair*
Panel on Understanding Cross-National
Health Differences Among High-Income Countries

Contents

Figures, Tables, and Boxes

FIGURES

TABLES

BOXES

Summary

The United States is among the wealthiest nations in the world, but it is far from the healthiest. Although life expectancy and survival rates in the United States have improved dramatically over the past century, Americans live shorter lives and experience more injuries and illnesses than people in other high-income countries. A growing body of research is calling attention to this problem, with a 2011 report by the National Research Council confirming a large and rising international "mortality gap" among adults age 50 and older. The U.S. health disadvantage cannot be attributed solely to the adverse health status of racial or ethnic minorities or poor people, because recent studies suggest that even highly advantaged Americans may be in worse health than their counterparts in other countries.

As a follow-up to the 2011 National Research Council report and in light of this new evidence, the National Institutes of Health asked the National Research Council (NRC) and the Institute of Medicine (IOM) to convene a panel of experts to study this issue. The Panel on Understanding Cross-National Health Differences Among High-Income Countries was charged with examining whether the U.S. health disadvantage exists across the life span, exploring potential explanations, and assessing the larger implications of the findings.

THE INFERIOR HEALTH STATUS OF THE UNITED STATES

The panel's analysis compared health outcomes in the United States with those of 16 comparable high-income or "peer" countries: Australia, Austria, Canada, Denmark, Finland, France, Germany, Italy, Japan,

Norway, Portugal, Spain, Sweden, Switzerland, the Netherlands, and the United Kingdom. We examined historical trends dating back several decades, with a focus on the more extensive data available from the late 1990s to 2008.

Over this time period, we uncovered a strikingly consistent and pervasive pattern of higher mortality and inferior health in the United States, beginning at birth:

- For many years, Americans have had a shorter life expectancy than people in almost all of the peer countries. For example, as of 2007, U.S. males lived 3.7 fewer years than Swiss males and U.S. females lived 5.2 fewer years than Japanese females.
- For the past three decades, this difference in life expectancy has been growing, especially among women.
- The health disadvantage is pervasive—it affects all age groups up to age 75 and is observed for multiple diseases, biological and behavioral risk factors, and injuries.

More specifically, when compared with the average for peer countries, the United States fares worse in nine health domains:

1. **Adverse birth outcomes:** For decades, the United States has experienced the highest infant mortality rate of high-income countries and also ranks poorly on other birth outcomes, such as low birth weight. American children are less likely to live to age 5 than children in other high-income countries.
2. **Injuries and homicides:** Deaths from motor vehicle crashes, non-transportation-related injuries, and violence occur at much higher rates in the United States than in other countries and are a leading cause of death in children, adolescents, and young adults. Since the 1950s, U.S. adolescents and young adults have died at higher rates from traffic accidents and homicide than their counterparts in other countries.
3. **Adolescent pregnancy and sexually transmitted infections:** Since the 1990s, among high-income countries, U.S. adolescents have had the highest rate of pregnancies and are more likely to acquire sexually transmitted infections.
4. **HIV and AIDS:** The United States has the second highest prevalence of HIV infection among the 17 peer countries and the highest incidence of AIDS.
5. **Drug-related mortality:** Americans lose more years of life to alcohol and other drugs than people in peer countries, even when deaths from drunk driving are excluded.

6. **Obesity and diabetes:** For decades, the United States has had the highest obesity rate among high-income countries. High prevalence rates for obesity are seen in U.S. children and in every age group thereafter. From age 20 onward, U.S. adults have among the highest prevalence rates of diabetes (and high plasma glucose levels) among peer countries.

7. **Heart disease:** The U.S. death rate from ischemic heart disease is the second highest among the 17 peer countries. Americans reach age 50 with a less favorable cardiovascular risk profile than their peers in Europe, and adults over age 50 are more likely to develop and die from cardiovascular disease than are older adults in other high-income countries.

8. **Chronic lung disease:** Lung disease is more prevalent and associated with higher mortality in the United States than in the United Kingdom and other European countries.

9. **Disability:** Older U.S. adults report a higher prevalence of arthritis and activity limitations than their counterparts in the United Kingdom, other European countries, and Japan.

The first half of the above list occurs disproportionately among young Americans. Deaths that occur before age 50 are responsible for about two-thirds of the difference in life expectancy between males in the United States and peer countries, and about one-third of the difference for females. And the problem has been worsening over time; since 1980, the United States has had the first or second lowest probability of surviving to age 50 among the 17 peer countries. Americans who do reach age 50 generally arrive at this age in poorer health than their counterparts in other high-income countries, and as older adults they face greater morbidity and mortality from chronic diseases that arise from risk factors (e.g., smoking, obesity, diabetes) that are often established earlier in life.

The U.S. health disadvantage is more pronounced among socioeconomically disadvantaged groups, but even advantaged Americans appear to fare worse than their counterparts in England and some other countries. That is, Americans with healthy behaviors or those who are white, insured, college-educated, or in upper-income groups appear to be in worse health than similar groups in comparison countries.

Certain factors do *not* appear to be responsible for the U.S. health disadvantage. The United States has higher survival after age 75 than do peer countries, and it has higher rates of cancer screening and survival, better control of blood pressure and cholesterol levels, lower stroke mortality, lower rates of current smoking, and higher average household income. In addition, U.S. suicide rates do not exceed the international average. Finally, the nation's large population of recent immigrants is generally in better health than native-born Americans.

With these important exceptions, Americans under age 75 fare poorly among peer countries on most measures of health. This health disadvantage is particularly striking given the wealth and assets of the United States and the country's enormous level of per capita spending on health care, which far exceeds that of any other country.

POSSIBLE EXPLANATIONS FOR THE
U.S. HEALTH DISADVANTAGE

The panel's search for potential explanations revealed that important antecedents of good health—such as the quality of health care and the prevalence of health-related behaviors—are also frequently problematic in the United States. For example, the U.S. health system is highly fragmented, with limited public health and primary care resources and a large uninsured population. Compared with people in other countries, Americans are more likely to find care inaccessible or unaffordable and to report lapses in the quality and safety of care outside of hospitals.

In terms of individual behaviors, Americans are less likely to smoke and may drink less heavily than their counterparts in peer countries, but they consume the most calories per capita, abuse more prescription and illicit drugs, are less likely to fasten seatbelts, have more traffic accidents involving alcohol, and own more firearms than their peers in other countries. U.S. adolescents seem to become sexually active at an earlier age, have more sexual partners, and are less likely to practice safe sex than adolescents in other high-income countries.

Adverse social and economic conditions also matter greatly to health and affect a large segment of the U.S. population. Despite its large and powerful economy, the United States has higher rates of poverty and income inequality than most high-income countries. U.S. children are more likely than children in peer countries to grow up in poverty, and the proportion of today's children who will improve their socioeconomic position and earn more than their parents is smaller than in many other high-income countries. In addition, although the United States was once the world leader in education, students in many countries now outperform U.S. students. Finally, Americans have less access to the kinds of "safety net" programs that help buffer the effects of adverse economic and social conditions in other countries.

Although all of these differences are compelling and important, no single factor fully explains the U.S. health disadvantage, for example:

- Problems with the health care system might exacerbate illnesses and heighten mortality from certain diseases but cannot account for transportation-related accidents or violence.

- Individual behaviors may contribute to the overall disadvantage, but studies show that even Americans with healthy behaviors, for example, those who are not obese or do not smoke, appear to have higher disease rates than their peers in other countries.
- The problem is not confined to socially or economically disadvantaged Americans; as noted above, several recent studies have suggested that even Americans with high socioeconomic status may experience poorer health than their counterparts in peer countries.

Many conditions that might explain the U.S. health disadvantage—from individual behaviors to systems of care—are also influenced by the physical and social environment in U.S. communities. For example, built environments that are designed for automobiles rather than pedestrians discourage physical activity. Patterns of food consumption are also shaped by environmental factors, such as actions by the agricultural and food industries, grocery store and restaurant offerings, and marketing. U.S. adolescents may use fewer contraceptives because they are less available than in other countries. Similarly, more Americans may die from violence because firearms, which are highly lethal, are more available in the United States than in peer countries. A stressful environment may promote substance abuse, physical illness, criminal behavior, and family violence. Asthma rates may be higher because of unhealthy housing and polluted air. In the absence of other transportation options, greater reliance on automobiles in the United States may be causing higher traffic fatalities. And when motorists do take to the road, injuries and fatalities may be more common if drunk driving, speeding, and seatbelt laws are less rigorously enforced, or if roads and vehicles are more poorly designed and maintained.

The U.S. health disadvantage probably has multiple explanations, some of which may be causally interconnected, such as unemployment and a lack of health insurance. Other explanations may share antecedents, especially those rooted in social inequality. Still others may have no obvious relationship, as in the very distinct causes of high rates of obesity and traffic fatalities. The relationships between some factors may develop over time, or even over a person's entire life course, as when poor social conditions during childhood precipitate a chain of adverse life events. Turmoil and risk-taking in adolescence can lead to subsequent setbacks in education or employment, fomenting life-long financial instability or other stresses that inhibit healthy life-styles or access to health care. In some cases, the explanation may simply be that the United States is at the leading edge of global trends that other high-income countries will follow, such as smoking and obesity.

Given the pervasive nature of the low U.S. rankings—on measures of health, access to care, individual behaviors, child poverty, and social mobility—the panel considered the possibility that a common thread might link the multiple domains of the U.S. health disadvantage. Might certain aspects of life in modern America—including some of the choices that American society is making (knowingly or not)—be part of the explanation for the U.S. health disadvantage? There are no definitive studies on this subject, but the public health literature certainly documents the health benefits of strengthening systems for health and social services, education, and employment; promoting healthy life-styles; and designing healthier environments. These functions are not solely the province of government: effective policies in both the public and private sector can create incentives to encourage individuals and industries to adopt practices that protect and promote health and safety. In countries with the most favorable health outcomes, resource investments and infrastructure often reflect a strong societal commitment to the health and welfare of the entire population.

Because choices about political governance structures, and the social and economic conditions they reflect and shape, matter to overall levels of health, the panel asked whether some of these underlying societal factors could be contributing to greater disease and injury rates and shorter lives in the United States. And might these choices also explain the inability of the United States to keep pace with peer countries in other important health-related domains, such as education and child poverty? These are important questions for which further research is needed. It will also be important for Americans to engage in a thoughtful discussion about what investments and compromises they are willing to make to keep pace with health advances other countries are achieving. Before this can occur, the public must first be informed about the country's growing health disadvantage, a problem that may come as a surprise to many Americans.

NEXT STEPS

The evidence regarding the U.S. health disadvantage is considerable and growing, but many fundamental questions remain about its underlying causes, the complex causal pathways that link health determinants with health outcomes, and how these pathways differ for specific subgroups of people over time and place. New data and new analyses are needed to answer these questions and to uncover the best ways of improving health outcomes in the future.

The panel offers three research recommendations for the scientific community to better understand what is driving the U.S. health disadvantage and how it can be reduced (see Box S-1). The panel recommends work to harmonize the data that are currently collected in many countries and to add questions to existing surveys, both in the United States and elsewhere; to develop new measures of health outcomes and new analytic methods

BOX S-1
Recommendations Relating to Research

RECOMMENDATION 1 Acting on behalf of all relevant data-gathering agencies in the U.S. Department of Health and Human Services, the National Institutes of Health and the National Center for Health Statistics should join with an international partner (such as the OECD or the World Health Organization) to improve the quality and consistency of data sources available for cross-national health comparisons. The partners should establish a data harmonization working group to standardize indicators and data collection methodologies. This harmonization work should explore opportunities for relevant U.S. federal agencies to add questions to ongoing longitudinal studies and population surveys that include various age groups—especially children and adolescents—and to replicate validated questionnaire items already in use by other high-income countries.

RECOMMENDATION 2 The National Institutes of Health and other research funding agencies should support the development of more refined analytic methods and study designs for cross-national health research. These methods should include innovative study designs, creative uses of existing data, and novel analytical approaches to better elucidate the complex causal pathways that might explain cross-national differences in health.

RECOMMENDATION 3 The National Institutes of Health and other research funding agencies should commit to a coordinated portfolio of investigator-initiated and invited research devoted to understanding the factors responsible for the U.S. health disadvantage and potential solutions, including lessons that can be learned from other countries.

for determining how various factors affect these outcomes; and to adopt a long-term sustained commitment to support this research agenda.

While these efforts are under way, the panel urges that the nation not simply wait for more data before addressing the U.S. health disadvantage: evidence is already available to begin tackling this important problem and the lead conditions responsible for it. The strength of our findings—which was a surprise to us—led us to consider what public- and private-sector leaders can do to begin to catch up with the health advances that other countries are achieving. In the recommendations related to policy, listed in Box S-2 and explained in greater detail in Chapter 10, we encourage three avenues for action: pursuing established national health objectives, alerting

---

BOX S-2
Recommendations Relating to Policy

RECOMMENDATION 4 The nation should intensify efforts to achieve established national health objectives that are directed at the specific disadvantages documented in this report and that use strategies and approaches that reputable review bodies have identified as effective.

RECOMMENDATION 5 The philanthropy and advocacy communities should organize a comprehensive media and outreach campaign to inform the general public about the U.S. health disadvantage and to stimulate a national discussion about its implications for the nation.

RECOMMENDATION 6 The National Institutes of Health or another appropriate entity should commission an analytic review of the available evidence on (1) the effects of policies (including social, economic, educational, urban and rural development and transportation, health care financing and delivery) on the areas in which the United States has an established health disadvantage, (2) how these policies have varied over time across high-income countries, and (3) the extent to which these policy differences may explain cross-national health differences in one or more health domains. This report should be followed by a series of issue-focused investigative studies to explore why the United States experiences poorer outcomes than other countries in the specific areas documented in this report.

---

the public, and exploring innovative policy options. More specifically, the panel recommends

- **Pursuing National Health Objectives** The panel urges a strengthened national commitment to existing public health objectives that address the specific health disadvantages documented in this report. That commitment should include the application of effective strategies and policies, as identified by reputable review bodies, to reform the health system, promote healthy behaviors, and improve health-related social conditions and community environments.
- **Alerting the Public** The panel envisions a robust outreach effort to inform the public about the growing U.S. health disadvantage relative to other high-income countries and to stimulate a national discussion about the implications of this for future policy, practice, and research.

- **Identifying Innovative Policies** The panel believes that there is much to learn from a thorough examination of the policies and approaches that countries with better health outcomes have found useful and that may have application, with adaptations, in the United States. Also of value would be a series of issue-focused investigative studies to seek explanations for the specific health disadvantages documented in this report.

The life-course perspective adopted by the panel underscores the importance of early life, not only because children and youth are often the victims of the U.S. health disadvantage, as in the case of infant mortality and adolescent homicides, but also because early life is a critical developmental period that can shape health development trajectories throughout life. The seeds of illnesses that strike older adults are often planted before age 25, a period when adverse social and environmental exposures and the establishment of unhealthy behaviors and risk factors can lead to life-long consequences. The striking health and social disadvantages documented among U.S. infants, children, and adolescents emphasize the importance of child and family services, support for education, especially in early childhood, and social services that safeguard young people. At the same time, public health and social policy solutions that target middle-aged and older adults can produce important improvements in life expectancy and health, particularly because of the high prevalence of chronic diseases that afflict Americans at older ages.

COSTS OF INACTION

The consequences of not attending to the growing U.S. health disadvantage and reversing current trends are predictable: the United States will probably continue to fall further behind comparable countries on health outcomes and mortality. In addition to the personal toll this will take, the drain on life and health may ultimately affect the economy and the prosperity of the United States as other countries reap the benefits of healthier populations and more productive workforces. With so much at stake, especially for America's youth, the United States cannot afford to ignore its growing health disadvantage.

Introduction

The United States is one of the wealthiest nations in the world (World Bank, 2012a), and Americans pride themselves on the quality of their health care, physicians, hospitals, and academic medical centers. The United States also plays an international leadership role in biomedical and health services research and in developing cutting-edge medical technologies, pharmaceutical products, and treatment innovations.

Although Americans have achieved very high levels of health over the past century and are healthier than people in many other nations, a growing body of research suggests that the health of the U.S. population is not keeping pace with the health of people in other economically advanced, high-income countries. Multiple studies have documented more favorable health outcomes in the United Kingdom, continental Europe, Australia, New Zealand, Japan, and Canada (Banks et al., 2006; Crimmins et al., 2008, 2010; Martinson et al., 2011a; Meslé and Vallin, 2006; National Research Council, 2010, 2011; Rau et al., 2008). This research documents a growing U.S. health disadvantage[1]: the United States is losing ground in the control of diseases, injuries, and other sources of morbidity.

---

[1]The term "health disadvantage," which is used in the statement of task given to the panel and throughout this report, is defined here as a condition of relative inferiority, reflecting the unfavorable health outcomes in the United States compared with those in other high-income countries. The term is not meant to imply that the United States, among the wealthiest countries in the world, is *disadvantaged*, i.e., "lacking in the basic resources or conditions (as standard housing, medical and educational facilities, and civil rights) believed to be necessary for an equal position in society" (Merriam-Webster Dictionary, 2012).

The U.S. health disadvantage also appears to be costing lives: Americans are not living as long as their counterparts in other countries. According to a report from the Paris-based OECD (2011b), 27 countries now outperform the United States on life expectancy at birth.[2] At 78.2 years, the United States is well below the average of 79.5 years among OECD member nations and far below the life expectancy (83.0 years) of Japan, the top-ranking country. Looking at mortality trends only among adults over age 50, a recent National Research Council (2011) study found that the United States began losing ground relative to other high-income countries around 1980, falling from the middle of the group in 1980 to near the bottom by 2007. Between 1980 and 2007, life expectancy at age 50 increased by only 2.5 years in the United States compared with 6.4 years in Japan, 5.2 years in Italy, and an average of 3.9 years in nine high-income countries other than the United States (National Research Council, 2011).

THE ROLE OF DIVERSITY IN HEALTH DISADVANTAGES

Compared with many other high-income countries, the population of the United States is more racially and ethnically diverse, receives immigrants from multiple countries, and struggles with higher poverty rates (OECD, 2011e). In addition, unlike the populations of most comparable countries, many Americans—especially racial and ethnic minorities and socioeconomically disadvantaged groups—lack health insurance coverage (National Center for Health Statistics, 2012). The poor health of racial and ethnic minorities and socioeconomically disadvantaged groups is well documented (Agency for Healthcare Research and Quality, 2011). As a result, some might question whether the poor health status of the U.S. population in the aggregate reflects the adverse health status of minorities or the poor and whether affluent, white Americans are just as healthy as their counterparts in other countries.

A growing body of research is beginning to suggest that the U.S. health disadvantage is not limited to socioeconomically disadvantaged groups: even the most advantaged Americans are in worse health than their counterparts in other countries. For example, a study that purposely limited its analysis to non-Hispanic whites found that U.S. residents aged 55-64 were less healthy than their English counterparts (Banks et al., 2006). The authors added that "health insurance cannot be the central reason for the better health outcomes in England because the top SES [socioeconomic status] tier of the U.S. population has close to universal access but their health outcomes are often worse than those of their English counterparts" (p. 2,043). Likewise, a similar comparison with England found that "the patterns were similar when

---

[2]Life expectancy at birth is the average number of years a newborn can expect to live if, over a lifetime, he or she experienced the age-specific mortality rates that were reported in that year.

the sample was restricted to whites, the insured, nonobese, nonsmoking nondrinkers, and specific income categories and when stratified by normal weight, overweight, and obese weight categories" (Martinson et al., 2011a, p. 858).[3] Two other studies have confirmed that even affluent and highly educated Americans suffer from a health disadvantage relative to residents of other high-income countries (Avendano et al., 2009, 2010).[4]

FINDING ANSWERS IS CRITICAL

Why is the United States falling behind? Answering this question is not just an academic exercise. The answers could reveal one or more factors that threaten the health of Americans and their economic competitiveness relative to other countries (World Economic Forum, 2011). Understanding the complex factors responsible for the U.S. health disadvantage could improve understanding of the factors responsible for health itself and point toward more strategic policies to improve the health of the American public.

Cross-national comparisons can shed new light on a nation's health and what drives it, pointing to strengths and unrecognized weaknesses (Marmor et al., 2009). They can suggest critical directions for improvement by identifying what has been achieved elsewhere and suggest key priorities for research. The lessons learned from these comparisons may also be instructive to countries following in the footsteps of the United States. For example, the United States leads the world in the share of the population that is obese, but other countries are close behind (Finucane et al., 2011; OECD, 2011b). To the extent that obesity explains the growing health disadvantage of Americans, other countries can forecast similar threats to the health of their populations from increasing rates of obesity.

One potential explanation for the U.S. health disadvantage is the country's health care system. Many Americans understand that health care in the United States needs improvement (Pew Research Center, 2009), and indeed, national reform proposals often target specific weaknesses in the U.S. health care system. These weaknesses include fragmentation, duplication, inaccessibility of records, the practice of defensive medicine, misalignment of physician and patient incentives, limited access for many people, and excessively fast adoption of expensive technologies of uncertain efficacy and cost-effectiveness (Fineberg, 2012; Garber and Skinner, 2008). The Institute of Medicine and the National Research Council have issued a number of reports over the past several decades documenting and describing many of

---

[3]More recent analyses of these data are available in Martinson (2012).

[4]These data also show much more serious and larger cross-national health differences for people with lower levels of wealth.

these problems and their implications (see, e.g., Institute of Medicine, 2001, 2003c, 2007b, 2010).

But health care systems and the health services they deliver are not the only influences on population health. Life-styles and behaviors, social and economic circumstances, environmental influences, and public policies can also play key roles in shaping individual and community health. And a number of these factors may be critical to understanding why some high-income countries experience significantly better health outcomes than the United States.

As noted above, a recent National Research Council (2011) study of mortality trends began to explore potential explanations for the U.S. health disadvantage, including smoking, obesity, cardiovascular diseases, cancer, and deficiencies in health care. However, the panel for that study was charged with looking at mortality trends only among people age 50 and older. This report extends that study by examining whether a U.S. health disadvantage can also be found among children and young adults. We adopted a life-course approach and a broad social-ecological framework in this study of international differences in health.

STUDY CHARGE AND PANEL APPROACH

The Office of Behavioral and Social Sciences Research (OBSSR) at the National Institutes of Health (NIH)[5] gave the following charge to the National Research Council and the Institute of Medicine:

Appoint an ad hoc committee of experts to examine what is known about international differences among high-income countries in measures of health and disability over the life cycle, and what those findings imply for public health. The findings from this report could suggest the need for new data collection, an agenda for further research, or the opportunity to design more effective public health strategies in the future. More specifically, the committee was charged with the following tasks:

1. Describe the sources, purpose, and limitations of international health comparisons;
2. Describe the nature and the strength of the evidence that exists to support the conclusion of a health disadvantage of the U.S. population versus comparable industrialized nations;
3. Determine to what extent the reported health disadvantage in the U.S. population holds true across various diseases and conditions;

---

[5]Established in 1995, OBSSR is located in the Office of the Director of NIH. Its mission is to stimulate behavioral and social sciences research across NIH and to integrate this research more fully into ongoing areas of NIH research. Its ultimate goal is to improve the nation's understanding, treatment, and prevention of disease.

4. Determine to what extent the reported health disadvantage in the U.S. population holds true across various age ranges (infant, child, adolescent, adult, elderly). Can onset of the difference in health be traced to a single age group or does it develop over certain parts of the life course?

5. Propose alternative explanations, or potential causes of the reported health disadvantage going beyond previously tested explanations. This would include an examination of individual risk factors (e.g., diet, exercise, smoking, drugs, and alcohol); societal level factors (e.g., social organization of work and leisure, social and interpersonal relationships, and social networks); and other factors (e.g., access to health care) that may have a differential impact on health outcomes across countries; and

6. If insufficient evidence (data) exists currently to test new hypotheses, indicate the nature and extent of data that would be required.

The Panel on Understanding Cross-National Health Differences Among High-Income Countries began its work in 2011. Members were drawn from key disciplines relevant to the study charge, including demography, economics, epidemiology, medicine, public health, and sociology. The panel included several members from Europe, individuals working in academia and research, and those with experience in governmental agencies and nongovernmental organizations. It also included several members of the panel that produced the prior National Research Council (2011) study that inspired this report (including its cochairs), providing a useful link between the two studies. The panel produced this report over a period of 18 months and met in person four times between January 2011 and June 2012. Three of these meetings included public sessions during which the panel heard from the study sponsors as well as other leading experts and researchers from both the United States and overseas. Outside experts presented on their areas of expertise with a particular view toward this study's focus on cross-national health differences and the U.S. position relative to other high-income countries. The panel also commissioned several data analyses and research reviews on topics for which it sought additional information. This included an analysis of data on mortality, morbidity, and disease determinants in high-income countries, drawn from databases (e.g., the Human Mortality Database) from the World Health Organization (WHO), the OECD, and other major data repositories.[6] All of these analyses compared the health of the U.S. population with those in a peer group of comparable

---

[6]The Human Mortality Database provides detailed mortality and population data to researchers, students, policy analysts, and others interested in human longevity. The database is an outgrowth of earlier projects in the Department of Demography at the University of California, Berkeley, and at the Max Planck Institute for Demographic Research in Rostock, Germany, and can be accessed at http://www.mortality.org or http://www.humanmortality.de.

affluent countries, and some had a special focus on social factors (based on life-cycle events and living conditions) that might explain health differentials.

The panel began its work acknowledging both the importance of the study and the tremendous scope it might encompass. With less than 2 years to complete the project, it was critical to set some practical boundaries and realistic expectations that would drive the development of a useful report in a timely manner. The panel recognized that relevant scientific evidence was at once vast (e.g., for understanding the determinants of health) and scant (e.g., for establishing causality for the U.S. health disadvantage). As a result, it understood its charge to examine what is currently known and propose a research agenda for future work rather than to draw definitive conclusions on every aspect of this topic. The panel sought to establish a concise and compelling framework for the analysis, citing relevant systematic reviews of the literature from reputable sources.

The panel considered a broad range of potential explanatory factors in response to its charge to "propose alternative explanations or potential causes of the reported health disadvantage, going beyond previously tested explanations." Potential explanations for the U.S. health disadvantage range from those factors that are commonly understood to influence health (e.g., such health behaviors as diet, physical inactivity, and smoking, or inadequate access to physicians and high-quality medical care) to more "upstream" social and environmental influences on health (e.g., income, education, and the conditions in which people live and work). All of these factors, in turn, may be shaped by broader national contexts and public policies that might affect health and the determinants of health, and therefore might explain why one advanced country enjoys better health than another.

This broad perspective—spanning downstream, proximate determinants of health to more upstream, distal ones—has informed several other major health studies including the World Health Organization's Commission on the Social Determinants of Health (2008), the "Marmot Review" of health in the United Kingdom (Marmot, 2010), and the Robert Wood Johnson Foundation's Commission to Build a Healthier America in the United States (Braveman and Egerter, 2008). It has also been central to several other studies conducted by the Institute of Medicine (e.g., 2003a).

In addition to taking a social-ecological approach, both the study sponsor and the panel members recognized the value of adopting a life-course perspective on health development in order to understand international differences in health status. As detailed in Chapter 3, differences between countries in the health of older adults may have much to do with the different conditions they experienced as children, adolescents, and young adults. Furthermore, some diseases in late life, such as cardiovascular disease and many cancers, are attributable in large measure to unhealthy behaviors (e.g., tobacco use) and modifiable risk factors (e.g., obesity) that are established in adolescence and young adulthood (Tirosh et al., 2011). Finally, an adult's socioeconomic status—which also affects health—is shaped by childhood circumstances, in particular the adversities a family faces. For these many reasons, understanding why the health of Americans is not keeping pace with that of people in other countries must take into account their entire life experiences.

Given the breadth of factors encompassed in a behavioral and social science perspective, the panel had to be both systematic and selective in its approach. For each group of factors (e.g., health systems, health behaviors, social and environmental factors), the panel reviewed the available evidence that (1) the set of factors matters to health; (2) the set of factors is worse in prevalence or health impact in the United States compared with other high-income countries; and (3) this difference between the United States and other countries could contribute to the U.S. health disadvantage. More details on the methods of the systematic review are discussed in Chapter 3.

TOPICS BEYOND THE SCOPE OF THIS STUDY

The health disparities that exist between advantaged and disadvantaged populations within countries often eclipse the health disparities between countries. Health inequities are shaped less by the geographic boundaries that define nations, which are often arbitrary by-products of history and geopolitics, than by differences within populations (within and across borders) in demographic characteristics, socioeconomic resources, and environmental factors that affect health (Hans, 2009).[7] Most high-income countries report significant health gradients by income, education, social class, occupation, and other social factors, and in some countries the gradients are alarmingly steep (Mackenbach et al., 2008). In the United States, health status differs markedly for poor people, for people with low educational attainment, and for some minority populations, such as blacks and Native Americans (Agency for Healthcare Research and Quality, 2011; Bleich et al., 2012; Braveman et al., 2011b; Satcher et al., 2005; Woolf et al., 2004).

Although understanding and ameliorating health disparities is a priority in the United States and other countries, the panel focused on its charge of understanding why the *aggregate* health status of the United States is poorer than the *aggregate* health status of other countries. However, because of questions about the role of disparities, the panel did explore whether aggregate health status in the United States might be compromised by the large health inequities that exist within the population, which are

---

[7]In addition, as discussed further in Chapters 8 and 9, nations such as the members of the European Union cannot be considered fully independent from each other, from either a statistical or policy perspective.

discussed in Chapters 6 and 7. Moreover, the causal mechanisms for cross-national health differences and for health gradients within countries may differ in important ways that are only beginning to be understood. For example, studies suggest that some Scandinavian countries with superior aggregate health status have larger health disparities than other European countries with poorer aggregate health status (Bambra, 2007; Bambra and Eikemo, 2009; Dahl et al., 2006; Eikemo et al., 2008a; Huijts and Eikemo, 2009; Kunst et al., 1998; Lahelma and Lundberg, 2009; Mackenbach et al., 1997, 2008; Stirbu et al., 2010).

It is also important to note that, unlike the National Research Council (2011) report that examined trends in mortality above age 50 in high-income countries, this panel's work focused more on current cross-national differences and less on trends over time. Although current levels are certainly reflections of past trends and suggest the direction of future trends, both time constraints and inadequacies in available data limited our ability to explore changes over time. This work remains an important priority for future research.

THE REPORT

This report is structured to address three aims: (1) to document the nature and scope of the U.S. health disadvantage (Part I, Chapters 1-2), (2) to explore potential explanations for this disadvantage (Part II, Chapters 3-8), and (3) to propose next steps for the field (Part III, Chapters 9-10).

Part I

Part I reviews the current evidence on mortality and morbidity differences across high-income countries. This information establishes a preliminary evidence base on cross-national health differences spanning all ages of life:

- Chapter 1: Shorter Lives
- Chapter 2: Poorer Health Throughout Life

Part II

Part II of this report is devoted to exploring potential explanations for the U.S. health disadvantage, framed around the factors that are known to influence individual and population health. It is organized as follows:

- Chapter 3: Framing the Question
- Chapter 4: Public Health and Medical Care Systems

- Chapter 5: Individual Behaviors
- Chapter 6: Social Factors
- Chapter 7: Physical and Social Environmental Factors
- Chapter 8: Policies and Social Values

The wide variety of factors considered responds to the panel's charge, which stipulates that we "go beyond previously tested explanations." Part II also develops the argument that these potential explanations are inter-connected across chapters in "upstream/downstream" relationships. These relationships are important in two ways: they demonstrate causality, e.g., when policies affect education or obesity rates; and they offer a temporal, life-course perspective, e.g., when environmental conditions in childhood precipitate unhealthy behaviors and pathophysiological disease processes later in life.

Part III

The panel's conclusions and recommendations are in Part III of this report, which sets out priorities for research and action:

- Chapter 9: Research Agenda
- Chapter 10: Next Steps

Given the ambitious charge to the panel, the significance of the findings to emerge from Part I, and the breadth of the explanatory framework laid out in Part II, we see this report as laying the foundation for a challenging research agenda but also for immediate next steps the nation can take while awaiting the results of future studies.

Part I:
Documenting the U.S.
Health Disadvantage

The scientific evidence pointing to a "U.S. health disadvantage" relative to other high-income countries has been building over time. Several studies using comparable data sources in the United States and United Kingdom have reported that Americans have higher disease rates and poorer health than the British. In 2006, James Banks and colleagues (2006) compared the self-reported health and biological markers of disease among residents of the United States and England aged 55-64 and found that U.S. residents were much less healthy than their English counterparts. A few years later, Melissa Martinson and colleagues confirmed these findings but extended them to younger ages, noting that the cross-country differences were of similar magnitude across all age groups (Avendano and Kawachi, 2011; Martinson et al., 2011a, 2011b). The investigators also reported that this health disadvantage persisted even when the comparison was limited to people in the highest socioeconomic brackets, whites, the insured, and those without a history of tobacco use, drinking, or obesity.

Other studies have documented a higher prevalence of diseases and risk factors among Americans than among the Japanese (Crimmins et al., 2008; Reynolds et al., 2008), Western Europeans (Banks and Smith, 2011; Michaud et al., 2011; Thorpe et al., 2007), the populations of England and Western Europe (Avendano et al., 2009), and those of England, Western Europe, Canada, and Japan (Crimmins et al., 2010). Although life expectancy in high-income countries has been increasing for decades, the pace of the increase in the United States has fallen dramatically behind that of other high-income countries (Meslé and Vallin, 2006; National Research Council,

21

2011; Rau et al., 2008). Between 1980 and 2006, life expectancy in the United States grew more slowly than in almost every other high-income country (Meslé and Vallin, 2006; Rau et al., 2008).

Although the U.S. health disadvantage has been ably documented in prior research, much of the focus has been on adults age 50 and older. This age group has received attention because of its higher burden of disease and because comparable international surveys were often only available for older adults. The National Research Council (2011) study that inspired the establishment of this panel was sponsored by the National Institute on Aging and focused on documenting and explaining differences in life expectancy at age 50, not the U.S. health disadvantage at younger ages. The impetus for the present study included a strong interest in extending the earlier analysis to those under age 50.

Part I of this report is devoted to documenting the nature and strength of the evidence regarding a health disadvantage of the U.S. population. It draws on prior work in discussing the health and survival of older adults, but it focuses on documenting cross-national differences in the health and life expectancy of children, adolescents, and young adults.

Because death is such a defining event and because national mortality and cause of death data tend to be widely available and of high quality, the panel chose to begin its examination by looking at mortality. Chapter 1 provides an initial look at cross-national mortality differences in high-income countries, focusing on Americans' shorter life spans. The chapter begins by documenting the disparities in cause-specific mortality rates and then summarizes the existing literature on life expectancy differentials between the United States and other countries. So as not to duplicate the prior National Research Council (2011) study of mortality over age 50 and because of time constraints, the panel chose to focus on documenting differences between the United States and other high-income countries in life expectancy *under* age 50 and the causes of death responsible for years of life lost. The chapter features new analyses, commissioned by the panel, on cross-national life expectancy differentials before age 50. These analyses look at differences in life expectancy by sex, age group, and specific causes of death.

Chapter 2 moves beyond mortality and adopts a life-course perspective to examine health differences between the United States and other countries by age group. The chapter examines cross-national differences in the prevalence, severity, and mortality rates for specific diseases and injuries affecting each age group across the life course. As in Chapter 1, the panel's primary focus was on individuals under age 50, especially children. We were particularly interested in adolescence, which is both an important life stage in its own right (Gore et al., 2011; Patton et al., 2009) and may offer clues to understanding cross-national differences that emerge later in life. The chapter's examination of health under age 50 was supplemented with a review of the rich literature on health differences among individuals over age 50, a period of life when chronic illness becomes more prevalent. Given the breadth of material covered in Chapter 2, the panel had to be much more parsimonious in its approach to indicators and data sources.

1

Shorter Lives

D
o Americans live as long as people in other high-income countries? This chapter reviews one of the most reliable sources of information about cross-national health differences, vital statistics on deaths. Unlike measures obtained from survey data, these data pertain to an unambiguous indicator of health. High-quality vital statistics are available for nearly all deaths in high-income countries. Their continuous coverage permits the construction of accurate time series, and the data can be converted into meaningful popular indicators, such as life expectancy at birth, which is an intuitively appealing summary measure that is often used as the basis for evaluating overall health status. Data on mortality by cause of death can also provide important clues about the social and epidemiologic pathways that affect population health.

This chapter examines mortality from multiple perspectives to present a comprehensive picture of the evidence: we examine mortality rates (the number of deaths from particular causes per 100,000 persons), life expectancy at various ages, the probability of living to age 50, and years of life lost from particular causes. We present both the data and the United States' ranking on these data relative to other high-income countries.[1] In

---

[1]We report rankings to simplify comparisons across countries, but it is important to recognize that this is an ordinal measure that does not reflect the size of the difference between one rank and the next. Rankings can change when small differences in rates shift a country's rank.

this chapter we focus on 17 high-income countries:[2] Australia, Austria, Canada, Denmark, Finland, France, Germany, Italy, Japan, Norway, Portugal, Spain, Sweden, Switzerland, the Netherlands, the United Kingdom, and the United States.

MORTALITY RATES

For many years, global health statistics compiled by the OECD[3] and the World Health Organization (WHO) have documented higher mortality rates in the United States relative to other high-income countries. Among the 17 peer countries examined by the panel, Americans faced the second highest risk of dying from noncommunicable diseases in 2008 (418 per 100,000 persons) and the fourth highest risk of dying from communicable (infectious) diseases in 2008 (World Health Organization, 2011a) (see Figures 1-1 and 1-2).

Death rates from noncommunicable diseases, notably cardiovascular diseases, have declined everywhere but less so in the United States. As of 2009, ischemic heart disease mortality among males in the United States was 129 per 100,000, higher than the other 16 peer countries except Finland (OECD, 2011b).[4] Table 1-1 provides cause-specific mortality rates

---

[2]The panel selected these 17 as "peer countries" because they are most comparable to the United States. We set three criteria for designating peer countries: (1) high levels of development for a long period of time, (2) sufficient population size to ensure stability of estimates, and (3) data from the Human Mortality Database (2012) of suitable quality and availability for the time period used in our analysis, 2006-2008. Excluded countries did not meet one or more of these criteria. For example, data quality has been a problem in Belgium, and its latest year of available data was 2005; Greece and Korea were not included in the Human Mortality Database at the time of our analysis; and several other high-income countries are former Soviet satellites with atypical mortality experiences. For consistency, this report's documentation of the U.S. health disadvantage is based on comparisons with these 16 peer countries. The panel uses a more general term, "high-income countries," to refer to other groups of high-income countries. On occasion, we make comparisons with these other high-income countries and even emerging economies (e.g., Mexico, Russia) because data were available for this larger comparison group, because we cited studies that included these countries, and because for certain conditions (e.g., mortality rates, child poverty) comparisons with emerging economies underscore the United States' relative position.

[3]The OECD is a membership organization of 34 member countries that share a commitment to democratic government and the market economy. Well known for its publications and statistics, the work of the OECD covers both economic and social issues, including macroeconomics, trade, employment, education, health, and social welfare. The organization was established in 1961 when the United States and Canada joined the 18 former members of the Organization for European Economic Cooperation (established in 1947 for postwar reconstruction) to work together on shared economic development. The OECD's 34 members now include countries from North and South America, Europe, and the Asia-Pacific region, and it includes not only most advanced economies, but also such emerging economies as Chile, Mexico, and Turkey.

[4]U.S. mortality rates from ischemic heart disease are even higher than those of some emerging economies, such as Mexico and Slovenia (OECD, 2011b).

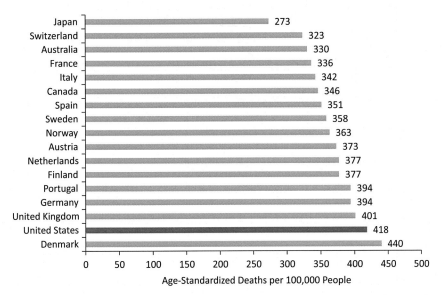

FIGURE 1-1 Mortality from noncommunicable diseases in 17 peer countries, 2008.
SOURCE: Data from World Health Organization (2011a, Table 3).

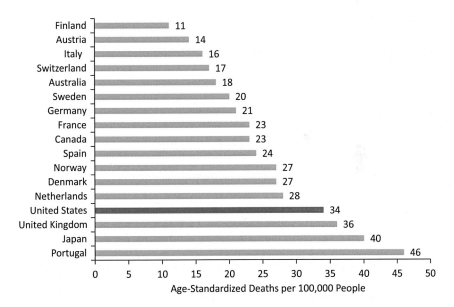

FIGURE 1-2 Mortality from communicable diseases in 17 peer countries, 2008.
SOURCE: Data from World Health Organization (2011a, Table 3).

TABLE 1-1 Mortality Rates in 17 Peer Countries, 2008

	Australia	Austria	Canada	Denmark	Finland	France	Germany	Italy	Japan	Netherlands	Norway	Portugal	Spain	Sweden	Switzerland	U.K.	U.S.
ALL CAUSES	378.0	420.5	401.2	500.8	446.6	397.7	440.6	383.0	349.3	427.3	425.3	467.7	397.7	409.8	371.2	462.1	504.9
Specific causes																	
Noncommunicable diseases	329.9	372.5	345.8	440.4	377.0	336.4	393.9	342.1	273.0	376.8	363.1	393.9	351.0	357.6	323.3	400.5	418.4
Cardiovascular diseases[a]	117.2	154.4	118.6	144.6	163.6	99.2	174.9	132.3	97.4	122.4	132.9	148.2	115.7	150.8	123.1	141.7	155.7
Malignant neoplasms[b]	118.8	123.6	125.8	157.7	106.5	138.4	127.5	124.2	115.1	147.1	130.3	134.3	122.0	116.4	112.7	137.0	123.8
Other neoplasms	2.9	3.5	2.4	3.6	2.1	5.2	2.8	5.3	3.3	3.6	2.5	2.9	3.7	3.2	2.8	2.9	2.9
Neuropsychiatric conditions[c]	26.2	19.6	28.2	38.8	48.4	34.0	19.5	18.8	7.2	32.7	32.7	16.1	28.2	34.2	33.5	31.7	39.2
Respiratory diseases	21.6	15.5	22.4	31.2	12.7	13.3	17.3	17.3	15.8	23.9	25.4	25.6	31.1	15.5	13.4	34.4	34.3
• Chronic obstructive pulmonary disease	12.8	12.9	15.4	26.8	9.1	5.4	11.9	11.5	4.0	19.0	20.8	10.5	12.6	11.7	9.9	21.5	24.3
• Asthma	1.2	0.9	0.5	0.8	0.8	0.8	1.0	0.4	0.7	0.3	1.2	0.6	0.8	0.6	0.5	1.1	0.9
Digestive diseases[d]	13.0	19.4	15.5	29.5	27.2	20.4	25.7	16.4	14.5	17.4	13.7	25.6	21.3	15.1	15.9	26.7	19.8
Diabetes mellitus	9.9	17.4	13.4	12.3	5.5	8.1	11.1	12.4	4.5	10.1	7.4	19.3	9.4	9.2	8.2	5.0	15.2

Genitourinary diseases[e]	8.2	5.4	7.6	8.1	3.0	5.7	7.1	5.7	8.0	9.3	7.7	12.2	9.2	4.6	4.0	9.1	12.3
Endocrine disorders	5.2	8.1	5.1	6.5	1.6	5.8	3.5	4.3	2.6	3.5	3.8	5.3	3.4	3.2	2.7	3.1	7.1
Congenital anomalies	3.2	3.9	3.7	4.0	3.6	2.8	2.8	2.8	2.6	3.3	3.4	2.7	2.9	2.7	3.8	3.7	4.3
Musculoskeletal diseases[f]	2.8	1.5	2.4	3.4	2.3	2.5	1.2	2.0	1.7	2.4	2.6	1.4	3.0	2.1	2.8	3.5	2.9
Skin diseases	0.8	0.3	0.5	0.6	0.3	0.9	0.3	0.5	0.3	1.1	0.7	0.1	1.0	0.5	0.5	1.5	0.8
Oral conditions	0.1	0.0	0.0	0.1	0.1	0.0	0.0	0.0	0.0	0.1	0.1	0.0	0.0	0.0	0.0	0.0	0.0
Injuries	*30.0*	*33.6*	*32.1*	*33.1*	*58.5*	*38.2*	*25.2*	*24.6*	*36.3*	*22.1*	*35.6*	*28.3*	*23.2*	*32.2*	*30.5*	*25.5*	*52.8*
Unintentional injuries	21.0	21.2	20.6	22.3	38.6	23.3	15.4	19.0	16.1	13.7	25.2	19.3	16.7	19.7	17.2	17.4	35.5
• Road traffic accidents	6.8	6.9	7.8	5.4	5.7	6.6	5.7	8.4	3.8	4.0	5.2	10.0	6.1	4.1	3.8	4.8	13.9
• Poisonings	3.0	0.2	2.1	3.8	13.9	1.6	0.9	0.7	0.6	0.9	6.8	0.2	1.0	3.2	1.9	2.8	8.9
• Falls	3.1	5.3	3.7	2.8	10.1	4.0	4.6	2.2	2.5	4.0	3.7	2.5	2.2	4.1	7.5	3.5	4.4
• Fires	0.3	0.3	0.6	0.9	1.2	0.6	0.3	0.2	0.7	0.3	1.0	0.5	0.3	0.7	0.1	0.5	0.9
• Drowning	0.9	0.8	0.8	0.8	1.8	1.3	0.5	0.6	2.4	0.4	0.8	0.7	0.9	0.9	0.6	0.4	1.2
• Other	6.9	7.8	5.6	8.6	6.0	9.3	3.3	7.0	6.2	4.1	7.7	5.4	6.1	6.7	3.3	5.5	6.2
Intentional injuries	9.0	12.3	11.5	10.8	19.9	14.8	9.9	5.6	20.2	8.4	10.4	9.0	6.5	12.6	13.3	8.0	17.3
• Self-inflicted injuries	7.8	11.8	9.9	10.1	17.7	13.6	9.1	4.5	19.8	7.4	9.7	7.5	5.6	11.7	12.6	6.9	10.3

continued

TABLE 1-1 Continued

	Australia	Austria	Canada	Denmark	Finland	France	Germany	Italy	Japan	Netherlands	Norway	Portugal	Spain	Sweden	Switzerland	U.K.	U.S.
• Violence	1.3	0.5	1.6	0.8	2.2	1.3	0.7	1.1	0.4	0.9	0.6	1.5	0.9	0.9	0.7	1.1	6.5
• War	—	—	—	—	—	0.0	—	—	—	0.1	0.0	0.0	—	—	—	—	0.4
Communicable, maternal, perinatal, and nutritional conditions	*18.1*	*14.4*	*23.3*	*27.3*	*11.1*	*23.1*	*21.5*	*16.3*	*40.0*	*28.4*	*26.6*	*45.5*	*23.6*	*20.0*	*17.3*	*36.1*	*33.7*
Respiratory infections[g]	7.0	5.7	9.0	13.9	4.0	8.4	10.1	4.7	29.7	16.8	13.8	23.6	9.3	8.9	7.4	23.8	9.7
Infectious and parasitic diseases[h]	5.9	4.7	7.6	7.6	4.4	9.0	7.4	6.8	8.1	6.9	9.2	17.5	9.8	7.8	4.7	6.7	15.4
Maternal conditions	0.1	0.0	0.1	0.2	0.1	0.1	0.1	0.0	0.1	0.1	0.1	0.1	0.1	0.1	0.0	0.1	0.4
Perinatal conditions	4.6	3.8	5.9	4.4	2.5	3.6	3.6	3.8	1.3	3.6	3.1	3.4	3.7	2.4	4.8	5.2	7.1
• Prematurity and low birth weight	1.2	1.9	2.0	2.7	0.9	0.6	2.0	1.5	0.4	0.8	0.4	0.9	0.9	0.4	1.4	3.4	3.2
• Birth asphyxia and birth trauma	0.6	0.5	1.1	0.6	0.4	0.8	0.5	1.1	0.4	0.7	0.5	0.8	0.7	0.5	0.7	0.6	0.9

• Neonatal infections and other conditions	2.8	1.4	2.7	1.1	1.2	2.2	1.1	1.1	0.5	2.0	2.1	1.7	0.9	2.1	1.4	2.7	1.2	3.0
Nutritional deficiencies[i]	0.5	0.2	0.7	1.2	0.1	2.0	1.0	1.0	0.8	1.0	0.4	0.9	0.7	0.9	0.4	0.3		1.0

NOTES: Age-adjusted mortality rates in the 17 peer countries show similar relationships when the age-adjusted data are examined separately by sex. However, for some conditions, cross-national patterns change slightly. For example, although the mortality rate for overall injuries (and unintentional injuries) is higher in aggregate in Finland than in the United States, among females only the United States has a higher rate. Similarly, among females, the U.S. mortality rate for infectious and parasitic diseases is higher than that of Portugal, the mortality rate for respiratory diseases is higher than that of the United Kingdom, and the mortality rate for musculoskeletal diseases is higher than that of Spain. Austrian females have a higher mortality rate for cardiovascular diseases, and Swedish and Swiss females have higher rates of mortality for intentional injuries. Among males, the U.S. mortality rate for communicable diseases is higher than that of the United Kingdom; Portugal has a higher mortality rate for noncommunicable diseases in general and genitourinary diseases in particular; Denmark has a higher mortality rate for neuropyshicatric disorders; both Denmark and Austria have higher mortality rates for congenital anomalies; Sweden has a slightly higher mortality rate for cardiovascular diseases; Spain has a higher mortality rate for respiratory diseases; and France and Switzerland have higher mortality rates for musculoskeletal diseases. Not listed here are conditions with relatively low mortality rates (e.g., less than 2 per 100,000), for which the standing of the United States compared with peer countries may differ more substantially by sex.

[a]Primarily includes ischemic heart disease but also includes higher death rates from hypertensive and inflammatory heart disease. U.S. death rates from cerebrovascular disease and rheumatic heart disease are at or below average.

[b]Includes cancers of the mouth and oropharynx, esophagus, stomach, colon and rectum, liver, pancreas, trachea, bronchus, lung, skin, breast, cervix uteri, corpus uteri, ovary, prostate, and bladder; lymphomas; multiple myeloma; and leukemia.

[c]Includes Alzheimer and other dementias, Parkinson disease, multiple sclerosis, and drug use disorders, for which the United States has above-average mortality rates, and the following neuropsychiatric disorders for which the United States has average or below-average mortality rates: unipolar depressive disorder, bipolar disorder, schizophrenia, epilepsy, alcohol use disorder, posttraumatic stress disorder, obsessive-compulsive disorder, panic disorder, insomnia, and migraine.

[d]Includes peptic ulcer disease, cirrhosis of the liver, and appendicitis.

[e]Primarily nephritis and nephrosis. U.S. death rates from benign prostatic hypertrophy are at or below average.

[f]Includes rheumatoid arthritis and osteoarthritis, but U.S. death rates from the latter are at or below average.

[g]Includes upper and lower respiratory infections and otitis media.

[h]Includes HIV/AIDS and other sexually transmitted infections, tuberculosis, diarrheal diseases, childhood-cluster diseases (e.g., pertussis, poliomyelitis), meningitis, hepatitis B and C, malaria, tropical-cluster diseases (e.g., schistosomiasis, leishmaniasis), leprosy, dengue, Japanese encephalitis, trachoma, and intestinal nematode infections.

[i]Includes protein-energy malnutrition, iodine deficiency, Vitamin A deficiency, and iron-deficiency anemia.

SOURCE: Adapted from World Health Organization (2011a, Table 3).

for the 17 peer countries and shows that the United States also experiences relatively high mortality rates for neuropsychiatric conditions, respiratory diseases, diabetes and other endocrine disorders, genitourinary disease, congenital anomalies, infectious diseases, and perinatal conditions. This pattern differs little when the data are examined separately by sex (see NOTES in Table 1-1). An interactive graph, which allows a more thorough examination of the data in Table 1-1, is located at http://nationalacademies. org/IntlMortalityRates.

Figure 1-3 shows that in 2008 the United States had the second highest death rate from injuries among the 17 peer countries, exceeded only by Finland (World Health Organization, 2011a, Table 3). Unintentional injuries are the leading cause of death among Americans, from ages 1-44 (National Center for Health Statistics, 2012).

An important contributor has been deaths related to transportation. In 2009, the United States had the highest death rate from transportation-related accidents among the 17 peer countries (and the third highest in the OECD, exceeded only by Mexico and the Russian Federation). The death rate from transportation-related accidents decreased by 42 percent in OECD countries between 1995 and 2009, but by only 11 percent in the United States (OECD, 2011b). Although there are more motorists and miles driven in the United States, calculations of fatality rates per vehicle-kilometer, which correct for this confounding variable, also show that the United States

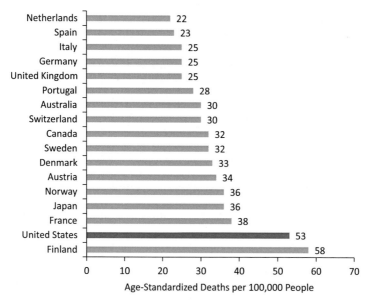

FIGURE 1-3 Mortality from injuries in 17 peer countries, 2008.
SOURCE: Data from World Health Organization (2011a, Table 3).

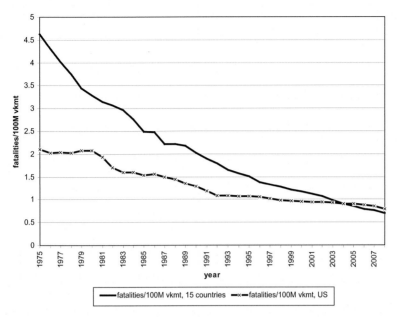

FIGURE 1-4 Motor vehicle fatalities in the United States and 15 other high-income countries, 1975-2008.

NOTE: The comparison set of countries in this analysis are Australia, Austria, Belgium, Denmark, Finland, France, Germany, Great Britain, Israel, Japan, the Netherlands, Norway, Slovenia, Sweden, and Switzerland.

SOURCE: Transportation Research Board (2011, Figure 2-2c).

has lost the advantage it once held over other countries. Figure 1-4 shows the trend over three decades. As the Transportation Research Board (2011, p. 40) explains:

> Fatality rates per vehicle kilometer have declined greatly in every high-income country in the past several decades, and the absolute disparity of rates among countries has lessened. A comparison of the U.S. experience with that of 15 other high-income countries for which 1975–2008 data are available shows that the U.S. fatality rate was less than half the aggregate rate in the other countries in 1975 but has been higher since 2005. Consequently, total annual traffic deaths in the 15 countries fell by 66 percent in the period, while U.S. deaths fell by only 16 percent. The U.S. fatality rate was among the best before 1990 but has been below the median rate of the group every year since 2001.

The United States also has dramatically higher rates of death from violent injuries, especially from firearms. In a study that compared 23 OECD countries in 2003, the U.S. homicide rate was 6.9 times higher than the other high-income countries and the rate of firearm homicides was 19.5 times higher. Although overall suicide rates were lower in the United States than in those countries, firearm suicide rates were 5.8 times higher than in other countries. Across the 23 countries in the study, 80 percent of all firearm deaths occurred in the United States (Richardson and Hemenway, 2011). This pattern is not new; data from the early 1990s showed similar results (Krug et al., 1998).

Although the incidence of AIDS has fallen since the early 1990s, the United States still has the highest incidence of AIDS among the 17 peer countries (and the third highest in the OECD, exceeded only by Brazil and South Africa) (OECD, 2011b). The incidence of AIDS in the United States (122 per million) is almost nine times the OECD average (14 per million).[5]

High mortality rates in the United States relative to other rich nations have been the subject of numerous research studies. A 2005 study reported that U.S. adults aged 15-59 had higher mortality rates than those in nine economically comparable nations: "Compared with other nations in the WHO's mortality database, in the United States 15-year-old girls rank 38th and 15-year-old boys rank 34th in their likelihood of reaching age 60" (Jenkins and Runyan, 2005, p. 291). These researchers noted that the higher mortality was true for both sexes and throughout the first five decades of life (Jenkins and Runyan, 2005).

The U.S. health disadvantage is not limited to death rates; the United States also has relatively high prevalence rates for disease and disability. Chapter 2 details this morbidity disadvantage *by age group*, but in Box 1-1 we briefly note the key findings that apply across the *entire* U.S. population.

The United States does enjoy some health advantages compared with other countries. In 2009, the United States had the third lowest mortality rate from stroke among the 17 peer countries (OECD, 2011b), despite its above-average mortality for ischemic heart disease.[6] As of 2009, the U.S. suicide rate (10.5 per 100,000 persons) was also below the average of the 16 peer countries (OECD, 2011b). Finally, although the U.S. incidence rate for cancer is the fourth highest of the 17 peer countries (OECD, 2011b),[7] mortality rates for certain cancers (e.g., cervical and colorectal cancer) are lower than most peer countries (World Health Organization, 2011a).

---

[5]The United States has the fifth highest prevalence of HIV infection among 40 OECD countries, exceeded only by Portugal, the Russian Federation, Estonia, and South Africa (OECD, 2011b), and the highest prevalence of HIV infection (for ages 15-49) among the 17 peer countries (World Health Organization, 2010).

[6]The reasons for this differential pattern are not entirely clear, but they may relate to cross-national differences in risk factors and treatment for cerebrovascular disease.

[7]The incidence of cancer may be skewed by the intensity of screening programs in the United States and may not accurately reflect the prevalence of the disease.

Table 1-2 lists other conditions for which the U.S. mortality rate is at or below the average of the 16 other peer countries.

CROSS-NATIONAL DIFFERENCES IN LIFE EXPECTANCY

Not surprisingly, higher mortality rates affect life expectancy in the United States. Perhaps the single most impressive achievement of the past century is the striking increase in longevity in nearly all parts of the world. At the turn of the 20th century, North American and Western European countries experienced life expectancies at birth of 40-50 years (Preston and Haines, 1991): 100 years later (in 2007), no country in these regions had a life expectancy of less than 75 years, and most had levels of more than 80 years (Human Mortality Database, 2012).

However, as shown in Table 1-3, there remain large differences in life expectancy at birth among high-income peer countries. In 2007, men in Switzerland and women in Japan enjoyed the longest life expectancies for their sexes. In contrast, the United States ranked last among males and next to last among females.[8] These differences with the top-performing countries amount to approximately 3.7 years for males and 5.2 years for females (Ho and Preston, 2011).[9]

We emphasize that these large cross-national differences are often eclipsed by even larger *within-country* disparities in life expectancy. As discussed in Box 1-2, such disparities are substantial in the United States (Agency for Healthcare Research and Quality, 2011; Bleich et al., 2012; Braveman et al., 2011a; Satcher et al., 2005; Woolf et al., 2004), and they may be part of the reason that the United States compares so unfavorably with its peers.

The U.S. disadvantage in life expectancy relative to other high-income countries is not a recent phenomenon (although the gap has grown over time), nor is this the first report to call attention to the problem. Jenkins and Runyan (2005) reported that U.S. survival rates for each of the five decades

---

[8]The life expectancy of females was lower in Denmark than in the United States in 2007. Life expectancy in aggregate (for males and females) has historically been lower in Denmark than in the United States, but not since 2005. These findings are from the Human Mortality Database (2012), which provides regularly updated detailed mortality and population data to researchers, students, and others interested in the history of human longevity. It is available at http://www.mortality.org.

[9]Ho and Preston's analysis for this panel is modeled on a similar analysis of mortality above age 50 that they conducted for the National Research Council (2011) panel and also published in Ho and Preston (2010). The current analysis draws on data from three sources: the Human Mortality Database, the WHO Mortality Database, and Statistics Canada. The data were downloaded July 2011, and, for each country, the latest year of data available between 2006 and 2008 was extracted. Dana Glei of Georgetown University provided the panel with a focused mid-project technical review of this analysis.

BOX 1-1
The U.S. Morbidity Disadvantage

As of 2010, the United States had the highest prevalence of diabetes (for adults aged 20-79) among the 17 peer countries (and among all OECD countries except Mexico). The U.S. obesity epidemic probably plays a major role in the prevalence of diabetes. The United States has the highest prevalence of adult obesity among the 17 peer countries (and all other OECD countries), a position it has held for decades. As of 2009, the prevalence of obesity in the United States (33.8 percent) was twice the OECD average (16.9 percent) (OECD, 2011b).

In a comparison of the health of Americans and the English across the life span, from birth to age 80, the United States had a higher prevalence of obesity, lipid disorders, diabetes, and asthma (Martinson et al., 2011a). Among females, the United States also had a higher prevalence of hypertension, heart attack and angina, and stroke. The differences were as large for young people as for old people. The researchers found that the English advantage persisted even when the samples were restricted to whites, people with health insurance, nonsmokers, nondrinkers, individuals of normal weight, or those in specific income categories (Martinson et al., 2011a, 2011b).

Studies of risk factors, rather than diseases, have yielded more mixed results. For example, some studies find that hypertension is less common in the United States than in other countries (Danaei et al., 2011a; Wolf-Maier et al., 2003), while others report the opposite (Banks et al., 2006; Martinson et al., 2011a). Similarly, some studies report that serum lipid levels are lower in the United States than in other countries (Farzadfar et al., 2011; Martinson et al., 2011a), but another study that compared biological risk factors in American and Japanese adults over age 20

---

between ages 15 and 59 were lower than those in nine economically comparable nations. Kunitz and Pesis-Katz (2005) reported that Americans have a shorter life expectancy than their neighbors in Canada. Meslé and Vallin (2006) and Rau et al. (2008) reported that, from 1980 to 2006, life expectancy in the United States grew more slowly than in almost every other high-income country.

The most extensive and recent analysis was a report by the National Research Council (2011). As noted in the Introduction of this report, that earlier report analyzed how life expectancy at age 50 had changed between 1980 and 2007, noting that it had increased by only 2.5 years in the United States compared with increases of 6.4 years in Japan, 5.2 years in Italy, and an average of 3.9 years in nine high-income countries other than the United

found that Americans had a higher summary risk score, including higher levels of serum lipids, glycosylated hemoglobin, and obesity, especially before age 50 (Crimmins et al., 2008).*

The percentage of American adults who describe their health as "good" or "very good" is the highest among people in high-income countries (OECD, 2011b), but this metric is subject to some limitations. Questions about self-rated health may be answered differently across countries due to cultural differences—such as differences in the likelihood or threshold for reporting good health—and may not always track well with objective health indicators. Danish residents, for example, are known to have shorter life expectancy than people in many other countries, but they are more likely to report their health as good or excellent (Oksuzyan et al., 2010). To some extent, this paradox may reflect attitudinal differences across cultures about the relative importance of physical health for a satisfying life. Self-rated health is influenced by mental health, which may differ in the United States from other countries. Finally, questions have been raised about the statistical validity of questions about self-rated health when presented to subjects of varied nationalities. The significance of the high self-rated health of Americans is therefore not entirely clear.

---

*Some of these inconsistencies may reflect cross-national differences in treatment patterns. Americans with hypertension and hyperlipidemia may be more likely than others to receive medication for these conditions (see Chapter 4), and this may account for lower levels of blood pressure and serum lipids observed in some studies.

States. As a result, the U.S. ranking in life expectancy at age 50 fell from the middle of the distribution for peer countries in 1980 to the bottom quartile by 2007. The drop was especially sharp for U.S. women (National Research Council, 2011).

Data on life expectancy at birth reveal an even more alarming pattern. Figure 1-5 shows that, among peer countries, male life expectancy at birth in the United States ranked near the bottom in 1980 and at the bottom in 2006. Figure 1-6 shows that female life expectancy, which had been near the median in 1979, ranked at the bottom in 2006. As documented in Box 1-3, the U.S. ranking on life expectancy has been deteriorating for decades and is now decades behind many peer countries.

A notable exception to this unfavorable pattern is the higher life expec-

TABLE 1-2 U.S. Death Rates Relative to 16 Peer Countries, 2008

| | Mortality Rate (per 100,000) | | |
| | | Peer Countries (N = 16) | |
Cause of Death	United States	Unweighted Mean	Range
	Death Rates Above Average		
Cardiovascular diseases[a]	**155.7**	133.6	97.4-174.9
Neuropsychiatric conditions[b]	**39.2**	28.1	7.2-48.4
Respiratory disease	**34.3**	21.0	12.7-34.4
Infectious and parasitic diseases	**15.4**	7.7	4.4-17.5
Diabetes mellitus	**15.2**	10.2	4.5-19.3
Genitourinary diseases[c]	**12.3**	7.2	3.0-12.2
Endocrine disorders	**7.1**	4.2	1.6-8.1
Congenital anomalies	**4.3**	3.3	2.6-4.0
Musculoskeletal diseases[d]	**2.9**	2.4	1.2-3.5
Nutritional deficiencies	**1.0**	0.7	0.1-2.0
Skin diseases	**0.8**	0.6	0.1-1.5
Maternal conditions	**0.4**	0.1	0.0-0.2
Perinatal conditions	**7.1**	3.7	1.3-5.9
Unintentional injuries[e]	**35.5**	20.4	13.7-38.6
Intentional injuries	**17.3**	11.4	5.6-20.2
	Death Rates at or Below Average		
Malignant neoplasms[f]	123.8	**127.3**	106.5-157.7
Digestive diseases[g]	19.8	**19.8**	13.0-29.5
Respiratory infections	9.7	**12.3**	4.0-29.7
Other neoplasms	2.9	**3.3**	2.1-5.3
Oral conditions[h]	0.0	0.0	0.0-0.1
Sense organ diseases[h]	0.0	0.0	0.0

NOTE: Higher death rates shown in bold.

[a]Primarily ischemic heart disease but also includes higher death rates from hypertensive and inflammatory heart disease. U.S. death rates from cerebrovascular disease and rheumatic heart disease are at or below average.

[b]The United States has average or below-average mortality rates for the following neuro-psychiatric disorders: unipolar depressive disorder, bipolar disorder, schizophrenia, epilepsy, alcohol use disorder, posttraumatic stress disorder, obsessive-compulsive disorder, panic disorder, insomnia, and migraine.

[c]Primarily nephritis and nephrosis. U.S. death rates from benign prostatic hypertrophy are at or below average.

[d]Includes rheumatoid arthritis and osteoarthritis, but U.S. death rates from the latter are at or below average.

[e]The U.S. rate for "other" unintentional injuries (excluding road traffic accidents, poisonings, fires, and drownings) is equivalent to the peer countries.

[f]With the exception of the following malignant neoplasms for which the United States has higher mortality rates: cancers of the trachea, bronchus, and lung; corpus uteri cancer; lymphomas and multiple myeloma; and leukemia.

[g]Includes peptic ulcer disease, cirrhosis of the liver, and appendicitis.

[h]Low mortality rates round to 0.0 per 100,000.

SOURCE: Adapted from World Health Organization (2011a, Table 3).

TABLE 1-3 Life Expectancy at Birth in 17 Peer Countries, 2007

Males			Females		
Country	LE	Rank	Country	LE	Rank
Switzerland	79.33	1	Japan	85.98	1
Australia	79.27	2	France	84.43	2
Japan	79.20	3	Switzerland	84.09	3
Sweden	78.92	4	Italy	84.09	3
Italy	78.82	5	Spain	84.03	5
Canada	78.35	6	Australia	83.78	6
Norway	78.25	7	Canada	82.95	7
Netherlands	78.01	8	Sweden	82.95	7
Spain	77.62	9	Austria	82.86	9
United Kingdom	77.43	10	Finland	82.86	9
France	77.41	11	Norway	82.68	11
Austria	77.33	12	Germany	82.44	12
Germany	77.11	13	Netherlands	82.31	13
Denmark	76.13	14	Portugal	82.19	14
Portugal	75.87	15	United Kingdom	81.68	15
Finland	75.86	16	United States	80.78	16
United States	75.64	17	Denmark	80.53	17

NOTE: LE = life expectancy at birth (years), or e^0.
SOURCE: Ho and Preston (2011, Table 1).

tancy of very old Americans. Manton and Vaupel (1995) demonstrated that life expectancy from ages 80-95 was higher in the United States than in England, France, Japan, and Sweden.[10] Fifteen years later, Ho and Preston (2010) observed the same pattern for adults over age 65. U.S. older adults had among the lowest age-specific mortality rates of 17 peer countries: this pattern has been observed in every decade since 1960, suggesting that the underlying cause is not a recent phenomenon. Researchers are unclear whether this phenomenon reflects unusually aggressive efforts in the United States to identify and treat chronic diseases or a selection process, that is, healthier adults surviving to old age (Finch and Crimmins, 2004; Ho and Preston, 2010; Janssen et al., 2005a).

---

[10]The one exception was for males in Japan, but the difference was not statistically significant.

BOX 1-2
Disparities in Life Expectancy in the United States

The range of life expectancies across U.S. states, which is 7.0 years for males and 6.7 years for females, exceeds the cross-national range between the United States and peer countries, which is 3.7 years for males and 5.2 years for females (see Table 1-3). Although life expectancy is very low in Appalachia and the Deep South, some states in the northern Plains and along the Pacific coast and Eastern seaboard easily outrank many peer countries. For example, males in Minnesota and Hawaii have a higher life expectancy than those in nine peer countries. Females in Hawaii have a life expectancy of 84.8 years, higher than that of every peer country except Japan (Measure of America, 2012).

Even greater disparities are found at the county level. For example, a recent study estimated that life expectancy in 2007 varied dramatically across U.S. counties for both males (from 65.9 to 81.1 years) and females (from 73.5 to 86.0 years), although most U.S. counties still compared unfavorably with the best-performing high-income countries. Merging 3,147 counties into 2,357 county clusters suitable for statistical analysis, researchers found that only 33 counties had a male life expectancy that exceeded the average of 10 leading countries, and only 8 counties had a higher female life expectancy (Kulkarni et al., 2011).

Of course, small geographic areas that compare unfavorably with the best countries can be found anywhere, but this may occur more often in the United States than in other high-income countries. For example, in a comparison of U.S. counties with small geographic units in three peer countries—Japan, Canada, and the United Kingdom—researchers found that 17 percent of U.S. counties had a male life expectancy that was more than 30 years behind that of the top 10 countries, whereas the same was only true for 2 percent of Canadian health areas, for 0.2 percent of British local authorities, and for no Japanese municipalities (Kulkarni et al., 2011).

The large geographic health disparities in the United States, which have been documented to the level of census tracts and neighborhoods (Bay Area Regional Health Inequities Initiative, 2008; Center on Human Needs, 2012b; Ezzati et al., 2008; Krieger et al., 2008), are less about geography than the characteristics of the local population and environment (Hans, 2009): see Chapters 6 and 7. For example, black infants in the United States are more than twice as likely as white infants to die before their first birthday (National Center for Health Statistics, 2012). Among black males with less than 12 years of education, life expectancy in 2008 was 14.2 years shorter than for white males with 16 or more

---

BOX 1-2 Continued

years of education (Olshansky et al., 2012). Many other poignant examples of major health disparities by race, ethnicity, and socioeconomic
status in the United States have been documented (see, e.g., Agency for
Healthcare Research and Quality, 2011; Braveman and Egerter, 2008;
Institute of Medicine, 2003d; Murray et al., 2006; National Center for
Health Statistics, 2012). Disadvantaged Americans living in disadvantaged regions of the United States have very unfavorable health outcomes when compared with other countries. One study found that black
and Native Americans in some regions had mortality rates that were
almost twice that of the OECD countries with the *highest* mortality rates
(Murray et al., 2006).

Large within-country health disparities in the United States may contribute in important ways to the nation's overall health disadvantage
relative to other high-income countries. Although studies reviewed in this
report suggest that the health disadvantage relative to peer countries
persists even when the U.S. data are limited to non-Hispanic whites or
upper-income populations, the U.S. health disadvantage is clearly far
greater among the large proportion of Americans who live amid unfavorable health conditions. And as discussed in Chapters 6 and 7, the health
of the entire population may be affected by the conditions that more
severely compromise the health of disadvantaged groups.

The National Research Council (2011) study explored potential explanations for the relatively poor U.S. performance at age 50 and older and
concluded that the long history of heavy cigarette smoking in the United
States accounted for a substantial share of the shortfall. Other contributors to the U.S. disadvantage included a rapid growth in obesity, significant
socioeconomic inequalities, and a lack of health insurance for large segments of the population (National Research Council, 2011). These (and
other) explanations are examined in Part II of this report.

SURVIVAL TO AGE 50

The panel was charged with looking at "health and disability over the
life-cycle" and therefore extended the prior report's analysis of U.S. health
conditions (above age 50) to younger Americans, from birth to age 50, to

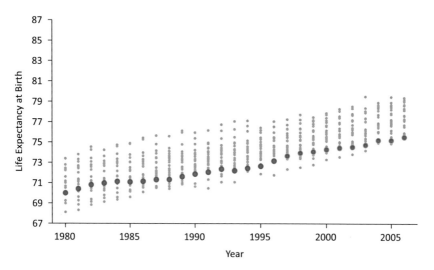

FIGURE 1-5 U.S. male life expectancy at birth relative to 21 other high-income countries, 1980-2006.
NOTES: Red circles depict newborn life expectancy in the United States. Grey circles depict life expectancy values for Australia, Austria, Belgium, Canada, Denmark, Finland, France, Iceland, Ireland, Italy, Japan, Luxembourg, the Netherlands, New Zealand, Norway, Portugal, Spain, Sweden, Switzerland, the United Kingdom, and West Germany.
SOURCE: National Research Council (2011, Figure 1-3).

determine if there is a similar U.S. health disadvantage in life expectancy.[11] A cross-national analysis of mortality data from birth to age 50 reveals a U.S. health disadvantage at ages under 50 that is more serious than what has been found for those over age 50 (Ho and Preston, 2011) (see Figures 1-7 and 1-8), which show the probabilities of survival from birth to age 50 since 1980. The United States has clearly fallen far behind its peer countries in life expectancy—both under and over age 50.

Is higher mortality in the United States concentrated at specific ages or is it more general? Figure 1-9 displays the rank order of U.S. mortality relative to the 16 peer countries listed in Table 1-3, stratified by age group. Remarkably, the U.S. rank for either sex is never better than 15th at any age below 75. The United States has the worst ranking in most age groups, especially in the long span that stretches between birth and age 55. In short, in terms of sheer physical survival, people living in the United States fare

---

[11]Chapter 2 examines the U.S. disadvantage on health measures other than life expectancy, including the prevalence of diseases and injuries across the life course.

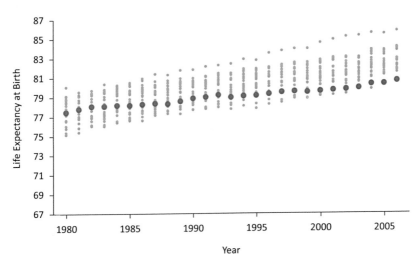

FIGURE 1-6 U.S. female life expectancy at birth relative to 21 other high-income countries, 1980-2006.
NOTES: Red circles depict newborn life expectancy in the United States. Grey circles depict life expectancy values for Australia, Austria, Belgium, Canada, Denmark, Finland, France, Iceland, Ireland, Italy, Japan, Luxembourg, the Netherlands, New Zealand, Norway, Portugal, Spain, Sweden, Switzerland, the United Kingdom, and West Germany.
SOURCE: National Research Council (2011, Figure 1-4).

worse than their counterparts in peer countries except at the very oldest ages.

This finding is not simply a reflection of the racial and ethnic diversity of the U.S. population. When the analysis was limited to non-Hispanic whites in the United States, the poor U.S. ranking hardly changed (see Figure 1-10). At no age below 55 do U.S. non-Hispanic whites rank better than 16th out of 17 countries (for either sex). Therefore, the overall poor position of the United States cannot be attributed to any particular minority group because the disadvantage is observed even among non-Hispanic whites.

Another distinct aspect of the United States is the shape of its survival curve: higher mortality rates among young Americans and the increased survival of the elderly have produced a wider spread[12] in the age of death

---

[12]The U.S. survival curve is less "rectangular" (Wilmoth and Horiuchi, 1999) than in other countries. In most peer countries, mortality is declining at a faster rate among younger ages than older ages. This phenomenon appears to be more pronounced in other countries than in the United States.

BOX 1-3
How Many Years Behind Is the United States?

One focus of this report is the U.S. health disadvantage among younger adults. One summary measure of mortality in this age group is the probability that a 15-year-old will die before reaching age 50 (given current age-specific mortality rates). Demographers refer to this measure as $_{35}q_{15}$, or the probability of dying in the 35 years following one's 15th birthday. For females in the 16 peer countries, $_{35}q_{15}$ was around 2 percent in 2007 but was approximately twice as high—4 percent—in the United States. This means that the probability of a 15-year-old U.S. female dying within 35 years was double the average for 16 peer high-income countries.

In all high-income countries, including the United States, $_{35}q_{15}$ has been declining for more than half a century. But the relative position of the United States has deteriorated since the late 1950s, when it was near the average of its peers. These countries, on average, had reduced their $_{35}q_{15}$ for females to the U.S. 2007 level of 4 percent almost 40 years earlier. In this sense, one can say that, in 2007, the United States was 40 years behind the average of its peers (and 50 years behind the leading peer country).

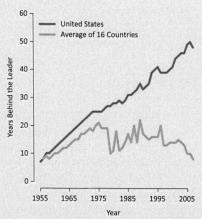

FIGURE 1-3a Number of years behind the leading peer country for the probability of dying between ages 15 and 50 among females, 1958-2007.
NOTES: The figure plots $_{35}q_{15}$, or the probability that a 15-year-old female will die before age 50. The y-axis measures how many years earlier the $_{35}q_{15}$ values of the 17 peer countries had been achieved by the country with the lowest $_{35}q_{15}$ value for that year.
SOURCE: Verguet and Jamison (2011, Figure 6).

continued

BOX 1-3 Continued

This concept of "years behind" provides a useful indicator of how well a given country is keeping pace with other countries. Figure 1-3a, from Verguet and Jamison (2011), plots years behind the leader for both the United States and the average of its peers for the period 1958-2007. It shows that the United States has fallen further behind the leader, while its peer countries began to "catch up" with the leader (albeit at an uneven pace) beginning in the mid-1970s. The net result of these uneven trends has been a steady decline of the U.S. position from near average to far below average.

Figure 1-3b also plots "years behind the leader" for both the United States and the average of its peers, but in this case it shows female mortality rates, by 5-year age groups, for a single year, 2007. These results confirm a U.S. mortality gap for females across the life span. It is most pronounced between the ages of 15 and 50 and diminishes somewhat for women above age 60.

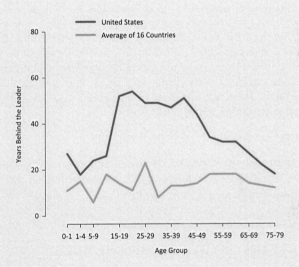

FIGURE 1-3b Number of years behind the leading peer country for female mortality by 5-year age group, 2007.
NOTE: The figure shows how many years earlier female mortality in the 17 peer countries had been achieved by the country with the lowest age-specific mortality for that year.
SOURCE: Verguet and Jamison (2011, Figure 7).

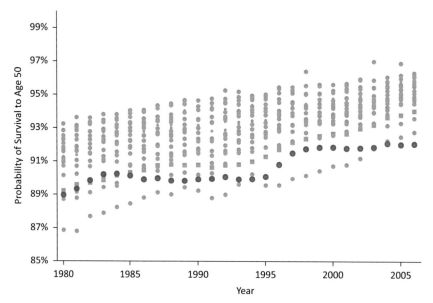

FIGURE 1-7 Probability of survival to age 50 for males in 21 high-income countries, 1980-2006.
NOTES: Red circles show the probability a newborn male in the United States will live to age 50. Grey circles show the probability of survival to age 50 in Australia, Austria, Belgium, Canada, Denmark, Finland, France, Iceland, Ireland, Italy, Japan, Luxembourg, the Netherlands, New Zealand, Norway, Portugal, Spain, Sweden, Switzerland, the United Kingdom, and West Germany.
SOURCE: National Research Council (2011, Figure 1-5).

than in other countries with a similar overall life expectancy (Shkolnikov et al., 2003). As a result, Americans lose more years of life than do those in other high-income countries (Shkolnikov et al., 2011), a topic we address in the next section.

YEARS OF LIFE LOST BEFORE AGE 50

At the turn of the 20th century, an individual born in Western Europe or North America could expect to live no more than 34 years between birth and age 50 (Keyfitz and Flieger, 1990), a loss of 16 years of life. In many countries today, a newborn can expect to live more than 49 of the first 50 years of life.[13] These remarkable gains are the result of major reductions in infectious diseases among infants and young children, as well as declines

---

[13]Data are from the Human Mortality Database (2012).

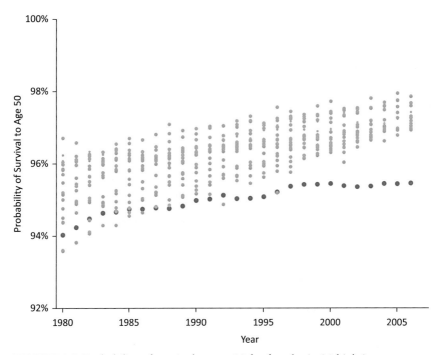

FIGURE 1-8 Probability of survival to age 50 for females in 21 high-income countries, 1980-2006.
NOTES: Red circles show the probability a newborn female in the United States will live to age 50. Grey circles show the probability of survival to age 50 in Australia, Austria, Belgium, Canada, Denmark, Finland, France, Iceland, Ireland, Italy, Japan, Luxembourg, the Netherlands, New Zealand, Norway, Portugal, Spain, Sweden, Switzerland, the United Kingdom, and West Germany.
SOURCE: National Research Council (2011, Figure 1-6).

in maternal mortality, the virtual elimination of infectious diseases among adolescents and middle-aged adults (particularly respiratory tuberculosis), and more recently, lower death rates from cardiovascular diseases from age 35 onward (Cutler and Miller, 2004; Riley, 2001).

Although the United States has shared in these improvements, it still forfeits the most years of potential life before age 50. Figures 1-11 and 1-12 show the number of years lost before age 50 by males and females, respectively, in the 17 peer countries. U.S. male and female newborns can expect to lose about 1.4 years and 0.8 years of life, respectively, before age 50. In the best performing country, Sweden, the corresponding losses are only 0.7 and 0.4 years, respectively. This mortality gap has also grown significantly over time. In 1990, U.S. females and males lost approximately 35 percent more years of life before age 50 than did those in other high-

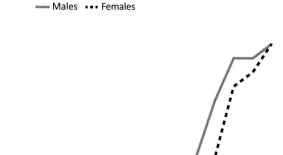

FIGURE 1-9 Ranking of U.S. mortality rates, by age group, among 17 peer countries, 2006-2008.

NOTES: The top rank is number 1, indicating the lowest death rate, and the bottom rank is number 17, indicating the highest death rate. Rankings are based on all-cause mortality rates for 2006-2008. Data for this figure were drawn from (1) the Human Mortality Database, 2011, University of California, Berkeley (USA), and Max Planck Institute for Demographic Research (Germany), available at http://www.mortality.org or http://www.humanmortality.de (data downloaded July 18, 2011) and (2) Arias, Elizabeth, 2011, United States Life Tables, 2007. *National Vital Statistics Reports*, 59(9), Hyattsville, MD: National Center for Health Statistics.
SOURCE: Adapted from Ho and Preston (2011, Figure 1).

income countries, but by 2009 this figure had grown to nearly 75 percent (Palloni and Yonker, 2012).

CAUSES OF PREMATURE DEATH

What causes of death are responsible for this excess loss of life in the United States? Because deaths in high-income countries are assigned to various causes of death by medical certifiers using internationally accepted criteria, it is possible to examine how life expectancy varies cross-nationally by cause. In this section, the panel presents an analysis of years of life

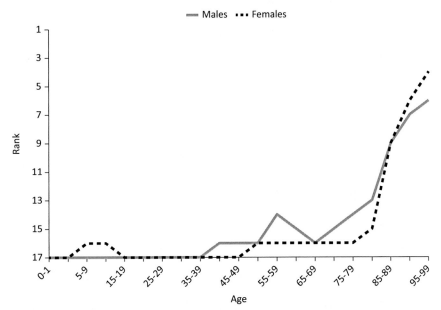

FIGURE 1-10 Ranking of U.S mortality rates for non-Hispanic whites only, by age group, among 17 peer countries, 2006-2008.
NOTES: The top rank is number 1, indicating the lowest death rate, and the bottom rank is number 17, indicating the highest death rate. Rankings are based on all-cause mortality rates for 2006-2008. Data for this figure were drawn from (1) the Human Mortality Database, 2011, University of California, Berkeley (USA), and Max Planck Institute for Demographic Research (Germany), available at http://www.mortality.org or http://www.humanmortality.de (data downloaded on July 18, 2011) and (2) Arias, Elizabeth, 2011, United States Life Tables, 2007, *National Vital Statistics Reports, 59*(9), Hyattsville, MD: National Center for Health Statistics.
SOURCE: Adapted from Ho and Preston (2011, Figure 2).

lost by cause for age groups under age 50, as reported by Ho and Preston (2011).[14] Years of life lost before age 50 is a measure that combines the intensity of a particular cause of death with its age incidence. It indicates how many potential years of life below age 50 are claimed by a particular cause of death—that is, how many additional years would be lived before age 50 if deaths from a particular cause were eliminated.

To facilitate comparisons with the United States, Ho and Preston (2011) created a composite of the other 16 peer countries by calculating an unweighted average of the age- and cause-specific death rates across these countries and grouping causes of death into the major categories used in the

---

[14]For deaths by cause after age 50, see National Research Council (2011).

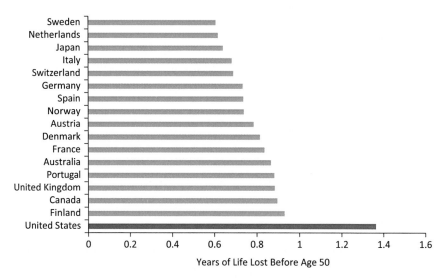

FIGURE 1-11 Years of life lost before age 50 by males in 17 peer countries, 2006-2008.
NOTE: Data for this figure come from the Human Mortality Database (downloaded July 18, 2011, last updated July 13, 2011); the WHO Mortality Database (downloaded July 18, 2011, last updated March 25, 2011); and Statistics Canada (downloaded July 22, 2011, data released February 23, 2010).
SOURCE: Adapted from Ho and Preston (2011, Figure 3).

Global Burden of Disease Study (Mathers et al., 2006). As shown in Figures 1-13 and 1-14, the results show that the United States loses a larger number of years of life to *all* of the major disease and injury groupings than do the other peer countries. Although communicable diseases and nutritional conditions are no longer a leading cause of premature deaths in most high-income countries, the United States still fares poorly in this category. The gap in years of life lost from noncommunicable diseases—which includes heart disease, cancer, and other conditions not caused by infections—is also large. For both males and females, cardiovascular disease and congenital anomalies together account for more than half of the U.S. excess mortality from noncommunicable diseases. Diabetes, digestive diseases, and respiratory diseases also contribute to the gap.

Intentional and unintentional injuries are also major contributors to the excess years of life lost by Americans before age 50. Intentional injuries—homicide and suicide—are particularly important causes of early deaths among U.S. males. Ho and Preston (2011) found that 69 percent of all U.S. homicide deaths in 2007 (73 percent of homicides before age 50) involved firearms (for both sexes combined), compared with a mean of 26 percent

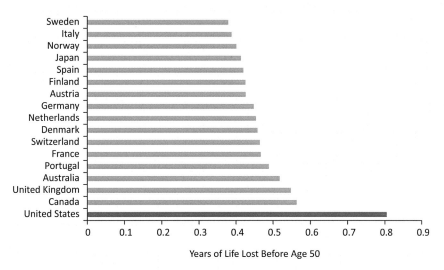

FIGURE 1-12 Years of life lost before age 50 by females in 17 peer countries, 2006-2008.

NOTE: Data for this figure come from the Human Mortality Database (downloaded July 18, 2011, last updated July 13, 2011); the WHO Mortality Database (downloaded July 18, 2011, last updated March 25, 2011); and Statistics Canada (downloaded July 22, 2011, data released February 23, 2010).

SOURCE: Adapted from Ho and Preston (2011, Figure 4).

in the other countries. Both males and females in the United States lose an equivalent number of years of life to unintentional injuries, such as motor vehicle accidents, falls, poisonings, fires, and drowning. Unintentional poisonings are the largest contributor to nontransportation-related accidents: in 2007, they accounted for 64 percent and 72 percent of nontransportation-related U.S. deaths under age 50 among males and females, respectively.

Drug-related deaths are another category in which the United States loses more years of life than other countries. Drug-related deaths include both drug- and alcohol-induced deaths, which account for 76 percent and 24 percent, respectively, of all drug-related deaths before age 50. This category includes deaths from medical conditions or complications induced by alcohol or drugs, selected mental and behavioral disorders due to alcohol or the use of psychoactive substances, accidental or intentional alcohol or drug poisonings or overdoses, and deaths with measurable blood levels of alcohol or other addictive drugs (e.g., opiates, cocaine, hallucinogens, psychotropic drugs). This category does not include deaths from drunk driving or other accidents, homicides, or other deaths that may be indirectly related to alcohol or drug use.

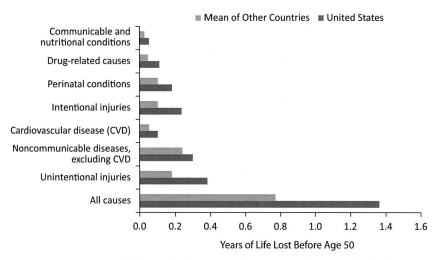

FIGURE 1-13 Years of life lost before age 50 due to specific causes of death among males in 17 peer countries, 2006-2008.

NOTES: Drug-related and other causes are not always mutually exclusive. The largest areas of overlap occur between drug-related causes and noncommunicable diseases and injuries. For example, drug-related digestive diseases (e.g., alcoholic liver disease) and drug-related neuropsychiatric disorders are also included in the noncommunicable disease category. Suicide and homicide by drugs fall under both drug-related causes and intentional injuries, and accidental drug overdoses fall under both drug-related causes and unintentional injuries. The drug-related causes category is included to illustrate the excess years of life lost from drug-related causes of death in the United States relative to other countries. Data for this figure come from the Human Mortality Database (downloaded July 18, 2011, last updated July 13, 2011); the WHO Mortality Database (downloaded July 18, 2011, last updated March 25, 2011); and Statistics Canada (downloaded July 22, 2011, data released February 23, 2010).

SOURCE: Adapted from Ho and Preston (2011, Figure 5).

The specific conditions responsible for the extra years of life lost in the United States are shown in Figures 1-15 and 1-16. Three causes—homicide, motor vehicle accidents, and nontransportation-related injuries—each contribute between 16 and 19 percent of the U.S. shortfall for males, and suicide contributes another 4 percent. Thus, deaths from injury of one form or another contribute the majority, 57 percent, of the excess mortality among American males under age 50. This is especially noteworthy given that mortality below age 50 accounts for the bulk of the U.S. male disadvantage in longevity. Noncommunicable diseases among men are also not trivial, accounting for 18 percent of the U.S. excess in years of life lost, with 8 percent coming from cardiovascular disease and 10 percent from all other noncommunicable diseases.

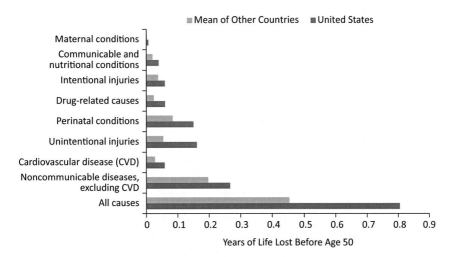

FIGURE 1-14 Years of life lost before age 50 due to specific causes of death among females in 17 peer countries, 2006-2008.
NOTES: Drug-related and other causes are not always mutually exclusive. The largest areas of overlap occur between drug-related causes and noncommunicable diseases and injuries. For example, drug-related digestive diseases (e.g., alcoholic liver disease) and drug-related neuropsychiatric disorders are also included in the noncommunicable disease category. Suicide and homicide by drugs fall under both drug-related causes and intentional injuries, and accidental drug overdoses fall under both drug-related causes and unintentional injuries. The drug-related causes category is included to illustrate the excess years of life lost from drug-related causes of death in the United States relative to other countries. Data for this figure come from the Human Mortality Database (downloaded July 18, 2011, last updated July 13, 2011); the WHO Mortality Database (downloaded July 18, 2011, last updated March 25, 2011); and Statistics Canada (downloaded July 22, 2011, data released February 23, 2010).
SOURCE: Adapted from Ho and Preston (2011, Figure 6).

The causes of the excess years of life lost by U.S. females are more diverse. Homicide, motor vehicle accidents, and nontransportation-related accidents also play an important role, contributing a total of 37 percent of the excess years of life lost in the United States. For intentional injuries, the excess comes from homicide alone, because U.S. women lose fewer years of life to suicide than women in other peer countries. Noncommunicable diseases are more important for women, contributing 29 percent of the U.S. excess in years of life lost compared with 18 percent for men. Of this 29 percent, 9 percent comes from cardiovascular disease and the remaining 20 percent from all other noncommunicable diseases. Perinatal conditions also affect females more than males: they contribute 19 percent to the U.S. excess in years of life lost among females and 13 percent among males (Ho and Preston, 2011).

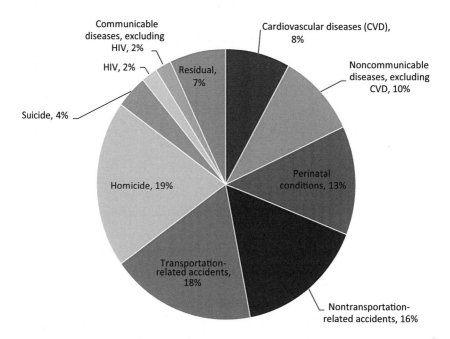

FIGURE 1-15 Contribution of cause-of-death categories to difference in years of life lost before age 50 between the United States and the mean of 16 peer countries, males, 2006-2008.

NOTES: Because of the overlap with other cause-of-death categories, drug-related causes are not included as a separate category in this figure, which shows mutually exclusive contributions of specific causes of death (see NOTES in Figure 1-13). Data for this figure come from the Human Mortality Database (downloaded July 18, 2011, last updated July 13, 2011); the WHO Mortality Database (downloaded July 18, 2011, last updated March 25, 2011); and Statistics Canada (downloaded July 22, 2011, data released February 23, 2010).

SOURCE: Adapted from Ho and Preston (2011, Figure 7).

INFLUENCE OF EARLY DEATHS ON LIFE EXPECTANCY AT BIRTH

It is useful to consider how much of the U.S. disadvantage in life expectancy at birth is attributable to deaths before or after age 50. That is: does the average American newborn have a shorter life expectancy than peers in other countries because of the diseases of old age or because of threats to health earlier in life? Answering this question involves a calculation that combines the actual years of life lost before age 50 (as shown in Figures 1-11 and 1-12) with the probability of surviving to age 50 (as shown in Figures 1-7 and 1-8). The latter is included because individuals who die before age 50 obviously forfeit all years of life beyond that age, but the forfeiture

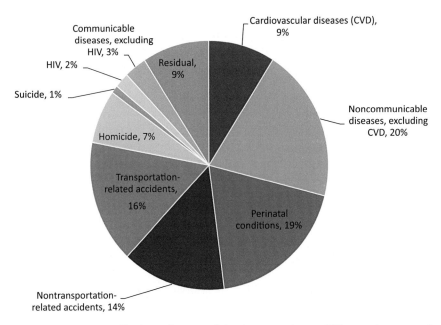

FIGURE 1-16 Contribution of cause-of-death categories to difference in years of life lost before age 50 between the United States and the mean of 16 peer countries, females, 2006-2008.
NOTES: Because of the overlap with other cause-of-death categories, drug-related causes are not included as a separate category in this figure, which shows mutually exclusive contributions of specific causes of death (see NOTES in Figure 1-14). Data for this figure come from the Human Mortality Database (downloaded July 18, 2011, last updated July 13, 2011); the WHO Mortality Database (downloaded July 18, 2011, last updated March 25, 2011); and Statistics Canada (downloaded July 22, 2011, data released February 23, 2010).
SOURCE: Adapted from Ho and Preston (2011, Figure 8).

is attributable to death before age 50. The calculation reveals that about two-thirds of the U.S. shortfall in life expectancy in 2007 relative to France and Japan—two very high-performing countries—were attributable to high U.S. mortality after age 50 (Ho and Preston, 2011).

A somewhat different picture emerges when the results are separated by sex and the comparison is made with the composite of the other 16 peer countries. Deaths after age 50 contributed to 58 percent of the U.S. shortfall in newborn life expectancy among females but only to 32 percent of the shortfall among newborn males (Ho and Preston, 2011). That is, most of the life expectancy difference among males is attributable to high U.S. mortality *before* age 50. This finding also implicates intentional and unintentional injuries, discussed above, which together account for a majority of the excess in U.S. male mortality before age 50.

CONCLUSIONS

Vital statistics paint a definitive and vivid portrait of the relative position of the United States in cross-national health comparisons. On nearly all indicators of mortality, survival, and life expectancy, the United States ranks at or near the bottom among high-income countries. Its poor performance pertains to both sexes, to all ages below 75, to white non-Hispanics as well as to the population as a whole, and to the most important causes of death.

Although the poor ranking of U.S. life expectancy at birth is partly attributable to relatively higher mortality rates after age 50, that is not the entire story: the United States compares unfavorably on mortality rates up to age 75. U.S. performance is particularly poor from birth to age 50, ranking near the bottom among peer countries. These findings and those from previous research, including the prior National Research Council (2011) report, suggest that throughout the life course people living in the United States fare worse than their peers, except at the oldest ages.

The data reported here highlight specific threats to health early in life, beginning in infancy: the United States has the lowest life expectancy at birth of the 17 peer countries the panel examined. Accidents (unintentional injuries), many of which involve adolescents and young adults, claim about 30 percent of the years lost before age 50, and suicides and violence also contribute to deaths in this age group. Noncommunicable diseases become more of a factor after age 30.

In summary, there is a growing mortality gap between the United States and comparable high-income countries. If the United States experienced the same rates of mortality due to unintentional injuries and noncommunicable diseases as do other peer countries, then almost two-thirds of the excess losses in years of life lost before age 50 would be eliminated (Palloni and Yonker, 2012). To add to the analysis in this chapter, which focuses on life expectancy, the next chapter examines how the United States compares with other countries in terms of quality of life, specifically, health status, the prevalence of disease, and the incidence of injuries.

2

Poorer Health Throughout Life

The previous chapter documented that life expectancy in the United States is shorter than in other high-income countries and identified the principal causes of death that account for this difference. However, health involves much more than staying alive. The goal of a healthy life is freedom from illness and injury: "Health is a state of complete physical, mental, and social well-being and not merely the absence of disease or infirmity" (World Health Organization, 1948). Thus, health is measured not only by mortality, but also morbidity and quality of life.

This chapter looks at cross-national comparisons of physical and psychological illnesses, injuries, and biological risk factors across the life course, from infancy to old age, with a special focus on childhood and adolescence. The panel chose to focus on youth for four reasons:

1. Previous research has concentrated on understanding the U.S. health disadvantage after age 50 (the age group for which most data are available), not on the question of whether there is a similar disadvantage for younger Americans. This is hardly an academic question. Young adults are among the most productive members of modern economies, and children and young adults will lead the next generation and determine the future strength and well-being of the nation. A health disadvantage early in life has profound implications for everyone.

2. The health problems facing children, adolescents, and young adults are often quite different from those affecting other age groups. As documented in Chapter 1, the U.S. mortality disadvantage is driven

by different causes at different ages: before age 50, these causes include conditions in infancy, nonintentional injuries, violence, and diseases of the heart and circulatory system at ages 30-50 (Palloni and Yonker, 2012).

3. The health and well-being of young people may help explain health disadvantages that emerge later in life. For example, approximately 80 percent of adults who are regular users of cigarettes or alcohol start these behaviors as adolescents (Kalaydjian et al., 2009; Oh et al., 2010; Viner, 2012). High rates of obesity among children and adolescents can track into adulthood and shed light on patterns of heart disease and diabetes among older people (Tirosh et al., 2011). Exposure to these risk factors, among others, and the pathophysiological damage they inflict generally occurs over many years before the disease processes they induce reach the point of producing clinical symptoms, yet they may play a vital role in understanding disease outcomes later in life.

4. The causal pathways that link early life health risks with subsequent diseases may involve many aspects of development that are seemingly unrelated. For example, health challenges in early life can disrupt intellectual and emotional development, impede physical growth, and limit education and employment opportunities (Fletcher and Richards, 2012), which in turn may set up a lifetime of socioeconomic disadvantage (see Chapter 6). Conditions of austerity may restrict access to health care (see Chapter 4) and limit opportunities to pursue healthy behaviors (see Chapter 5) or live in neighborhoods that promote good health (see Chapter 7).

This chapter compares, for each age group, a set of health indicators in the United States with those in a comparable group of other high-income or "peer" countries. The chapter focuses on illnesses and injuries, not unhealthy behaviors or other modifiable risk factors (health behaviors such as smoking and unhealthy diets are examined in Chapter 5). Unlike the previous chapter, which examined a relatively precise outcome, death, this chapter investigates a more general one, health. Health is more challenging to measure and quantify because it is multidimensional, and it reflects the culmination of a complex set of factors that include exposure to risk (or protective) factors and susceptibility (or resistance) to illness and injury. It is especially difficult to assess cross-national differences in health because of inconsistent data and metrics, conflicting findings, and in many cases, the absence of comparable data. This chapter critically examines the data that are currently available, and Chapter 9 outlines research priorities that will produce a stronger empirical basis for future work on this topic.

HEALTH ACROSS THE LIFE COURSE

The panel commissioned two special analyses of health under age 50. For the first analysis, Palloni and Yonker (2012) collated data for the same 17 peer countries discussed in Chapter 1 and examined the results across four age groups: infancy and early childhood (ages 0 to 4), late childhood and adolescence (ages 5-19), early adulthood (ages 20-34), and middle adulthood (ages 35-49).[1] The analysis drew on data from the World Health Organization (WHO), the Global Burden of Diseases, Injuries, and Risk Factors Study (GBD 2010 Study), UNICEF, and statistics that OECD has compiled from various national sources and household surveys conducted in its member countries. The second analysis, by Viner (2012), examined mortality rates by cause in the United States and 26 other high-income countries[2] for five age groups—1-4, 5-9, 10-14, 15-19, and 20-24—and then reanalyzed the data for the 17 peer countries of interest to the panel. These data were drawn from the WHO World Mortality Database.

The panel also commissioned a third analysis of data from the National Health and Nutrition Examination Survey in the United States (NHANES) and the Survey of Health, Ageing and Retirement in Europe (SHARE), in which Crimmins and Solé-Auró (2011) compared the cardiovascular risk profile of adults at age 50 in the United States and other high-income countries.

The data presented in this chapter are subject to important limitations. As already noted, there is no single measure of health status across the life course, and comparable data on many important health measures are not available across all countries. The available data often cover a very narrow (recent) time period and do not extend far enough back in time to capture health determinants that may explain current patterns. Ideally, longitudinal data would be used to examine changes in health conditions over time. Comparisons of national indicators also mask important within-country health disparities: a country's low ranking on any indicator may be a reflection of a health disadvantage in certain segments of the population or some geographic regions. Finally, despite any adjustments or harmonization, most data were not originally collected for the purposes of this report. The

---

[1]Age groups cannot always be defined consistently across data sources. Dividing the life course into age groups is largely a matter of convenience, although each phase of life presents different health challenges and opportunities, developmental milestones, and crucial transitions.

[2]These 27 countries are Australia, Austria, Canada, Chile, Denmark, Estonia, Finland, France, Greece, Hungary, Iceland, Ireland, Israel, Italy, Japan, Luxembourg, Mexico, the Netherlands, New Zealand, Norway, Poland, Portugal, Spain, Sweden, Switzerland, the United Kingdom, and the United States.

inferences we draw are dependent on the validity of the ranking methodology and the quality of the source data on which they are based.

Table 2-1 presents data on health outcomes from the analysis by Palloni and Yonker (2012). For each indicator, the table presents the range (and average) of peer country values, along with the U.S. value and ranking[3] relative to the other countries. The final column presents a composite ranking for the United States that standardizes and combines the distributions of each age-specific measure.[4]

CHILDREN AND ADOLESCENTS

Between 2001 and 2006, mortality rates for children aged 1-19 were higher in the United States than in all peer countries except Portugal. Most OECD countries had mortality rates between 15 and 25 deaths per 100,000 children; the U.S. rate was 32.7 (National Center for Health Statistics, 2011). For decades, the mortality disadvantage experienced by U.S. youth relative to their peers in other countries has followed a U-shaped pattern: the United States has exhibited higher infant mortality rates than the OECD average, comparable mortality rates at ages 1-9, and higher mortality rates at older ages, especially after age 15 (Viner, 2012).

Two recent international reports have also found that the overall well-being of U.S. children is quite low relative to other rich nations, at least according to the composite measures used in the studies. In a report by OECD (2009a), the United States ranked 24 out of 30 countries on children's well-being for health and safety.[5] A UNICEF (2007) report also found that children in the United States ranked poorly (21st out of 21 countries) based on selected indicators of child well-being.[6]

---

[3]Rankings are calculated from a standardized distribution that includes all OECD countries in the comparison set. The principal unit of analysis is the individual in the designated age group.

[4]The "composite rank" is the rank order of the averaged z-scores across indicators: see footnotes in Table 2-1 for details on how this was calculated.

[5]The OECD report examined six areas of child well-being: material well-being, housing and environment, educational well-being, health and safety, risk behaviors, and quality of the school day (OECD, 2009a).

[6]The UNICEF score was based on country statistics for low birth weight, infant mortality, breastfeeding, vaccinations, physical activity, mortality, and suicides (UNICEF, 2007).

TABLE 2-1 Health Indicators by Age Group, Range, and Rank of the United States Among 17 Peer Countries

Age Group (years)	Measure	Range (average)	U.S. Data	U.S. Rank[a] By Indicator	Composite for Age Group
0-4	Stillbirths[b]	2.0-3.9 (2.9)	3.0	11	17
	Low birth weight[c]	4.2-9.6 (6.6)	8.2	16	
	Perinatal mortality[d]	3.0-11.9 (5.7)	6.6	13	
	Neonatal mortality[e]	1.3-4.4 (2.7)	4.4	17	
	Infant mortality[f]	2.5-6.7 (3.9)	6.7	17	
	Days of life lost,[g] females	0.05-0.12 (0.07)	0.12	17	
	Days of life lost,[g] males	0.05-0.15 (0.09)	0.15	17	
5-19	Overweight,[h] girls	13.1-35.9 (22.6)	35.9	17	17
	Overweight,[h] boys	12.9-35.0 (22.4)	35.0	16	
	Dental caries[i]	0.7-2.1 (1.1)	1.2	11	
	Good health,[j] females	45.3-96.6 (86.4)	96.6	1	
	Good health,[j] males	49.8-97.1 (89.8)	97.1	1	
	Youth HIV,[k] females	0.0-0.5 (0.2)	0.3	14	
	Youth HIV,[k] males	0.1-0.7 (0.3)	0.7	16	
	Adolescent births[l]	5-41 (12)	41	17	
	Adolescent suicides[m]	2.4-12.2 (6.5)	7.7	11	
	Days of life lost,[g] females	0.10-0.21 (0.12)	0.21	17	
	Days of life lost,[g] males	0.16-0.37 (0.21)	0.37	17	
20-34	Good health,[j] females	41.1-93.4 (80.5)	92.6	2	17
	Good health,[j] males	42.8-93.9 (83.4)	93.9	1	
	Average BMI,[n] females aged 20-24	20.1-25.6 (22.9)	25.6	17	
	Average BMI,[n] males aged 20-24	22.2-26.3 (24.1)	26.3	17	
	Average BMI,[n] females aged 25-34	20.9-27.1 (24.0)	27.1	17	
	Average BMI,[n] males aged 25-34	23.3-27.8 (25.8)	27.8	17	
	Diabetes,[o] females	0.9-2.3 (1.6)	2.3	17	
	Diabetes,[o] males	1.6-4.1 (2.9)	4.1	17	
	Average FPG,[p] females	4.6-5.1 (4.9)	5.1	17	
	Average FPG,[p] males	4.9-5.4 (5.2)	5.4	17	

continued

TABLE 2-1 Continued

Age Group (years)	Measure	Range (average)	U.S. Data	U.S. Rank[a] By Indicator	Composite for Age Group
20-34	Average BP,[q] females	106.5-115.9 (111.3)	107.9	3	
	Average BP,[q] males	118.8-129.8 (124.0)	118.8	1	
	Average cholesterol,[r] females	4.4-4.8 (4.7)	4.8	11	
	Average cholesterol,[r] males	4.7-5.1 (4.9)	4.8	6	
	Maternal mortality[s]	2.0-14.9 (6.0)	13.7	16	
	Days of life lost,[g] females	0.20-0.43 (0.26)	0.43	17	
	Days of life lost,[g] males	0.34-0.77 (0.47)	0.77	17	
35-49	Average BMI,[n] females aged 35-44	21.9-28.8 (25.6)	28.8	17	17
	Average BMI,[n] males aged 35-44	24.0-29.0 (27.0)	29.0	17	17
	Average BMI,[n] females aged 45-54	22.7-29.9 (26.9)	29.9	17	
	Average BMI,[n] males aged 45-54	24.1-29.5 (27.7)	29.5	17	
	Diabetes,[o] females aged 35-44	1.7-4.7 (3.1)	4.5	16	
	Diabetes,[o] males aged 35-44	3.2-7.1 (5.1)	7.1	17	
	Diabetes,[o] females aged 45-54	3.6-9.7 (6.2)	9.3	16	
	Diabetes,[o] males aged 45-54	6.4-14.1 (9.9)	14.1	17	
	Average FPG,[p] females aged 35-44	4.8-5.4 (5.1)	5.4	16	
	Average FPG,[p] males aged 35-44	5.1-5.7 (5.4)	5.7	17	
	Average FPG,[p] females aged 45-54	5.0-5.7 (5.4)	5.7	16	
	Average FPG,[p] males aged 45-54	5.4-6.0 (5.7)	6.0	17	
	Average BP,[q] females aged 35-44	110.7-120.5 (115.9)	112.4	2	
	Average BP,[q] males aged 35-44	119.7-130.9 (126.1)	119.7	1	

TABLE 2-1 Continued

Age Group (years)	Measure	Range (average)	U.S. Data	U.S. Rank[a] By Indicator	Composite for Age Group
35-49	Average BP,[q] females aged 45-54	119.0-129.7 (124.1)	120.3	3	
	Average BP,[q] males aged 45-54	123.3-135.7 (131.1)	123.3	1	
	Average cholesterol,[r] females aged 35-44	4.7-5.2 (5.0)	5.0	7	
	Average cholesterol,[r] males aged 35-44	5.1-5.6 (5.3)	5.2	4	
	Average cholesterol,[r] females aged 45-54	5.2-5.7 (5.5)	5.4	3	
	Average cholesterol,[r] males aged 45-54	5.3-5.8 (5.5)	5.3	1	
	Days of life lost,[g] females	0.01-0.04 (0.02)	0.04	17	
	Days of life lost,[g] males	0.37-0.80 (0.47)	0.80	17	

NOTE: Data are for the most recent year available.

[a]Rankings are calculated from a standardized distribution that includes all 16 OECD countries in the comparison set. Rankings are from best (1) to worst (17), and all comparisons are for the 17 peer countries listed in the text. The "composite rank" is the rank order of the averaged z-scores across indicators. It was calculated in two stages: (1) the value of each indicator was converted into a z-score based on observed means and standard deviations and (2) the z-scores were averaged across indicators and then rank ordered. For consistency, some z-scores were reverse coded to preserve the meaning of high and low ranks across all indicators. Data were not always available to rank all 17 countries.

[b]Stillbirths are rate per 1,000 total births (2009). Stillbirth rates and numbers use the WHO definition of birth weight of at least 1,000 grams or a gestational age of at least 28 weeks (third-trimester stillbirth). Data from WHO at http://www.who.int/reproductivehealth/topics/maternal_perinatal/stillbirth/en/.

[c]Low birth weight is the number of live births weighing less than 2,500 grams as a percentage of the total number of live births. Values (if present) averaged over 2005-2009. Data from OECD at http://stats.oecd.org/index.aspx?DataSetCode=HEALTH_STAT.

[d]Perinatal mortality is the ratio of deaths of children within 1 week of birth (early neonatal deaths) plus fetal deaths with a minimum gestation period of 28 weeks or a minimum fetal weight of 1,000 grams, expressed per 1,000 births. Values (if present) averaged over 2005-2009. Data from OECD at OECD_Health_MaternalAndInfantMortality_2653d1a3-bb42-4768-9cfb-d405eb5050a1.xls.

[e]Neonatal mortality is the number of deaths of children under 28 days of age in a given year, expressed per 1,000 live births. Values (if present) averaged over 2005-2009. Data from OECD_Health_MaternalAndInfantMortality_2653d1a3-bb42-4768-9cfb-d405eb5050a1.xls.

[f]Infant mortality is the number of deaths of children under 1 year of age that occurred in a given year, expressed per 1,000 live births. Values (if present) averaged over 2005-2009.

continued

TABLE 2-1 Continued

Data from OECD at OECD_Health_MaternalAndInfantMortality_2653d1a3-bb42-4768-9cfb-d405eb5050a1.xls.

[g]Days of life lost is the number of potential days of life lost due to mortality in the designated age range. Calculated for 2009 from life tables. Data from WHO at http://www.who.int/whosis/whostat/2011/en/.

[h]Overweight is defined variously by country. Source years also vary by country. Data from International Association for the Study of Obesity at http://www.iaso.org/resources/world-map-obesity/?map=children.

[i]Dental caries is the weighted average of the number of decayed, missing, or filled teeth (DMFT) among 12-year-olds (2004). Data from WHO at http://apps.who.int/ghodata/.

[j]Good health is the percentage of the population who report their health as "good" or "better." Values (if present) averaged over 2005-2009. Data available for ages 15-24 are provided here for ages 5-19, and data available for ages 25-44 are provided here for ages 20-34. Data from OECD at http://stats.oecd.org/index.aspx?DataSetCode=HEALTH_STAT.

[k]Youth HIV is the percentage of the population infected with HIV (2007). Data available for ages 15-24 are provided here for ages 5-19. Data from United Nations Human Development Report 2010 at http://hdr.undp.org/en/statistics/.

[l]Adolescent births is the number of births to women ages 15-19 per 1,000 women (2010). Data from United Nations Human Development Report 2010 at http://hdrstats.undp.org/en/indicators/36806.html.

[m]Adolescent suicides is suicides per 100,000 people ages 15-19. Three-year averages of data from most recent years available. Data from OECD at http://stats.oecd.org/index.aspx?DataSetCode=HEALTH_STAT.

[n]Average BMI is average body mass index (kg/m^2) (2008). Obesity is a BMI above 30 kg/m^2. Data available for ages 20-24 and ages 25-34 are provided here for ages 20-34 and data for ages 35-44 and ages 45-54 are provided for ages 35-49. Data from Global Burden of Metabolic Risk Factors of Chronic Diseases Collaborating Group at http://www1.imperial.ac.uk/publichealth/departments/ebs/projects/eresh/majidezzati/healthmetrics/metabolicriskfactors/.

[o]Diabetes is the percentage of the population diagnosed with diabetes (2008). Data available for ages 25-34 are provided here for ages 20-34, and data for ages 35-44 and ages 45-54 are provided here for ages 35-49. Data from Global Burden of Metabolic Risk Factors of Chronic Diseases Collaborating Group at http://www1.imperial.ac.uk/publichealth/departments/ebs/projects/eresh/majidezzati/healthmetrics/metabolicriskfactors/.

[p]Average FPG is average fasting plasma glucose (mmol/L) (2008). Data available for ages 20-24 and ages 25-34 are provided here for ages 20-34, and data for ages 35-44 and ages 45-54 are provided for ages 35-49. Data from Global Burden of Metabolic Risk Factors of Chronic Diseases Collaborating Group at http://www1.imperial.ac.uk/publichealth/departments/ebs/projects/eresh/majidezzati/healthmetrics/metabolicriskfactors/.

[q]Average BP is average systolic blood pressure (mm Hg) (2008). Data available for ages 20-24 and ages 25-34 are provided here for ages 20-34, and data for ages 35-44 and ages 45-54 are provided here for ages 35-49. Data from Global Burden of Metabolic Risk Factors of Chronic Diseases Collaborating Group at http://www1.imperial.ac.uk/publichealth/departments/ebs/projects/eresh/majidezzati/healthmetrics/metabolicriskfactors/.

[r]Average cholesterol is average total serum cholesterol (mmol/L) (2008). Data available for ages 20-24 and ages 25-34 are provided here for ages 20-34, and data for ages 35-44 and ages 45-54 are provided here for ages 35-49. Data from Global Burden of Metabolic Risk Factors of Chronic Diseases Collaborating Group at http://www1.imperial.ac.uk/publichealth/departments/ebs/projects/eresh/majidezzati/healthmetrics/metabolicriskfactors/.

[s]Maternal mortality is the ratio of maternal deaths from all causes per 100 000 live births. Values (if present) averaged over 2005-2009. Data from OECD at http://stats.oecd.org/index.aspx?DataSetCode=HEALTH_STAT.

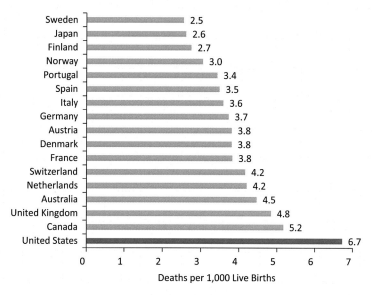

FIGURE 2-1 Infant mortality rates in 17 peer countries, 2005-2009.
NOTE: Rates averaged over 2005-2009.
SOURCE: Data from OECD (2012c).

Infancy and Early Childhood: Ages 0-4

Birth Outcomes

Infant mortality rates in the United States have been stagnant in the past decade and are now higher than in most high-income countries (Congressional Budget Office, 1992; MacDorman and Mathews, 2009; OECD, 2011b). From 2005-2009, the United States had the highest infant mortality rate (6.7 per 1,000 live births) of the 17 peer countries and the 31st highest in the OECD (OECD, 2011b) (see Figure 2-1).[7] The U.S. ranking on birth outcomes—including stillbirths, infant mortality, and low birth weight—can be seen in Table 2-1. Across these indicators, the United States has the lowest composite rank. The U.S. rate for stillbirths and perinatal mortality

---

[7]Among 40 OECD countries in 2009, the only countries with a higher infant mortality rate than the United States were Brazil, Chile, China, India, Indonesia, Mexico, the Russian Federation, South Africa, and Turkey (OECD, 2011b). Cross-national variation in infant mortality rates is partly affected by differences in how countries register preterm births. In the United States, Canada, and Nordic countries, preterm neonates who often have low probabilities of survival are registered as live births, thereby increasing the mortality rate relative to countries that do not include preterm neonates among live births.

ranks among the highest in the peer countries, but the most glaring difference is low birth weight and neonatal and infant mortality.

The United States ranks poorly on low birth weight, prematurity, and maternal health. Figure 2-2 shows that the proportion of low-birth-weight babies in the United States (8.2 percent for 2005-2009) is the second highest among the peer countries. Among the 17 peer countries examined in Chapter 1, the United States had the 14th highest rate of preterm deaths before age 5 in 2008 (World Health Organization, 2010). Figure 2-3 shows the results of a recent analysis of data from 184 countries (Blencowe et al., 2012), which found that the rate of preterm births in the United States (12 percent) was comparable to that of sub-Saharan Africa. Two important antecedents of infant and child health—adolescent pregnancies and maternal health—also show a clear U.S. disadvantage (see Figure 2-2). Taken together, these measures indicate that U.S. children often enter life under unfavorable health conditions.

The high rate of adverse birth outcomes in the United States does not appear to be a statistical artifact, such as a difference in coding practices for very small infants who die soon after birth (MacDorman and Mathews,

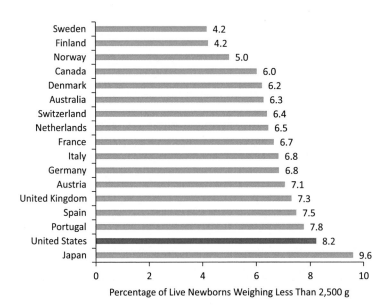

FIGURE 2-2 Low birth weight in 17 peer countries, 2005-2009.
NOTE: Values (if present) averaged over 2005-2009.
SOURCE: Data from OECD (2012l), OECD.StatExtracts: Health Status (database).

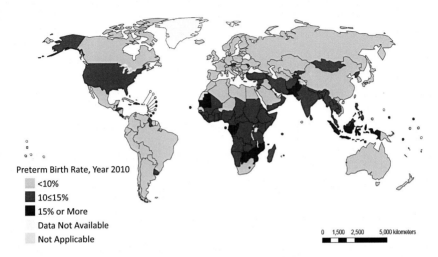

Preterm Birth Rate, Year 2010
- <10%
- 10≤15%
- 15% or More
- Data Not Available
- Not Applicable

0   1,500 2,500   5,000 kilometers

FIGURE 2-3 Global prevalence of preterm births, 2010.
SOURCE: Blencowe et al. (2012, Figure 3).

2009). Indeed, country rankings remained identical even when Palloni and Yonker (2012) recalculated the rates to exclude preterm births (less than 22 weeks of gestation).

Infant mortality and low birth weight are both markers of unhealthful *in utero* and postnatal conditions, findings that are also supported by related indicators. From birth through age 4, U.S. children lose more years of life than children in the other 16 peer countries (Palloni and Yonker, 2012). Infants who die during the first year of life, particularly preterm infants, are those at the lower tail of the distribution of newborns by health status, and low-birth-weight babies are at the extreme end of this tail. Both infant mortality and low birth weight are, in turn, influenced by maternal characteristics, including health-related behaviors (e.g., smoking, drinking, diet, and breastfeeding practices), marital and family status, maternal education, access to health care, and household and family conditions. The U.S. excess in infant mortality and the prevalence of low-birth-weight babies probably reflect both individual and societal contextual factors. For example, the United States fares poorly with respect to adolescent pregnancy and child poverty (see Chapter 6), which are, respectively, proximate and distal determinants of low birth weight and infant mortality.

Data are available to track trends over time in U.S. infant mortality and birth weight relative to other countries. By the early 1960s, the average infant mortality rate among peer countries had dropped to the U.S. rate, and by the 1970s the United States began to develop a disadvantage in infant mortality (Viner, 2012) (see Figure 2-4). Although U.S. infant

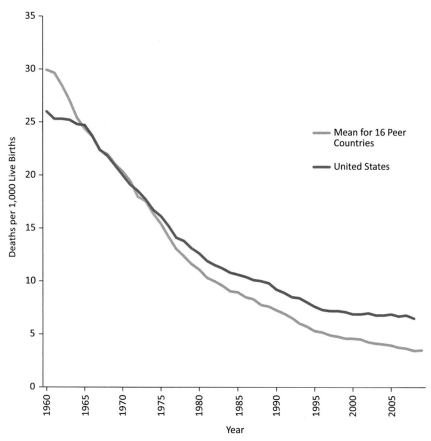

FIGURE 2-4 Infant mortality rates in the United States and average of 16 peer countries, 1960-2009.
NOTE: The average is calculated for the 16 peer countries examined in Chapter 1.
SOURCE: Viner (2012, supplemental analysis).

mortality declined by 20 percent between 1990 and 2010, other high-income countries experienced much steeper declines and halved their infant mortality rates over those two decades (Palloni and Yonker, 2012). Mortality rates after infancy followed a similar pattern: as of 1955, mortality rates at ages 1-5 were much lower in the United States than the OECD mean, but the latter fell dramatically and since 1980 have mirrored the United States (Viner, 2012).

Although birth outcomes are much worse for particular groups in the United States, such as black women (David and Collins, 1997; National Center for Health Statistics, 2012), the generally poor birth outcomes in the United States cannot be explained by the racial or ethnic diversity of the U.S. population. The U.S. infant mortality rate for non-Hispanic whites is also higher than the infant mortality rate of other countries (Palloni and Yonker, 2012).[8] Nor is this problem limited to infants from poor and disadvantaged families. The U.S. infant mortality rate among mothers with 16 or more years of education is still higher than the infant mortality rate in most high-income countries (Mathews and MacDorman, 2007).

Other Pediatric Health Outcomes

Children who survive infancy face other disadvantages in the United States relative to children in other high-income countries. For example, one study reported that the prevalence of asthma at ages 0-3 was higher in the United States than in England (Martinson et al., 2011a). The probability of children dying before age 5 in the United States is approximately 8 per 1,000, the highest rate of the 17 peer countries examined in Chapter 1 (World Health Organization, 2010). Injuries (primarily motor vehicle injuries, drowning, and fires and burns) are the leading cause of death among U.S. children in this age group (Bernard et al., 2007). In 2004, 11 percent of U.S. deaths under age 5 were from injuries, the largest proportion of the 17 peer countries except Japan (World Health Organization, 2010). In the United States, violence has been a long-standing cause of injury deaths in this age group, with homicide being the third leading cause of death for children aged 1-4 (National Center for Health Statistics, 2012). In the United States, infants are nearly four times more likely to be killed (usually by a parent) than children ages 1-4 (Karch et al., 2011). In 2006, the United States had the highest rate of child deaths due to negligence, maltreatment, or physical assault among the 17 peer countries (and among other OECD countries, including emerging economies) (OECD, 2012k). The rate of violent deaths in the United States among boys aged 1-4 has exceeded the OECD average since the late 1960s, and the same has been true for unintentional injury deaths in boys and girls (Viner, 2012).

The added health challenges that young children in the United States face may be especially significant because of the important developmental milestones that occur at this age. The period between ages 1 and 4 is not only a time of dramatic physical growth and development, but also a critical phase during which language and mathematical skills are first established, and children's experiences at these ages shape personal-

---

[8]The rate of low-birth-weight babies among non-Hispanic whites has been about average. Comparing non-Hispanic whites to the aggregate population rates of other countries is of limited validity without excluding similar subgroups in other countries that also have significant disadvantaged minority populations. Furthermore, excluding certain ethnic groups in the United States (e.g., Hispanics) removes a population with favorable birth outcomes due to the "Hispanic paradox" (see Chapter 6).

ity features and social connections—all of which can later affect health outcomes. Whether a child's developmental trajectory follows a healthy, risky, or delayed course between infancy and age 5 appears to depend both on the degree of exposure to health risks and on access to counterbalancing protective factors that are of crucial importance during various phases of development. For example, parental education, literacy, and emotional health may be especially important in the first 12 months of a child's life, whereas parenting behaviors (e.g., reading and discipline) and other inputs (e.g., health services and preschool) may be more important for toddlers. (See Chapter 6 for a discussion of the role of the family and household environment in explaining the U.S. health disadvantage.)

Childhood and Adolescence: Ages 5-19

Preadolescence and adolescence cover a vulnerable period during which positive experiences and setbacks can affect developmental accomplishments, such as consolidating self-image, discipline, and the capacity for abstract thinking and future planning. Developmental theorists have long identified adolescence as a key period of dramatic central nervous system, pubertal, and social development, second only to early childhood in the rate and breadth of development (Viner, 2012).

In addition, young people are highly susceptible to socialization processes and behaviors that affect health, such as sensation seeking and experimentation with alcohol, drugs, sex, and smoking. "Adolescence contributes significantly to adult burden of disease through initiation of major disease risk factors in adolescence, each of which tracks into adult life" (Viner, 2012, p. 54). This stage of life includes critical transition periods that modify childhood trajectories relating to health and well-being (Viner, 2012; World Bank, 2007). This stage sets the foundations of adult performance and influence educational attainment, future employment status, and career opportunities—factors that strongly predict health status and life expectancy (see Chapter 6).

The available evidence indicates that the U.S. health disadvantage is clearly evident at this age too, especially after age 15. Due mainly to the high prevalence of overweight and obesity, HIV infection, and adolescent pregnancy, the United States has the lowest composite rank for ages 5-19 (see Table 2-1). Although the United States and OECD countries have equivalent child mortality rates after infancy, the United States develops a clear mortality disadvantage in the adolescent and young adult years (Viner, 2012). Among adolescents aged 15-19[9] in 2005, the United States had the

---

[9]No all-cause mortality disadvantage was observed for U.S. adolescents aged 10-14 (Viner, 2012).

highest all-cause mortality rate among peer countries (Viner, 2012), a pattern that is decades old: high mortality rates at ages 15-24 (relative to the OECD mean) have existed among U.S. males since the 1950s and among females since the 1970s (Viner, 2012). A U.S. health disadvantage at this age has been documented in particular for obesity, chronic illness, adolescent pregnancy, sexually transmitted infections, mental illness, and injuries.

Obesity

The increasing prevalence of childhood overweight and obesity is a global phenomenon, but obesity is especially prevalent among children in the United States (Institute of Medicine, 2011c; Ogden et al., 2012a). As of 2005, the prevalence of obesity among adolescents aged 12-17 was more than twice the OECD mean (Viner, 2012). By 2011, 35.9 percent of girls and 35.0 percent of boys aged 5-17 were overweight or obese in the United States, the highest rate among the 17 peer countries (OECD, 2011b) (see Figure 2-5).[10]

As with infant mortality, the problem is not restricted to children of color or disadvantage. The prevalence of obesity among U.S. children aged 5-13 who are non-Hispanic whites is lower than that of other U.S. children, but still higher than the OECD average for ages 5-19 (a comparison group likely to experience higher obesity rates simply because it is older than the U.S. benchmark population). Results are similar among U.S. children born to mothers with more than a college education: their obesity rate also exceeds the OECD average (Han, 2011).

Higher rates of childhood obesity may have both immediate and extended consequences. There is some speculation and evidence that obesity may affect educational achievement and labor market success, thereby influencing health (see Chapter 6) (Crosnoe, 2007; Han, 2011; Sabia, 2007; Sarlio-Lahteenkorva et al., 2004; von Hinke Kessler Scholder et al., 2012). Moreover, the physiological metabolic effects of childhood obesity may contribute to diabetes, heart disease, and other obesity-related conditions that claim excess years of life in the United States after age 30 (see Chapter 1). The evidence linking childhood obesity to adult health outcomes is far from perfect, but it suggests that the acceleration of childhood obesity may precipitate excess mortality in the United States when today's children reach older ages. Obesity among women during their child-bearing years may also compromise birth outcomes (Aliyu et al., 2010; Cnattingius et al., 1998; McDonald et al., 2010; Siega-Riz et al., 2006).

---

[10]The prevalence of childhood obesity in the United States is the highest of 34 OECD countries except Greece (OECD, 2011b).

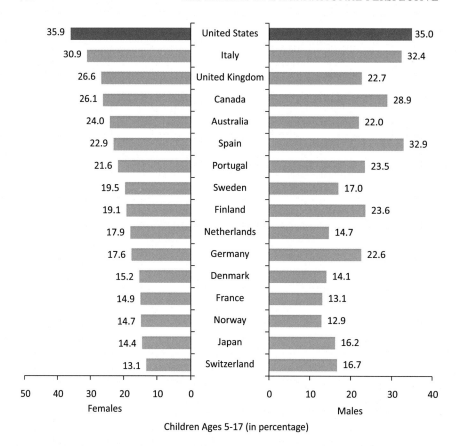

FIGURE 2-5 Prevalence of overweight (including obese) children in 17 peer countries, latest available estimates.
NOTES: Definitions of overweight and obese vary among countries. Prevalence rates are for the most current year available.
SOURCE: OECD (2011b, Figure 2.4.1).

Chronic Illnesses

The rise in childhood diabetes has mirrored the increase in obesity. Among the 17 peer countries in 2010, the United States had the fifth highest prevalence of diabetes among children ages 0-14, led only by Finland, Norway, Sweden, and the United Kingdom (OECD, 2011b). U.S. children also appear to be experiencing higher rates of other chronic illnesses. For example, one study found that the prevalence of asthma in children and adolescents was significantly higher in the United States than in England (Martinson et al., 2011a). Oral and dental health is important at all ages

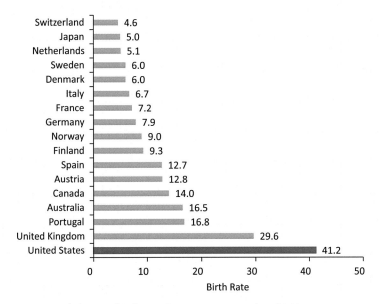

FIGURE 2-6 Adolescent birth rate in 17 peer countries, 2010.
NOTE: Adolescent birth rate is births per 1,000 girls aged 15-19.
SOURCE: Adapted from United Nations Development Programme (2011).

and especially among children (Institute of Medicine, 2011a): as of 2006, the average number of decayed, missing, or filled teeth in 12-year-olds was higher in the United States than in 11 peer countries (and three nonpeer countries) (OECD, 2009c).

Adolescent Pregnancy

Adolescent pregnancy is not strictly a health outcome, the focus of this chapter, but it often has adverse social implications for adolescent mothers, their children, and the family of the adolescent, implications that can eclipse any health risks associated with such pregnancies (see Chapter 6). As shown in Figure 2-6, the United States has the highest rate of adolescent pregnancy among peer countries (Palloni and Yonker, 2012; Viner, 2012), a trend that is consistent with its high prevalence of low-birth-weight babies and infant mortality.[11] The U.S. adolescent pregnancy rate in 2010 was 41 per 1,000 girls aged 15-19, nearly 3.5 times the average of the comparison countries (United Nations Development Programme, 2011). The high adolescent pregnancy rate in the United States is not a new problem. Surveys

---

[11]Early-age childbirth is a leading risk factor for low birth weight and infant mortality.

in the 1990s showed that U.S. adolescents aged 15-19 had a higher rate of pregnancies, births, and abortions than did their peers in four comparison countries: Canada, France, Sweden, and the United Kingdom (Darroch et al., 2001).

Sexually Transmitted Infections

Panchaud et al. (2000) compiled statistics from 16 developed countries that included a number of high-income countries. Among adolescents age 15-19, the U.S. prevalence of infections with syphilis (6.4 per 100,000), gonorrhea (571.8 per 100,000),[12] and chlamydia (1,131.6 per 100,000) was higher in the United States than in any other high-income country that provided comparison data.[13] A subsequent study based on 1996 data[14] from five countries also found that rates of infection with the same three organisms were markedly higher among U.S. adolescents than their peers in Canada, France, Sweden, and the United Kingdom (Darroch et al., 2001). The high prevalence of HIV infection in the United States (see Chapter 1) begins in adolescence. The prevalence of HIV infection at ages 15-24 is higher in the United States than in other high-income countries (Viner, 2012).

Mental Illness

Mental illness affects people of all ages and is discussed more fully toward the end of this chapter. However, we note it here because many mental illnesses first appear in late adolescence and young adulthood (Merikangas et al., 2010). One study reported that 75 percent of lifetime mental health problems start by age 24: more specifically, ages 12-24 is the usual age of onset for depression, anxiety disorders, psychoses, and eating and personality disorders (Kessler et al., 2005, 2007; Patel et al., 2007). Unfortunately, few comparative data exist to compare the mental health of adolescents in the United States with those elsewhere. A study that administered a common questionnaire in the United States and Europe found that

---

[12]In all comparison countries, the gonorrhea infection rate among adolescents (ages 15-19) ranged between 0.6 and 76.9 per 100,000. The Russian Federation was the only country with a higher rate of gonorrhea (596.5 per 100,000) than the United States (Panchaud et al., 2000).

[13]An important caveat noted by the authors is the deficiencies and inconsistencies in countries' reporting systems. Data on sexually transmitted infections are lacking in many countries, and even countries with somewhat reliable reporting systems are thought to underestimate the true incidence (Panchaud et al., 2000).

[14]The panel could not find more current data to make cross-national comparisons of sexually transmitted infection rates.

the prevalence of psychological disorders among adolescents was higher in the United States and the United Kingdom than in 11 other European countries (Viner, 2012). Another study suggested that young people aged 0-19 used more psychotropic medication[15] in the United States than in the Netherlands and Germany (Zito et al., 2008), but medication use may be a poor proxy for disease prevalence. (Alcohol and other drug use are discussed in Chapter 5.)

Injuries

The high all-cause mortality rate among U.S. adolescents results principally from injuries, the leading cause of death at these ages (National Center for Health Statistics, 2012). Among adolescents aged 15-19 in 2005, the United States had the highest injury mortality rate of the 17 peer countries (Viner, 2012). Many of these deaths occur in motor vehicle accidents: drivers aged 16-20 are more than twice as likely to be involved in fatal crashes (per licensed driver in the age group) than drivers over age 35 (Transportation Research Board, 2011). For decades, young people in the United States have been more likely to die in motor vehicle accidents than their peers in other countries. Since the 1950s, the mortality rate from transportation-related injuries at ages 15-24 has been higher in the United States than the mean for the 17 peer countries,[16] with more dramatic differences among males than females (see Figures 2-7a and 2-7b). Since the 1960s, the United States has also had a higher death rate from nontransportation-related injuries among children aged 5-9 and especially for males aged 10-19 (Viner, 2012).

Another long-standing trend is the high rate of deaths from youth violence in the United States. Homicide is the second leading cause of death among U.S. adolescents and young adults aged 15-24 (National Center for Health Statistics, 2012), and approximately 80 percent of these involve firearms (Bernard et al., 2007). Since the 1950s, a very large disparity has persisted in the rate of violent deaths among males aged 15-24 in the United States and their counterparts in peer countries (see Figures 2-8a and 2-8b). U.S. males aged 15-19 are five times more likely to die from violence than those in other OECD countries.[17] Violent deaths in this age group increased dramatically in the United States during the 1990s, but not elsewhere in

---

[15]Psychotropic drugs included antidepressants, antipsychotics, alpha-agonists, anxiolytics, hypnotics, lithium, antiparkinsonian agents, anticonvulsant-mood stabilizers, and stimulants.

[16]There was some equalization as OECD countries "caught up" in the 1970s, but the OECD subsequently experienced a steeper decline in the 1990s and the gap with the United States widened again.

[17]Although the United States is actively engaged in military conflicts, the risk of death from violence among young people in the United States far exceeds the number of deaths that occur in the military.

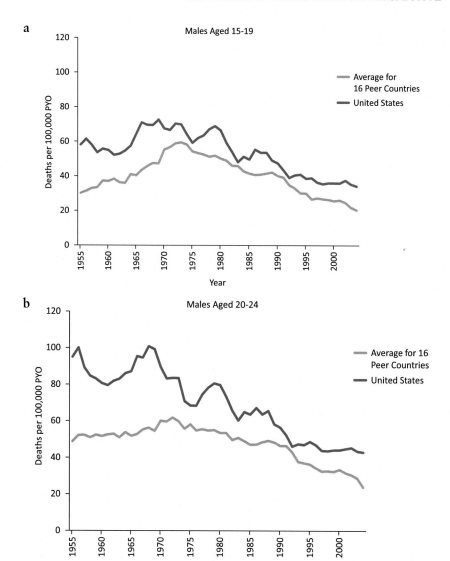

FIGURE 2-7 Transportation-related mortality among adolescent and young adult males in the United States and average of 16 peer countries, 1955-2004.
NOTES: PYO = person years of observation. The average is calculated for 16 peer countries (see Chapter 1).
SOURCE: Viner (2012, supplemental analysis).

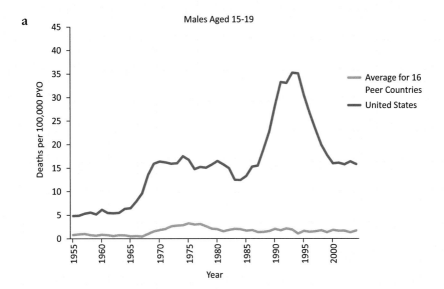

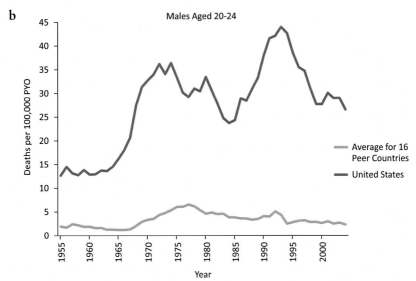

FIGURE 2-8 Violent mortality among adolescent and young adult males in the United States and average of 16 peer countries, 1955-2004.

NOTES: PYO = person years of observation. The average is calculated for 16 peer countries (see Chapter 1).

SOURCE: Viner (2012, supplemental analysis).

the peer countries, thereby widening the adolescent violence gap. Violent deaths in the United States declined thereafter but remained considerably higher than the peer country mean (see Figures 2-8a and 2-8b). Violent deaths among adolescent females have also been more common in the United States than in other countries, but the size of the disparity has been smaller (Viner, 2012).

Suicide

Suicide is the third leading cause of death in the United States for ages 10-24 (National Center for Health Statistics, 2012). The suicide rate among U.S. adolescents is currently close to the median among peer countries, but it is worth noting that suicide (like any cause of death at this age) claims many years of life lost before age 50 (see Chapter 1). From the 1960s to the 1990s, suicides were more common among U.S. males ages 15-24 than elsewhere in the OECD (Viner, 2012).[18]

ADULTS

Young Adulthood: Ages 20-34

The period between ages 20 and 34 has direct bearing on employment and family formation and development. The roots of unhealthy behaviors are often already entrenched by this age. Although young adults in the United States have average cholesterol and self-reported blood pressure levels and above average self-reported health (see caveats about self-reports in Chapter 1), they have higher rates of obesity and diabetes (Palloni and Yonker, 2012). High rates of overweight and obesity in the United States can be found in every age group above age 25 (Crimmins and Solé-Auró, 2012). In 2008, U.S. adults age 25 and older had the highest average body mass index of the 17 peer countries (see Figures 2-9a-c). Figures 2-10a-c show that U.S. adults also had the highest prevalence of diabetes. Among the same countries, the highest average fasting plasma glucose levels were in the United States and Spain. As noted above, indicators of maternal health in the United States are also less favorable than in other affluent countries. For example, for 2005-2009, the U.S. maternal mortality ratio was 13.7 per 100,000, the second highest of the 17 peer countries.

Unintentional injuries remain the leading cause of death in the United States for ages 25-34 (National Center for Health Statistics, 2012). Since

---

[18]See Chapter 5 for a discussion of the role of firearms, which are used in 52 percent of U.S. suicides (Karch et al., 2011) and increase the lethality of suicide attempts (Miller et al., 2011a).

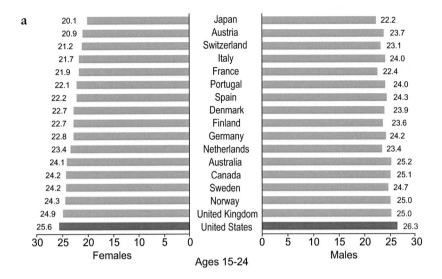

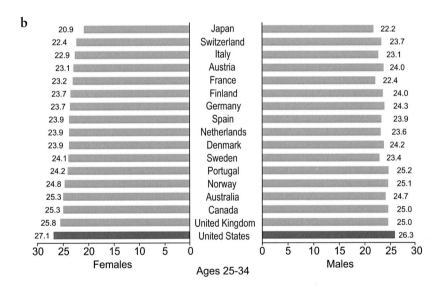

FIGURE 2-9 Average body mass index (BMI), by age and sex, in 17 peer countries, 2008.
SOURCE: Data from the Global Burden of Metabolic Risk Factors of Chronic Diseases Collaborating Group (2012), BMI by country.

continued

c

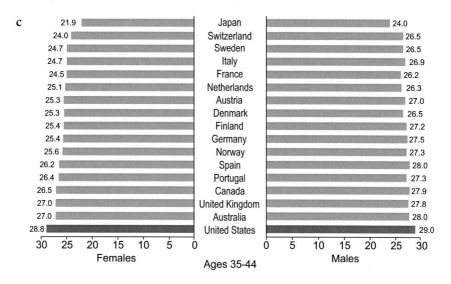

FIGURE 2-9 Continued.

the 1950s, deaths at these ages from transportation-related and non-transportation-related injuries have been markedly higher in the United States than in other OECD countries (Viner, 2012). In the United States, the second and third leading causes of death in this age group are, respectively, suicide and homicide (Karch et al., 2011). For U.S. males aged 20-24, the risk of dying from violence is nearly seven times higher than in other OECD countries, a trend that dates back to the 1950s.

Middle Adulthood: Ages 35-49

The symptoms of chronic illnesses often first appear between ages 35 and 49. At this age, cardiovascular diseases claim the largest fraction of years lost to disability (Palloni and Yonker, 2012). Continuing the pattern observed at younger ages, Americans who reach middle adulthood rank poorly on measures of obesity and diabetes (see Figures 2-9a-c and 2-10a-c) and have relatively high fasting plasma glucose levels. However, unintentional injuries remain the leading cause of death at this age (National Center for Health Statistics, 2012).

Health at Age 50

On average, Americans reach age 50 in significantly poorer health than their peers in other high-income countries. The panel commissioned

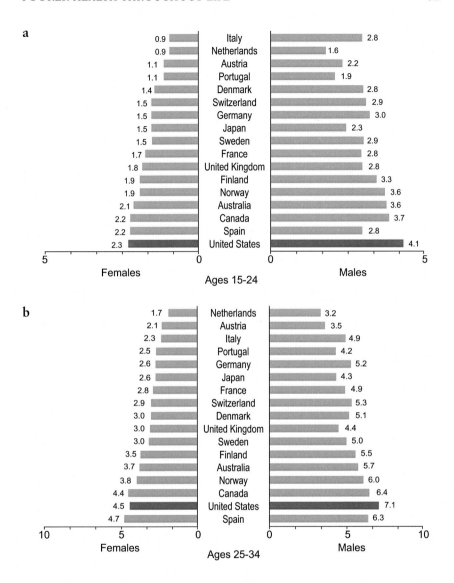

FIGURE 2-10 Self-reported prevalence of diabetes, by age and sex, in 17 peer countries, 2008.
SOURCE: Data from the Global Burden of Metabolic Risk Factors of Chronic Diseases Collaborating Group (2012), diabetes by country.

continued

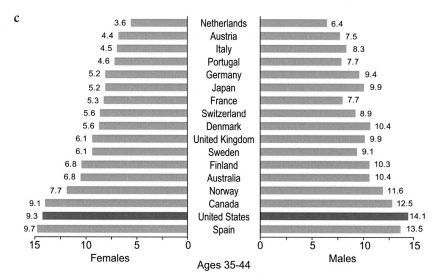

FIGURE 2-10 Continued.

an analysis of cardiovascular risk factors among adults aged 50-54 in the United States and 10 European countries. This age group was selected because it is the beginning of the age range when cardiovascular mortality, the leading cause of death, begins to be important, but mortality has not yet affected the population through selection.

Data obtained from NHANES and SHARE indicate that U.S. adults aged 50-54 report a higher prevalence of heart disease, stroke, diabetes, hypertension, and obesity than do their counterparts in 10 European countries. Crimmins and Solé-Auró (2012) calculated a cardiovascular risk score by adapting the method of Gaziano et al. (2008), which includes the above-mentioned risk factors along with smoking. The proportion of U.S. adults aged 50-54 with at least a 20 percent risk of having a fatal or nonfatal cardiac event within 5 years was higher in the United States than in the European countries (see Table 2-2). The percentage of U.S. adults at high risk exceeded the European percentage by 34 percent for men and by more than 159 percent for women (Crimmins and Solé-Auró, 2012). That is, Americans reach age 50 with significantly higher levels of cardiovascular risk than their European counterparts.

Maturity: Age 50 and Older

As noted in Chapter 1, a previous National Research Council (2011) report documented that life expectancy at age 50 is lower in the United

TABLE 2-2 Distribution of Cardiovascular Risk for Adults Aged 50-54 Among 11 High-Income Countries

Males	Country	Level of Risk			
		Very High	High	Moderate	Low
	Austria	0.70	11.37	38.55	49.38
	Belgium	0.25	13.35	42.06	44.33
	Denmark	2.41	10.24	39.76	47.59
	France	0.00	12.48	34.19	53.32
	Germany	0.52	14.30	37.73	47.45
	Greece	0.43	10.40	50.14	39.04
	Italy	0.00	12.28	41.56	46.16
	Netherlands	1.89	8.43	42.23	47.46
	Spain	2.37	15.58	42.84	39.21
	Sweden	0.33	7.88	29.46	62.33
	Pooled values (10 countries)	**0.68**	**12.93**	**38.98**	**47.40**
	United States	1.86	16.37	38.68	43.09
Females					
	Austria	0.00	3.90	26.42	69.68
	Belgium	0.00	5.65	22.86	71.49
	Denmark	0.00	5.70	26.58	67.72
	France	0.00	5.10	17.58	77.32
	Germany	0.29	3.60	24.79	71.32
	Greece	0.00	4.66	29.29	66.06
	Italy	0.60	4.08	22.64	72.68
	Netherlands	1.85	4.34	24.83	68.98
	Spain	0.00	3.29	19.52	77.19
	Sweden	1.08	6.65	22.71	69.57
	Pooled values (10 countries)	**0.32**	**4.22**	**22.41**	**73.05**
	United States	1.34	10.37	30.69	57.60

NOTES: Data for European countries are from the Survey of Health, Ageing and Retirement in Europe (SHARE) and data for the United States are from the National Health and Nutrition Examination Survey (NHANES). Data reflect estimated probability of a cardiovascular event (fatal or nonfatal) within 5 years, including low risk (10 percent or less), moderate risk (10-20 percent), high risk (20-30 percent), and very high risk (greater than 30 percent). Probabilities were determined by applying the risk charts in Gaziano et al. (2008) to data from NHANES for 2001-2006 and from the first wave of SHARE in 2004. SHARE data include 10 countries included in the first wave (Switzerland is omitted because of a low response rate). Samples are weighted in the analyses to be representative of national populations. People with disease are automatically categorized as high risk. For those without disease, age- and sex-specific risk is based on having indicators of high blood pressure, diabetes, body mass index, and current smoking.
SOURCE: Crimmins and Solé-Auró (2011, Table 3).

States than in other high-income countries. Research also shows that U.S. adults aged 50 and older have a higher prevalence of cardiovascular and other chronic diseases. Comparing the results of national population surveys in the United States and England, one of the first studies of this issue found that U.S. adults aged 55-64 reported higher rates of diabetes, hypertension, heart disease, myocardial infarction, stroke, lung disease, and cancer (Banks et al., 2006). To account for reporting biases, the researchers also compared the prevalence of biomarkers, and the biomarker data followed the same pattern. Controlling for health behaviors, such as smoking and alcohol use, did not explain the disparity. Health disparities between the United States and England were largest among individuals at the lowest socioeconomic levels, but the U.S. health disadvantage persisted even at the highest socioeconomic levels (Banks et al., 2006). Similarly, Reynolds et al. (2008) found that U.S. adults age 65 and older were more likely than their Japanese counterparts to report being overweight or obese and having heart disease, diabetes, arthritis, and activity limitations.

Other studies have demonstrated that older Americans have higher rates of disease than Europeans. In one study that compared U.S. adults age 50 and older with those in 10 European countries, the United States had a higher prevalence of heart disease, hypertension, high cholesterol, cerebrovascular disease, diabetes, chronic lung disease, asthma, arthritis, and cancer (Thorpe et al., 2007). Similarly, another study that compared the health of U.S. adults aged 50-74 with those in England and 10 European countries reported that Americans had higher rates of heart disease, stroke, hypertension, diabetes, cancer, lung disease, and limited activity than their European counterparts (Avendano et al., 2009). This study also found that Americans at all socioeconomic levels were less healthy, but the disparity was greatest among low-income groups. Oral disease is an important public health problem in older adults (Griffin et al., 2012; Institute of Medicine, 2011a), but there is no evidence of a U.S. disadvantage: one study found that older U.S. adults did not appear to have a greater prevalence of dental disease than Europeans (Crocombe et al., 2009).

A study that compared adults age 50 and older in Canada, Denmark, England, France, Italy, Japan, the Netherlands, Spain, and the United States found that Americans had the highest prevalence of heart attacks, strokes, diabetes, cancer, and activity limitations, and twice the risk of multiple comorbidities (Crimmins et al., 2010). The U.S. health disadvantage was generally larger among those aged 50-64 than among older adults, perhaps because Americans experienced the onset of disease at earlier ages. Nonobese Americans were less healthy than non-obese people in other countries, suggesting that obesity alone did not explain the U.S. health disadvantage (Crimmins et al., 2010). Still another study that compared the health of adults aged 50-53 in the United States and eight European countries found a higher prevalence of heart disease, diabetes, stroke, lung, disease, cancer, hypertension, and limited activity among Americans (Michaud et al., 2011). In a study that compared the health of adults aged 55-64 in the United States and 12 European countries, Banks and Smith (2011) found that the United States had the highest rates of cancer, diabetes, lung disease, and stroke.

Mental Health

Reports about the health of populations often ignore mental health, yet mental illness may act as both a cause and a manifestation of the U.S. health disadvantage. People's emotional and neuropsychiatric health can affect diseases and injurious behavior that claim lives, and of course mental illness is itself an important health outcome (U.S. Department of Health and Human Services, 2001a). Depression, alcohol-use disorders, schizophrenia, and bipolar disorders are among the six leading causes of years lived with disability in the United States (World Health Organization, 2003). The years of life affected by mental illness can be substantial because these disorders often first appear in adolescence and young adulthood (Kessler et al., 2007).

In an analysis commissioned by the panel, Palloni and Yonker (2012) calculated the per capita years of life before age 60 that were afflicted by disability (or incapacitation) in 2002-2006. In terms of the number of years lived with a disability, the United States ranked in the bottom quartile of peer countries (i.e., living more years with a disability), and mental illness and other neuropsychiatric disorders accounted for a large proportion of these years, especially among youth. Neuropsychiatric disorders claimed approximately 75 percent of days lost to disability at ages 15-29 and approximately 50 percent of days lost at ages 30-44 (Palloni and Yonker, 2012).

Individuals with serious mental illnesses, such as depression, face a higher risk of physical illnesses such as diabetes, hypertension, and heart disease (Newcomer, 2007) and, in the United States at least, may die many years earlier than the general population (Felker et al., 1996; Parks et al. 2006). The higher mortality rates of people with mental illness (Thornicroft, 2011; Wahlbeck et al., 2011) could contribute to the U.S. health disadvantage through multiple causal pathways. For example, people with mental illness may turn to cigarettes, alcohol, or drugs to cope with their condition. Schroeder and Morris (2010) found that people with mental illness consume 44 percent of all cigarettes in the United States. Fully 45 percent of U.S. suicides are precipitated by mental illness (e.g., depression, dysthymia, bipolar disorder) (Karch et al., 2011). Although people with mental illness are more likely to be victims rather than perpetrators of violence (Eisenberg, 2005; McNally, 2011), it is also the case that people struggling with antisocial behavior, paranoid schizophrenia, or bipolar disorders may commit homicide or other violent acts. People with serious mental illness, especially those being treated with second-generation antipsychotics, are known to have higher rates of cardiovascular, pulmonary, and infectious diseases (Parks et al., 2006). Anxiety and stress may affect the brain and the endocrine system, alter the behavior of the immune system, and damage end organs (McEwen and Gianaros, 2010). Finally, by affecting work productivity, absenteeism, employability, and social roles (e.g., social isolation, interpersonal tensions, marital disruptions) (Kessler, 2007), mental health can also influence social and economic determinants of health (see Chapter 6).

Whether mental illness (in its various forms) is more prevalent in the United States than in other high-income countries is still unclear. Cross-national studies of mental illness are limited because of inconsistencies in diagnostic criteria and disease classifications,[19] survey administration methods, and measured covariables (e.g., demographic characteristics, risk factors, treatment) (Kessler, 2007).[20] Differences in the prevalence of mental illnesses may be confounded by differences in awareness, detection, diagnosis, treatment approaches, and comorbidities. With all these caveats, however, several studies suggest that the prevalence of mental illness may be higher in the United States than in other countries. For example, a WHO study (Demyttenaere et al., 2004) conducted in 14 countries found the highest prevalence of mental illness in the United States: 26 percent of Americans reported having a mental health disorder in the past 12 months. The United States had the highest rates of depression (18 percent) and of mood (10 percent) and impulse-control (7 percent) disorders, and only Ukraine exceeded the U.S. prevalence rate for substance abuse disorders (Demyttenaere et al., 2004). Other studies from the same project, involving a longer list of countries, also found a high prevalence of depression in the United States relative to other high-income countries (Andrade et al., 2003; Bromet et al., 2011). Further research is needed, however, to conclude that mental illness is more prevalent in the United States than in peer countries.

---

[19]Mental illnesses are classified differently under different editions of the Diagnostic and Statistical Manual of Mental Disorders (DSM)—editions III, IIIR, IV—and under the International Classification of Diseases (ICD).

[20]Over time, such changes as the development of the Diagnostic Interview Schedule in the early 1980s and the WHO Composite International Diagnostic Interview in 1990 and its translation into multiple languages, have helped stimulate more comparable cross-national data on mental illness. The WHO International Consortium in Psychiatric Epidemiology helped launch the World Mental Health Survey Initiative, which will continue to yield rich data on mental health (see Demyttenaere et al., 2004).

CONCLUSIONS

The evidence reviewed in this chapter, together with the findings in Chapter 1, clearly point to a pervasive pattern of poorer health, more injuries, and shorter lives in the United States than in other high-income countries. These data show that Americans have shorter life expectancy than people in almost all other high-income countries—at birth and at age 50—and they are less likely to survive to age 50. This mortality disadvantage has been growing for the past three decades, especially among females.

The U.S. health disadvantage is pervasive: it affects all age groups up to age 75 and is observed for multiple diseases, biological and behavioral risk factors, and injuries. More specifically, when compared with the average for other high-income countries, the United States fares worse in nine health domains: adverse birth outcomes; injuries, accidents, and homicides; adolescent pregnancy and sexually transmitted infections; HIV and AIDS; drug-related mortality; obesity and diabetes; heart disease; chronic lung disease; and disability.

1. **Adverse birth outcomes:** For decades, the United States has experienced the highest infant mortality rate of high-income countries. It also ranks poorly on other birth outcomes (e.g., low birth weight) and measures of child health. American children are less likely to live to age 5 than children in other peer countries.
2. **Injuries, accidents, and homicides:** Injuries are a major cause of premature death in the United States, claiming 53 and 37 percent, respectively, of the excess years of life lost before age 50 by U.S. males and females. Deaths from motor vehicle crashes, nontransportation-related injuries, and violence occur at much higher rates in the United States than in other countries and are a leading cause of death in children, adolescents, and young adults. Since the 1950s, U.S. adolescents and young adults have died at higher rates from traffic accidents and homicide than their counterparts in other countries.
3. **Adolescent pregnancy and sexually transmitted infections:** Among high-income countries, U.S. adolescents have the highest rate of pregnancies and are more likely to acquire sexually transmitted infections, such as gonorrhea, syphilis, and chlamydia.
4. **HIV and AIDS:** HIV and other communicable diseases claim approximately 5 percent of the excess years of life lost in the United States before age 50. The United States has the second highest prevalence of HIV infection among the 17 peer countries and the highest incidence of AIDS. The United States has a high prevalence of HIV infection beginning at age 5.

5. **Drug-related mortality:** Americans lose more years of life to alcohol and other drugs than people in peer countries, even when deaths from drunk driving are excluded.
6. **Obesity and diabetes:** For decades, the United States has had the highest obesity rate among high-income countries. High prevalence rates for obesity are seen in U.S. children and in every age group thereafter. Beginning at age 20, Americans have among the highest prevalence rates of diabetes (and high plasma glucose levels) among people in all OECD countries.
7. **Heart disease:** The U.S. death rate from ischemic heart disease is the second highest among the 17 peer countries. Americans reach age 50 with a less favorable cardiovascular risk profile than their peers in Europe, and adults over age 50 are more likely to experience and die from cardiovascular disease than are older adults in other high-income countries.
8. **Chronic lung disease:** Lung disease is more prevalent and associated with higher mortality in the United States than in the United Kingdom and several other European countries.
9. **Disability:** Older U.S. adults report a higher prevalence of arthritis and activity limitations than their counterparts in Europe, Japan, and the United Kingdom.

The evidence for some other health indicators is less clear or mixed. Some studies suggest that the United States may also have higher rates of mental illnesses and asthma than comparable countries. The prevalence of cancer is also higher in the United States than in other countries, but this finding may reflect more intensive cancer screening practices. The United States also has a higher prevalence of strokes, which may reflect better survival rates with treatment.

The first half of the above list occurs disproportionately among young Americans. Deaths that occur before age 50 are responsible for about two-thirds of the difference in life expectancy between males in the United States and other high-income countries and about one-third of the difference for females. Americans reach age 50 in poorer health than their counterparts in other high-income countries, and as older adults they face greater morbidity and mortality from chronic diseases that arise from risk factors (e.g., smoking, obesity, diabetes) that are often established earlier in life. These findings underscore the importance of early life interventions, but there is also considerable evidence to support the importance of mid- and late-life interventions (such as smoking cessation) in addressing the U.S. health disadvantage.

The U.S. health disadvantage is more pronounced among socioeconomically disadvantaged groups, but several studies have found that even the most advantaged Americans may be faring worse than their counterparts in other countries. In comparisons with England and some other countries, Americans with healthy behaviors or those who are white, insured, college-educated, or in upper-income brackets are in worse health than similar groups in other countries.

The United States enjoys some health advantages, including a suicide rate that is at or below the OECD average, low cancer mortality rates, and possibly greater control of blood pressure and serum lipids. And despite objective measures to the contrary, American adults are more likely than others to rate their health as good.

The disappointing U.S. rankings across the life course suggest the existence of "tracking" between early life behaviors and risk factors: that is, patterns related to obesity, self-reported diabetes, and mental disorders appear to carry forward into successive age groups. Furthermore, the conditions that affect one generation may affect the next, as when poor maternal health or adolescent pregnancy leads to low-birth-weight newborns and infant mortality. Current research is attempting to delineate and quantify the interaction variables in these relationships (see, e.g., Gavin et al., 2011). Although statistical associations may represent an artifact of the available data, a more likely explanation is that they reflect the influence of common underlying conditions on all stages of the life course or causal connections across ages.

The health disadvantages that exist in the United States relative to other countries are all the more remarkable given the size and relative wealth of the U.S. economy and the nation's enormous spending on health care. National health expenditures in the United States have grown from an annual $256 billion (9.2 percent of gross domestic product [GDP]) in 1980 to $2.6 trillion (17.9 percent of GDP) in 2010 (Martin et al., 2012). No other country in the world spends as much on health care, and per capita spending on health care is also much higher in the United States than in any other country (Squires, 2011).

Although this report focuses on health and not economics, it bears noting that the health disadvantage of the United States does have implications for other domains, such as the economy and national security. For example, military and national security experts have warned that rising rates of obesity and illness in young adults are making it harder to recruit healthy soldiers (Cawley and Maclean, 2010). Major corporations are also concerned about the effects of obesity on workforce productivity and competitiveness in the international marketplace (see, e.g., Heinen, 2006).

The economic costs of higher rates of illness and premature death may be substantial. As LaVeist and colleagues (2009, p. 2) explain:

[D]isparities in health and health care impose costs on many parts of society, including individuals, families, communities, health care organizations, employers, health plans, and government agencies, including, of course, Medicare and Medicaid. These costs include direct expenses associated with the provision of care to a sicker and more disadvantaged population as well as indirect costs such as lost productivity, lost wages, absenteeism, family leave to deal with avoidable illnesses, and lower quality of life. Premature death imposes significant costs on society in the form of lower wages, lost tax revenues, additional services and benefits for families of the deceased, and lower quality of life for survivors.

Estimating direct medical costs and indirect costs to the economy and to individuals is methodologically complex (Russell, 2011) and beyond the focus of this report, but the existing literature hints at its scale (Cutler et al., 1997; Gold et al., 1996; LaVeist et al., 2009; Waidmann, 2009). For example, one analysis that focused only on the health disadvantage experienced by U.S. blacks, Asians, and Hispanics relative to non-Hispanic whites found that the combined costs of health inequalities and premature death between 2003 and 2006 was $1.24 trillion (LaVeist et al., 2009). This estimate relies on certain assumptions, but it suggests that the costs associated with the entire population having a health disadvantage relative to other high-income countries are also very high. Although current thinking on this topic is still evolving (e.g., see Acemoglu and Johnson, 2007), some economists have found a strong positive correlation between life expectancy and economic growth (Bloom et al., 2004). Quantifying the net effects of longevity and illness on economic growth and productivity is an area of ongoing research, but it seems likely that the economic costs of the U.S. health disadvantage are substantial.

The pervasive U.S. health disadvantage documented in this and the preceding chapter could arise from problems with health care, individual behaviors, social factors, the environment, or various policies. In Part II of the report, we explore these issues in an effort to explain why, compared with their counterparts elsewhere, Americans face shorter lives and greater illness at almost all ages.

Part II:
Explaining the U.S. Health Disadvantage

In Part I of this report, the panel reviewed the available evidence regarding cross-national differences in health and concluded that the United States has experienced dramatic improvements in health over the past century but still appears to have a major health disadvantage compared with other high-income countries. The research literature shows that this disadvantage has actually existed for many decades and appears to be growing, especially for women. On almost every measure of life expectancy, the United States ranks at or near the bottom compared to other high-income countries. Each year, other high-income countries are improving their health at a much faster rate than the United States, and the United States currently ranks lowest on a variety of health measures.

The evidence reviewed in Part I also makes clear that this disadvantage is pervasive: the United States ranks at or near the bottom on multiple measures of mortality and morbidity, in all age groups up to age 75, in males and females alike, and in virtually all other subgroups of the population. Furthermore, the disadvantage does not appear to be simply a reflection of lower levels of health among Americans who are uninsured and/or poor, as important as these are. Even advantaged Americans seem to be less healthy than their peers in other high-income countries. This pervasiveness also suggests the need to not only look at specific health conditions such as heart disease or other causes of morbidity and mortality, such as injuries, but also to pursue overarching, multisystemic explanatory factors at play. It is these potential explanations that are the focus of this second part of the report.

Our approach to this task was informed by our charge, which was to "propose alternative explanations or potential causes of the reported health disadvantage, going beyond previously tested explanations." We began by adopting a social-ecological and life-course perspective to frame the question (Chapter 3), which led to our decision to systematically consider a broad range of factors that might influence individual- and population-level health: public health and medical care systems (Chapter 4), individual behaviors (Chapter 5), social factors, such as education and income (Chapter 6), environmental factors (Chapter 7), and policies and social values (Chapter 8).

Dividing these topics by chapter is an editorial device: the reality is that these influences are deeply interconnected. Rarely do these factors influence health in isolation, and a reductionist approach can miss interrelationships that affect health outcomes. For example, a U.S. health disadvantage with respect to diabetes might result partly from inadequate medical care (Chapter 4), but also from the obesity epidemic, a product of unhealthy diets and sedentary behavior (Chapter 5), and an obesogenic environment (Chapter 7). The latter disproportionately affects households that face financial stress (Chapter 6), because assistance programs to buffer the impact of this stress are limited (Chapter 8).

The editorial device of separating these topics into distinct chapters should therefore not obscure the complex, dynamic interrelationships between these factors and the different roles they play over the life course as health disadvantages evolve over time. While all of these disparate factors may play a role, it would be a mistake to assume that the topics in each chapter can be decomposed into independent risk categories that "add up" to the U.S. health disadvantage. The dynamic and synergistic interactions between causal factors, only some of which are fully understood, are central to the many issues we review in Part II.

Out of necessity, the panel was selective and systematic in its approach to these complex and comingled influences. In each of the chapters in Part II, the panel focused on three key questions to understand the U.S. health disadvantage:

1. Does the set of factors matter to health?
2. Does the set of factors have greater prevalence or health effects in the United States than in other high-income countries?
3. Could this difference between the United States and other countries contribute to the U.S. health disadvantage?

Large bodies of research, at various stages of evolution and quality, have been devoted to the first question in this three-stage logical sequence and have been ably reviewed elsewhere. Rather than presenting this research in great detail, and because this was not the panel's primary focus, the panel provides a concise summary of this evidence and refers readers to comprehensive research reviews and landmark studies.

The chapters that follow focus instead on the second and third questions. For example, the second question entails not only demonstrating whether particular risk factors are more common in the United States than elsewhere, but also whether they have different effects on health outcomes. Countries with the same levels of hypertension (in untreated populations) or the same levels of poverty may experience different health outcomes if, respectively, one country performs better in controlling blood pressure or has a stronger safety net to help poor people avoid health complications. Although we did not systematically examine these differential effects, we did consider them when we knew there was some evidence available. For example, Chapter 6 reviews evidence that the lack of a college degree may have greater health consequences in the United States than elsewhere.

3

Framing the Question

The chapters that follow in this part of the report (Chapters 4-8) present a systematic examination of potential explanations for the pervasive U.S. health disadvantages documented in Part I. A number of factors distinguish the United States from other countries, but their contribution to the nation's health is unclear. For example, the United States has a very large economy and a large geographic footprint. At 3.7 million square miles, its land mass is much larger than any other high-income country but Canada (United Nations, 2012b) and it encompasses large rural expanses, although comparable cross-national data on the urban-rural mix are limited.[1] The United States is also a much younger nation than most

---

[1]According to the OECD, fully 78 percent of the U.S. land mass is "predominately rural" (defined as more than 50 percent of the population living with fewer than 150 inhabitants per square kilometer). An even larger proportion of the land mass is predominately rural in some other peer countries, including Austria (79 percent), Norway (84 percent), Australia (85 percent), Sweden (90 percent), Finland (93 percent), and Canada (96 percent). Compared with Americans, larger proportions of the populations of Austria, Sweden, Denmark, Finland, and Norway live in predominately rural areas. Among 13 peer countries for which data were available, the United States had the third lowest percentage (3.8 percent) of people living in "remotely rural" areas, defined as having to drive more than 60 minutes to reach a locality with more than 50,000 inhabitants (OECD, 2011d). These definitions differ from those commonly used in the United States to define rural areas, such as those developed by the U.S. Office of Management and Budget and the U.S. Census Bureau, which themselves are not entirely consistent (Aron, 2006).

of its wealthy European counterparts. However, there is little empirical evidence to link these distinctive conditions to adverse health outcomes.[2]

The panel recognized the need to identify and organize the leading factors that could plausibly contribute to cross-national health differences, which naturally led to the question of what factors affect health in the first place. This chapter begins by examining the determinants of health and then turns to two issues that helped us frame our approach in looking for potential explanations for the U.S. health disadvantage: the need for a social-ecological perspective that reflects both upstream and downstream influences on health and the need for a life-course perspective that considers influences over time. These concepts were instrumental in persuading the panel to map out a systematic approach to examining the role of health systems (Chapter 4), individual behaviors (Chapter 5), social factors (Chapter 6), the environment (Chapter 7), and policies and social values (Chapter 8).

THE DETERMINANTS OF HEALTH

Attempting to explain the cause(s) of the U.S. health disadvantage leads one to the question of what causes health and disease, a topic that social epidemiologists and other social scientists have studied for decades. Some factors are innate biological characteristics, such as age, sex, and genes, that generally cannot be modified.[3] Age and other standard sociodemographic factors are highly predictive of health outcomes but, as documented in Part I, the U.S. health disadvantage persists even after adjusting for these factors.

Both the general public and policy makers often assume that health is determined primarily by health care (Robert and Booske, 2011). Thus, it is reasonable to wonder if the U.S. health disadvantage reflects a deficiency in the U.S. health care system. For example, in contrast to many other countries, a large proportion of the U.S. population is uninsured (National Center for Health Statistics, 2012). But even if health care plays some role, decades of research have documented that health is determined by far more than health care. The seminal article by McGinnis and Foege (1993) highlighted the important role of health behaviors. By some estimates, approximately 40 percent of all deaths in the United States are associated

---

[2]Subsequent parts of this report do discuss the role of rural conditions in the United States in contributing to health disadvantages, such as access to medical care.

[3]Although it is possible that differences in population gene pools or other innate biological characteristics contribute to observed cross-national health differences, these and other non-modifiable risk factors receive little emphasis in this report due to their unlikely contributory role. However, Chapter 6 does address the important topic of gene-environment interactions and the potential role of epigenetics as a causal pathway for social and environmental influences on health.

with four health behaviors: tobacco use, unhealthy diet, physical inactivity, and problem drinking (Mokdad et al., 2004, 2005).

Yet health care and health-related behaviors are still not the whole story and raise a bigger question: Why are adverse behaviors or deficiencies in health care more common in the United States than in peer countries? Analytic disciplines often draw from different theoretical traditions to answer this question. For example, the behavioral sciences have developed elegant theoretical frameworks for understanding the complex influences on human behavior (Glanz et al., 2008), and economists have proposed human capital models to explain the tradeoffs people make to optimize their "health capital" (Galama and Kapteyn, 2011).

The fact remains that the U.S. health disadvantage is pervasive, cutting across multiple population subgroups and diverse health outcomes (as different as chronic diseases and injuries). This pervasiveness highlights the need to look beyond individual behaviors and choices by examining systemic processes that may influence multiple health outcomes through various specific and often interrelated pathways. Two conceptual models that are useful in framing these more distal, or upstream, health influences are the social-ecological framework and a life-course framework.

THE SOCIAL-ECOLOGICAL FRAMEWORK

Extensive research documents the behavioral and biological consequences of income, occupation, education, and social and physical environments (see, e.g., Adler and Stewart, 2010; Braveman et al., 2011b; Commission on the Social Determinants of Health, 2008; Kawachi et al., 2010). In a knowledge economy, education is the main pathway for economic security, and socioeconomic conditions influence the ability of individuals and families to make healthier choices and to live in healthier neighborhoods. The environment influences one's ability to engage in healthy behaviors, receive health care, and protect oneself from direct environmental threats. Enhanced recognition that "place matters" has permeated both research and public policy discourse (see California Newsreel, 2008). Television, advertising, and other media also influence health and health-related behaviors, negatively or positively, for example, by promoting (un)healthy foods and (un)safe sex (Center on the Developing Child at Harvard University, 2010; Grier and Kumanyika, 2008; Harris et al., 2009).

All the determinants of health noted above—health care, health behaviors, neighborhood conditions, education, and income—are shaped by both public- and private-sector policies. Decisions made by government officials, business leaders, voters, and other stakeholders affect access to health care, the location of supermarkets, school lunch menus, crime rates, public transportation, toxic waste sites, employment opportunities, and the quality of schools (Institute of Medicine, 2011d). Environmental factors and policies can affect everyone in the population and can influence multiple disease processes, thereby having consequences for multiple outcomes and offering a potential explanation for the recurring patterns observed in Part I.

The multifactorial nature of social-ecological influences on health has long been recognized (see Evans and Stoddart, 1990; Dahlgren and Whitehead, 1991). Many diagrammatic representations of the social-ecological model have appeared in reports of the Institute of Medicine (e.g., 2003, 2011e), in the conceptual framework developed by the World Health Organization (WHO) Commission on Social Determinants of Health, and in a schematic for public outreach developed by the Robert Wood Johnson Foundation Commission to Build a Healthier America (Braveman and Egerter, 2008). Figure 3-1 shows one such model, which was adopted for *Healthy People 2020*, the federal government's national health objectives (U.S. Department of Health and Human Services, 2012a). All such models show (with nesting concentric circles or boxes with multiple arrows) connections between individuals, families, and communities, and the biological, physical, social, and economic environments that surround them. They all emphasize the existence of proximate, or "downstream," health

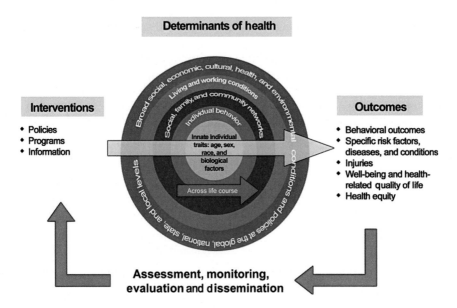

FIGURE 3-1 Model to achieve *Healthy People 2020* overarching goals.
SOURCE: Secretary's Advisory Committee on Health Promotion and Disease Prevention Objectives for 2020 (2008, p. 7).

influences (e.g., smoking) that are shaped by distal, or "upstream," factors (e.g., social norms regarding smoking, tobacco regulations). The models all attempt to highlight societal policies, cultures, and values as the important upstream context for conditions that promote, or undermine, good health downstream.[4]

Adverse conditions may also affect some population groups more than others,[5] thereby contributing to health disparities among vulnerable groups and, possibly, to the health disadvantage experienced in aggregate by the U.S. population (Bleich et al., 2012). The panel focused on its charge to examine and explain cross-national differences in health, not the causes of health differences (or disparities) within countries. Yet we recognize that the two might be interconnected in ways that the field is only beginning to understand. Within-country health disparities may contribute to aggregate cross-national health differences if the former are more pronounced in one country than in another; however, they cannot explain all of the U.S. health disadvantage. The evidence reviewed in Part I suggests that differences persist even among similarly advantaged groups. Understanding why advantaged populations in the United States appear to be less healthy than their counterparts in other countries led the panel to explore upstream processes that affect everyone, including national policies and other aspects of American life.

A LIFE-COURSE PERSPECTIVE

The complex relationships just described are in play at any given moment of time, but they also play out *over* time. For example, the absence of green space today may be the product of zoning decisions two decades ago. Such influences also extend over a person's lifetime: that is, the upstream-downstream continuum can also be a temporal experience for an individual. An individual's struggle through middle age with exertional angina from coronary artery disease may have originated in adolescence with the adoption of cigarette smoking, perhaps as a coping mechanism for a stressful childhood (Richter et al., 2009) or simply because the family lived in a poor neighborhood where smoking was the norm. In turn, the family's move into that poor neighborhood may have resulted from financial setbacks that occurred before the child was born. Health trajectories unfold not only over a lifetime, but also across generations as people are

---

[4]Policy is typically portrayed as the outer ring of models with concentric circles or as the far left-hand starting point of left-to-right logic models of influences on health outcomes.

[5]Such effects would be heightened if adverse conditions are more prevalent, more adverse, or have stronger effects in these populations.

subject to changing health influences stemming from family, neighborhood, and public policies.

This continuum of influences offers a potentially important systemic explanation for the higher prevalence of disorders (e.g., obesity, cardiovascular disease), violence (e.g., homicide), and risk factors documented in Part I. Over the past several decades, a great deal of social and biomedical sciences research has been devoted to looking across individuals' life spans in an effort to better understand developmental and health trajectories over time and, more specifically, how characteristics and experiences early in life may influence health and biobehavioral pathways much later in life (Billari, 2009; Braveman and Barclay, 2009; Braveman et al., 2011b; Gluckman and Hanson, 2006; Halfon and Hochstein, 2002; Keating and Hertzman, 1999; Kuh and Ben-Shlomo, 2004; Lynch and Davey Smith, 2005; Mayer, 2009; Palloni et al., 2009; Pollitt et al., 2005; Power and Hertzman, 1997). The life-course framework casts health as a developmental process influenced by multiple nested social, environmental, and biological spheres that continually interact over the course of one's life and shape the quality and nature of each person's growth, health, and development (Halfon and Hochstein, 2002).

Biological, physical, and social influences over the life course can exert both positive and negative effects on health trajectories. As one study explained, health early in life "lay(s) out the foundation for a life time of well-being" (Center on the Developing Child at Harvard University, 2010, p. 1). Conversely, harmful experiences early in life "are built into our bodies and significant adversity early in life can produce biological 'memories' that lead to lifelong impairments in both physical and mental health" (p. 3). The same DNA can be expressed differently during development as an outcome of adaptive responses and may be modified over time.

Ontogeny—the development or course of development of an individual organism—involves the development of both organ systems and personality. "Developmental plasticity" refers to how the fetus, child, and adolescent are sensitive to the physical environment and to interaction with other human beings. As noted in Chapter 2, the acquisition of certain cognitive traits and abilities, such as social skills, impulse control, self-image, motivation, and self-discipline, may occur during critical "transition periods" (Viner, 2012). Although protective factors during these sensitive periods can foster resilience and healthy development, harmful exposures may cause an individual to adapt abnormally or in unhealthy ways (Gluckman and Hanson, 2006). These patterns may be relevant to understanding some of the health problems documented in Part I that are more common among American youth than among their peers in other countries. Although research in genetics, molecular biology, developmental neurobiology, psychology, economics, and sociology is only beginning to elucidate the causal pathways that link early experiences with later disease, at least five general causal pathways have emerged as potentially important.

First, much of the association between early and adult health conditions results from the persistence of material, social, and cultural conditions over time. Children who grow up in poverty not only endure immediate health risks such as poor nutrition and exposure to lead, allergens, and other pollutants, but they are also more likely to face poverty and its consequences as adults (Case and Paxson, 2010, 2011; Case et al., 2005; Currie and Widom, 2010; Delaney and Smith, 2012; Isaacs et al., 2008). To the extent that such exposures remain stable across the life course, disadvantages during childhood and adolescence will persist later in life. Accordingly, poor health status early in life is associated with poor health status during adulthood (Kuh and Ben-Shlomo, 2004; Power et al., 1991; Wadsworth, 1991).

Second, biological exposures during early life, notably to certain infectious diseases—from human papilloma virus to HIV—can produce clinical complications later in life (Elo and Preston, 1996; Finch, 2010; Fong, 2000).

Third, adverse experiences and stimuli can result in biological embedding[6] (Hertzman, 2000). For example, Barker and colleagues were among the first to hypothesize that nutritional status in utero and during infancy and early childhood can cause organ damage that is responsible for hypertension, heart disease, metabolic syndrome, insulin resistance, obesity, and diabetes much later in life (Barker, 1998, 1999; Barker et al., 1993; Godfrey, 2006; Hales et al., 1991; Ravelli et al., 1998). Similarly, other investigators have documented associations between early life conditions (e.g., social class, income, nutrition, infections) and subsequent mortality and morbidity from diabetes, cancer, disabilities, and drug use (Freedman et al., 2008; Galobardes et al., 2008; Kuh and Ben-Shlomo, 2004; Kuh et al., 1997; Maty et al., 2008; Mayer, 2009; Palloni, 2006; Palloni et al., 2009; Turrell et al., 2007; Wadhwa et al., 2009; Warner and Hayward, 2006). Research on developmental plasticity is examining the mechanisms through which social and environmental exposures become biologically embedded or reversed during critical periods (Gluckman and Hanson, 2006; Gluckman et al., 2008, 2010). For example, sustained stress is thought to affect the brain and the neuroendocrine and immune systems (see Chapter 6).

Fourth, the relatively new field of epigenetics (see Chapter 6) has documented gene-environment interactions, in which permanent changes in

---

[6]Embedding is a process that leaves signatures in the brain and other organ systems that may remain latent for many years and then become manifest and influence adult physical and mental health.

gene expression, triggered by environmental cues, are thought to precipitate chronic conditions (Sandoval and Esteller, 2012) and may be inherited by offspring (Gluckman and Hanson, 2006; Institute of Medicine, 2006b).

Fifth, adversities during childhood and adolescence may cultivate dysfunctional traits (such as poor self-control, limited social skills, lack of perseverance and resilience, and shortsightedness), maladjusted personalities, and susceptibility to antisocial behavior (Caspi, 2000; Caspi and Moffit, 1995; Center on the Developing Child at Harvard University, 2010; Dozier and Peloso, 2006; Elder and Shanahan, 2007; Felitti et al., 1998; Knudsen et al., 2006; Trzesniewski et al., 2006). Stress may alter gene expression in ways that predispose children to specific behavioral phenotypes (Caspi et al., 2002, 2003; Taylor et al., 2006). These traits may in turn influence physical activity, diet, and risky behaviors (e.g., smoking, substance abuse, unprotected sex) (Richter et al., 2009) and they may exacerbate propensities for self-abuse, mental illness,[7] suicide, violence, unintended pregnancy, obesity, and diabetes (Dong et al., 2004; Edwards et al., 2003; Felitti et al., 1998; Guyer et al., 2009; Hillis et al., 2004; Horwitz et al., 2001; National Research Council and Institute of Medicine, 2000; Schilling et al., 2007; Shonkoff et al., 2009).[8] As noted in Chapter 2, adolescence is especially critical in shaping cognitive and social skills that fuel success in education, careers, and parenting[9] (Bowles et al., 2005; Heckman, 2007).

Life circumstances provide an essential context to properly interpret bivariate correlations between health and observed predictors, such as education or income. For example, educated adults may not necessarily experience better health as a direct result of what they have learned in school: the benefits may come from early experiences that foster self-efficacy, a desire to succeed in various endeavors—education among them—and an interest in good health (Pampel et al., 2010; Ross and Wu, 1995). On similar grounds, these factors may be relevant in understanding the evidence in Part I regarding the inferior health status and higher risk exposure of young people in the United States compared with those in other high-income countries. Higher rates of obesity, adolescent pregnancy, motor vehicle deaths, and

---

[7]Mental illness has other innate causes and can only partially be attributed to childhood experiences.

[8]Adults with cardiovascular disease, lung disease, depression, alcoholism, and mental disorders are more likely to report a history of child abuse (Anda et al., 2006; Caspi and Moffitt, 1995; Danese et al., 2007; Moffitt et al., 1992). Longitudinal studies of humans in particular situations (orphaned children, children living in refugee camps) and animals (McEwen, 2000; Zhang et al., 2005) suggest that inadequate maternal attachment and disruptions in protective adult behavior can trigger responses like those observed with child abuse, albeit less extreme (Thompson et al., 2002).

[9]Thus, adverse conditions during early childhood can breed adverse conditions in subsequent generations.

homicide may have a common explanation in the life experiences of the victims.

In short, existing research points to a variety of pathways that could link early childhood exposures and late adult conditions (e.g., diabetes, cardiovascular disease). The same pathways might also explain correlations between health status across contiguous stages of the life cycle, spanning shorter time intervals, such as between infancy and early childhood or between adolescence and young adulthood.

However, early life experiences are hardly the only influences on health and mortality later in life. It is important to differentiate between the health trajectories experienced by individuals and the epidemiologic trends observed in populations over time. For the latter, "period effects" can play a major role in explaining cross-national differences. For example, increases in smoking during the first half of the 20th century and in obesity in the second half were largely period effects that touched adults of all age groups and eventually extended to adolescents and children as well. Similarly, U.S. tobacco control efforts designed to reduce smoking among adults, which achieved success from the 1960s onward (see Chapter 5), benefited people of all ages.

The panel was acutely aware of one of the most well-known and vexing challenges when studying nonmedical influences on health, that of describing and empirically demonstrating causal pathways between a given health factor and a biological health outcome. In other words, how do the conditions presented by family, community, and national environments get under a person's skin to affect health? And how do these conditions affect people differently? Even twins do not experience the same health development trajectories over time (Madsen et al., 2010). Unlike the study of clinical interventions or biological effects, research on social, environmental, and policy factors involves more multidisciplinary and varied methodological approaches, including different notions in the epistemology of what constitutes "evidence" (Anderson and McQueen, 2009; Braveman et al., 2011c; McQueen, 2009; Rychetnik et al., 2002; Victora et al., 2004).[10] As the Measurement and Evidence Knowledge Network (Kelly et al., 2006, p. 33) of the WHO Commission on the Social Determinants of Health noted:

> The data and evidence which relate to social determinants of health come from a variety of disciplinary backgrounds and methodological traditions. The evidence about the social determinants comprises a range of ways of knowing about the biological, psychological, social, economic and material worlds. The disciplinary differences arise because social history, economics, social policy, anthropology, politics, development studies,

---

[10]The limitations of randomized controlled trials are discussed in Chapter 6.

psychology, sociology, environmental science and epidemiology, as well as biology and medicine, may all make contributions. Each of these has its own disciplinary paradigms, arenas of debate, agreed canons and particular epistemological positions. Some of the contributions of these disciplines are highly political in tone and intent.

CONCLUSIONS

The panel used both a social-ecological framework and a life-course perspective in determining what factors to consider in search of an explanation for the U.S. health disadvantage. Many theories or constructs could be pursued, but the panel identified five domains of particular interest that set the agenda for the next five chapters (see Figure 3-2).

We begin with health systems and individual behaviors, both of which can have important proximate influences on health, and are addressed in Chapters 4 and 5, respectively. The social-ecological model emphasizes that interactions with health systems, individual behaviors, and disease processes themselves are shaped by social factors and the environment. Chapters 6 and 7, respectively, examine their potential contributory role in the U.S. health disadvantage. Finally, Chapter 8 addresses how all these factors might be influenced at the macro level by the societal context of life in America, ranging from life-style to policies, governance, and social values.

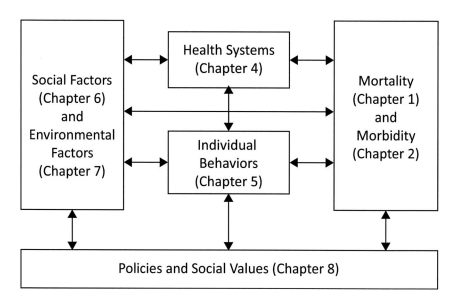

FIGURE 3-2 Panel's analytic framework for Part II.

Once again, we emphasize that the division of topics across the following chapters is not meant to imply that these categories of health influences operate independently or contribute in a simple additive fashion to the U.S. health disadvantage. The key dynamic trajectories of health, risk factors, socioeconomic circumstances, and physical and institutional environments are all integrally linked and cannot be decomposed in a reductionist fashion. As the following chapters make clear, these synergistic interactions are central to understanding the U.S. health disadvantage.

4

Public Health and Medical Care Systems

One explanation for the health disadvantage of the United States relative to other high-income countries might be deficiencies in health services. Although the United States is renowned for its leadership in biomedical research, its cutting-edge medical technology, and its hospitals and specialists, problems with ensuring Americans' access to the system and providing quality care have been a long-standing concern of policy makers and the public (Berwick et al., 2008; Brook, 2011b; Fineberg, 2012). Higher mortality rates from diseases, and even from transportation-related injuries and homicides, may be traceable in part to failings in the health care system.

The United States stands out from many other countries in not offering universal health insurance coverage. In 2010, 50 million people (16 percent of the U.S. population) were uninsured (DeNavas-Walt et al., 2011). Access to health care services, particularly in rural and frontier communities or disadvantaged urban centers, is often limited. The United States has a relatively weak foundation for primary care and a shortage of family physicians (American Academy of Family Physicians, 2009; Grumbach et al., 2009; Macinko et al., 2007; Sandy et al., 2009). Many Americans rely on emergency departments for acute, chronic, and even preventive care (Institute of Medicine, 2007a; Schoen et al., 2009b, 2011). Cost sharing is common in the United States, and high out-of-pocket expenses make health care services, pharmaceuticals, and medical supplies increasingly unaffordable (Commonwealth Fund Commission on a High Performance System, 2011; Karaca-Mandic et al., 2012). In 2011, one-third of American households reported problems paying medical bills (Cohen et al., 2012), a problem that seems to have worsened in recent years (Himmelstein et al., 2009). Health insurance premiums are consuming an increasing proportion of U.S. household income (Commonwealth Fund Commission on a High Performance System, 2011).

Apart from challenges with access, many Americans do not experience optimal quality when they do receive medical care (Agency for Healthcare Research and Quality, 2012), a problem that health policy leaders, service providers, and researchers have been trying to solve for many years (Brook, 2011a; Fineberg, 2012; Institute of Medicine, 2001). In the United States, health care delivery (and financing) is deeply fragmented across thousands of health systems and payers and across government (e.g., Medicare and Medicaid) and the private sector, creating inefficiencies and coordination problems that may be less prevalent in countries with more centralized national health systems. As a result, U.S. patients do not always receive the care they need (and sometimes receive care they do not need): one study estimated that Americans receive only 50 percent of recommended health care services (McGlynn et al., 2003).

Could some or all of these problems explain the U.S. health disadvantage relative to other high-income countries? This chapter reviews this question: it explores whether systems of care are associated with adverse health outcomes, whether there is evidence of inferior system characteristics in the United States relative to other countries, and whether such deficiencies could explain the findings delineated in Part I of the report.

DEFINING SYSTEMS OF CARE

The panel defines "health systems" broadly, to encompass the full continuum between public health (population-based services) and medical care (delivered to individual patients). As outlined in previous Institute of Medicine reports (e.g., 2011e), health systems involve far more than hospitals and physicians, whose work often focuses on tertiary prevention (averting complications among patients with known disease). Both public health and clinical medicine are also concerned with primary and secondary prevention.[1] The health of a population also depends on other public health services and policies aimed at safeguarding the public from health and injury risks (Institute of Medicine, 2011d, 2011e, 2012) and attending to the needs of people with mental illness (Aron et al., 2009). There

---

[1]Examples of primary prevention include smoking cessation, increased physical activity, administering immunizations to eliminate susceptibility to infectious diseases, and helping people avoid harmful environmental exposures (e.g., lead poisoning). Secondary prevention includes early detection of diseases and risk factors in asymptomatic persons (e.g., cancer and serum lipid screening).

is mounting evidence that chronic illness care requires better integration of professions and institutions to help patients manage their conditions, and that health care systems built on an acute, episodic model of care are ill equipped to meet the longer-term and fluctuating needs of people with chronic illnesses. Wagner and colleagues (1996) were among the first to document the importance of coordination in managing chronic illnesses. Many countries differ from the United States because public health and medical care services are embedded in a centralized health system and social and health care policies are more integrated than they are in the United States (Phillips, 2012).

The panel believes that the totality of this system, not just the health care component, must be examined to explore the reasons for differences in health status across populations. For example, a country may excel at offering colonoscopy screening, but ancillary support systems may be lacking to inform patients of abnormal results or ensure that they understand and know what to do next. Hospital care for a specific disease may be exemplary, but discharged patients may experience delayed complications because they lack coverage, access to facilities, transportation, or money for out-of-pocket expenses, and those with language or cultural barriers may not understand the instructions. The health of a population is influenced not only by health care providers and public health agencies but also by the larger public health system, broadly defined.[2]

Data are lacking to make cross-national comparisons of the performance of health systems, narrowly or broadly defined, in adequate detail. Only isolated measures are available, such as the 30-day case-fatality rate for a specific disease or the percentage of women who obtain mammograms. Nor is it clear what the ideal rate for a given health system measure (e.g., optimal wait times or density of physicians) should be for any given country. Out of necessity, this chapter focuses on the "keys under the lamppost"—the health system features for which there are comparable cross-national data—but the panel acknowledges that better data and measures are needed before one can properly compare the performance of national health care systems.

Based on the data that do exist, how well does the U.S. health care system prevent and treat injury and disease when compared with other high-income countries? As noted earlier, this chapter and the four that follow address three core questions. For this chapter, the three core questions are:

---

[2]The larger public health system includes not only public health agencies, but also public and private entities involved with food and nutrition, physical activity, housing and transportation, and other social and economic conditions that affect health (Institute of Medicine, 2011e). As discussed further in Chapter 8, public- and private-sector leaders are increasingly recognizing the health implications of "nonhealth" policies that relate to agriculture, transportation, land use, energy, housing, and other environmental conditions.

- Do public health and medical care systems affect health outcomes?
- Are U.S. health systems worse than those in other high-income countries?
- Do U.S. health systems explain the U.S. health disadvantage?

QUESTION 1:
DO PUBLIC HEALTH AND MEDICAL CARE SYSTEMS AFFECT HEALTH OUTCOMES?

As other chapters in this report emphasize, population health is shaped by factors other than health care, but it is clear that health systems—both those responsible for public health services and medical care—are instrumental in both the prevention of disease and in optimizing outcomes when illness occurs. The importance of population-based services is marked by the signature accomplishments of public health, such as the control of vaccine-preventable diseases, lead abatement, tobacco control, motor vehicle occupant restraints, and water fluoridation to prevent dental caries (Centers for Disease Control and Prevention, 1999, 2011b). Public health efforts are credited with much of the gains in life expectancy that high-income countries experienced in the 20th century (Cutler and Miller, 2005; Foege, 2004). The effectiveness of a core set of clinical preventive services (e.g., cancer screening tests) is well documented in randomized controlled trials (U.S. Preventive Services Task Force, 2012), as are a host of effective medical treatments for acute and chronic illness care (Cochrane Library, 2012). For example, gains in cardiovascular health have occurred with the adoption of evidence-based interventions including antiplatelet therapy, beta-blockers, and reperfusion therapy (Khush et al., 2005; Kociol et al., 2012).

Although some authors have questioned the impact of medical care on health (McKeown, 1976; McKinlay and McKinlay, 1977), others estimate that between 10-15 percent (McGinnis et al., 2002) to 50 percent (Bunker, 2001; Cutler et al., 2006b) of U.S. deaths that would otherwise have occurred are averted by medical care. Across various countries, medical care is credited with 23-47 percent[3] of the decline in coronary artery disease mortality that occurred between 1970 and 2000 (Bots and Grobbee, 1996; Capewell et al., 1999, 2000; Ford and Capewell, 2011; Ford et al., 2007; Goldman and Cook, 1984; Hunink et al., 1997; Laatikainen et al., 2005; Unal et al., 2005; Young et al., 2010).

Barriers to health care also influence health outcomes. Inadequate

---

[3]The same studies estimate that between 44 and 72 percent of the fall in mortality resulted from a reduction in cardiovascular risk factors (smoking, lipids, and blood pressure); see Chapter 5.

health insurance coverage is associated with inferior health care and health status and with premature death (Freeman et al., 2008; Hadley, 2003; Institute of Medicine, 2003b, 2009a; Wilper et al., 2009). Conversely, universal coverage has been associated with improved health, both in U.S. states (Courtemanche and Zapata, 2012) and in other countries (Hanratty, 1996). Two other barriers, inadequate numbers of physicians and a weak primary care system, are associated with higher all-cause mortality, all-cause premature mortality, and cause-specific premature mortality (Chang et al., 2011; Macinko et al., 2003, 2007; Or et al., 2005; Phillips and Bazemore, 2010; Starfield, 1996; Starfield et al., 2005).

Health is also affected by the quality of care. The Institute of Medicine (2000) estimated that medical errors claim 98,000 lives each year in the United States. Coordination of care also affects health outcomes because miscommunication, flawed handoffs, and confusion can result in lapses in patient safety and gaps and delays in the delivery of care (Institute of Medicine, 2007b).

Many of the specific causes of death discussed in Part I—such as transportation-related injuries, homicide, communicable diseases, and chronic diseases—have some connection to health professionals and medical care. For example, the survival of injury victims and their rehabilitation are dependent on emergency medical services and speedy, effective trauma care (Cudnick et al., 2009; Institute of Medicine, 2007a; MacKenzie et al., 2006). Medical care has obvious connections to other areas of the U.S. health disadvantage, such as infant mortality and other adverse birth outcomes, HIV infection, heart disease, and diabetes.

QUESTION 2:
ARE U.S. HEALTH SYSTEMS WORSE THAN THOSE IN OTHER HIGH-INCOME COUNTRIES?

The United States spends significantly more on health care than any other country (Anderson and Squires, 2010; Reinhardt et al., 2004; Squires, 2011). Median per capita spending among all OECD countries in 2009 was $3,223, which is less than half of the $7,960 per capita spent in the United States (OECD, 2011b). Such statistics have rallied interest in addressing the inefficiency of the health system and the causes of medical cost inflation (Berwick and Hackbarth, 2012; Fisher et al., 2011; Institute of Medicine, 2010; OECD, 2010b) and have sparked a campaign by medical organizations to discourage overutilization (Cassel and Guest, 2012).

Whether the high level of spending on health care contributes to the U.S. health disadvantage is not entirely clear. This spending, some of which reflects inefficiencies in health care delivery,[4] accounted for 17.9 percent of the nation's gross domestic product in 2010 (Martin et al., 2012). That spending carries a large opportunity cost: it could be diverting resources that might otherwise be applied to public health, education, social services, and the growth of businesses and the economy. The ramifications could include a deleterious effect on the health of Americans relative to their peers in other countries, but the panel found little empirical evidence to support this.

The panel did find some evidence comparing other characteristics of the health system—access and quality—that might explain the inferior health outcomes in the United States. This evidence is reviewed below.

Access to Public Health and Medical Care in the United States

Access to Public Health Services

Public health services in the United States are highly fragmented and are financed by a complex mixture of federal, state, local, and private sources that vary across communities, are earmarked for specific categorical disease priorities,[5] and fluctuate over time depending on budgets and separate appropriation decisions at the federal, state, and local level (Fielding and Teutsch, 2011; Institute of Medicine, 2012). The 2,565 local health departments in the United States operate under highly disparate resources and authorities (National Association of County and City Health Officials, 2011). In contrast, public health services in other countries are often coordinated by a central governmental body. It is estimated that the United States spends from 3 to 9 percent of its health budget on public health (Mays and Smith, 2011; Miller et al., 2008, 2012), and its model of specialized categorical program funding to subsidize public health activities does not always match well with the needs of catchment areas (Institute of Medicine, 2012).[6] However, there is no evidence that public health spending is higher per capita in other countries or that other countries are more effective in using public health investments to drive improvements in population health.

---

[4]Although a body of evidence suggests that a large proportion of health care spending in the United States is related to waste and inefficiency (Berwick and Hackbarth, 2012), the high consumption of health care resources may also be the product of the U.S. health disadvantage (reverse causality). Conversely, other evidence hints at an iatrogenic effect in which higher intensity of health care is associated with more unfavorable health outcomes (Fisher et al., 2003).

[5]Examples include maintaining programs in emergency preparedness, tuberculosis, HIV, maternal/child health activities, environmental sanitation, and hygiene.

[6]For example, on average, only 1.9 percent of the budget of the Centers for Disease Control and Prevention (CDC) and the budget of large metropolitan health departments is devoted to cardiovascular disease, the leading cause of death. State governments spend $1.22 per person on tobacco control, less than a quarter of the minimum level recommended by the CDC (Institute of Medicine, 2012).

Access to Medical Care

Access to medical care is limited for many people in the United States, a potentially important factor in understanding the U.S. health disadvantage relative to other countries. Americans seem less confident than people in other countries that the system will deliver the care they need. In a 2010 Commonwealth Fund survey, only 70 percent of U.S. adults reported being confident or very confident that they would receive the most effective treatments (e.g., drugs, tests) if they were seriously ill (Schoen et al., 2010). Patients in all countries but Norway and Sweden expressed greater confidence.

Health Insurance Coverage The large uninsured (and underinsured) population is a well-recognized problem in the United States. All other peer countries offer their populations universal or near-universal health insurance coverage. Only three OECD countries—Chile, Mexico, and Turkey—provide less coverage than the United States (OECD, 2011b).

Affordability Americans face greater financial barriers in accessing care—insurance deductibles, copayments, and out-of-pocket expenses—than do those in other high-income countries (Schoen et al., 2009b, 2010, 2011) (see Box 4-1). One out of three U.S. patients with a chronic illness or a recent need for acute care reports spending more than $1,000 per year in out-of-pocket costs (Schoen et al., 2011) (see Table 4-1). Higher medical costs could contribute to the U.S. health disadvantage if they cause patients to forgo needed care (Wendt et al., 2011). Even insured and higher-income Americans are more likely than their counterparts in other countries to report problems getting care (Huynh et al., 2006). Among insured adults in the United States under age 65, 25 percent reported serious difficulties paying medical bills, and approximately 40 percent reported access problems due to cost, out-of-pocket expenses exceeding $1,000, and gaps in care coordination (Schoen et al., 2011). In a comparison that looked specifically at adults with above-average incomes in 11 countries, only 74 percent of high-income respondents in the United States were confident that they would be able to afford needed care if they were to become seriously ill; in all comparison countries, the corresponding percentages were higher (Schoen et al., 2010).

Access to Clinicians For various reasons, U.S. patients are less likely to visit physicians than patients in other OECD countries. In 2009, annual consultations in the United States were 3.9 per capita, a lower rate than in all peer countries but Sweden and lower than the OECD average of 6.5 per capita (OECD, 2011b). However, physician consultation rates are an

BOX 4-1
Health Care Decommodification

"Health care decommodification" refers to the extent to which individuals' access to health care is independent of their financial resources or the market. To compare access based on resources across nations, the British social scientist Clare Bambra (2005) developed a health care decommodification index based on the following three variables: private health expenditure as a percentage of gross domestic product, private hospital beds as a percentage of total bed stock, and the percentage of the population covered by the health care system. She found that the United States had a lower decommodification score (9.0) than all the countries, including 14 peer countries (see Figure 4-1a). Bambra concluded that access to health care is much more market dependent in the United States than in other countries and therefore makes access to care more susceptible to the socioeconomic status of the patient.

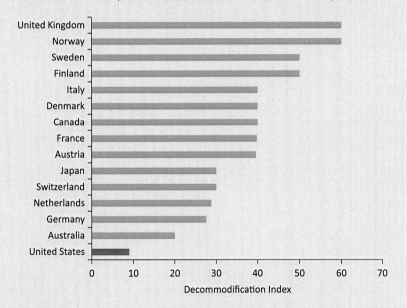

FIGURE 4-1a Access to health care independent of personal resources.
SOURCE: Adapted from Bambra and Beckfield (2012, Table 2).

TABLE 4-1 Cost-Related Access Problems in the Past Year Among U.S. Patients with Complex Chronic Conditions, 2011

Problem	Percentage of Respondents Reporting Access Problems in Selected Countries												
	Austria	Canada	France	Germany	Netherlands	New Zealand	Norway	Sweden	Switzerland	United Kingdom	United States
Difficulty paying or unable to pay medical bills	8	8	5	6	14	11	7	4	8	1	27
Cost-related access problems	30	20	19	22	15	26	14	11	18	11	42
Did not visit a doctor when had a medical problem	17	7	10	12	7	18	8	6	11	7	29
Did not get recommended test, treatment, or follow-up	19	7	9	13	8	15	7	4	11	4	31
Did not fill a prescription or skipped doses	16	15	11	14	8	12	7	7	9	4	30

SOURCE: Data from Schoen et al. (2011, Exhibit 1).

imperfect measure of access because they are confounded by many factors, such as policies that require an in-person physician visit for a referral or to refill a prescription.

Physician Density One reason for fewer physician visits in the United States may be a lower concentration of providers. According to the OECD, physician density (the number of practicing physicians per 1,000 population) in 2009 was 2.4 in the United States, lower than all peer countries but Japan (OECD, 2011b).[7] Physician density grew in the United States by only 0.5 per 1,000 people between 2000 and 2009, a lower growth rate in physician density than that reported by any peer country but France.[8] Access to physicians varies by geography, a particular problem in the United States with its large rural expanses.[9]

Primary Care Although the United States does well in providing access to many specialists, access to primary care physicians and a regular health care provider is more limited than in many other countries (OECD, 2011b; Schoen et al., 2009b, 2011; Starfield et al., 2005; World Health Organization, 2008b). According to the OECD, only 12.3 percent of U.S. physicians engage in primary care, the lowest proportion among 15 peer countries providing data (see Figure 4-1).[10] Macinko et al. (2003) applied 10 criteria to rank the primary care systems of 18 high-income countries (including Canada, Australia, Japan, and 14 European countries). The United States had the weakest primary care score of all the countries in 1975 and 1985 and the third weakest in 1995 (Macinko et al., 2003).

Continuity of care from a regular provider, which is important to effective management of chronic conditions (Liss et al., 2011), may be more tenuous in the United States than in comparable countries. Only slightly more than half (57 percent) of U.S. respondents to the 2011 Common-

---

[7]U.S. physician density was lower than that of 28 other countries, including all of Western and Eastern Europe (except Poland), Canada, Australia, New Zealand, and Russia (OECD, 2011b).

[8]In contrast, the density of nurses in the United States was 10.8 per 1,000 population in 2009, higher than the OECD average and the sixth highest nurse-to-physician ratio in OECD countries (OECD, 2011b).

[9]As of mid-2012, the Health Resources and Services Administration (HRSA) in the United States had formally designated 5,703 areas as having a primary care health professional shortage (U.S. Department of Health and Human Services, 2012c). The 54.5 million people living in these areas need another 15,168 health practitioners to meet their primary health care needs, assuming a population to practitioner ratio of 2,000:1. Almost 50 percent of U.S. counties had no obstetrician-gynecologists (National Center for Health Statistics, 2007).

[10]This percentage is less than half the OECD average (25.9 percent) and below the rates reported by such countries as Mexico, Turkey, and some Eastern European countries (e.g., the Czech Republic, Estonia, the Slovak Republic, Slovenia) (OECD, 2011b).

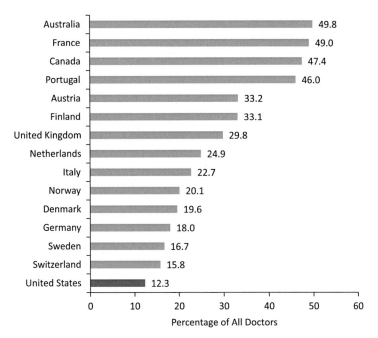

FIGURE 4-1 General practitioners as a proportion of total doctors in 15 peer countries, 2009.
SOURCE: Data from OECD (2011b, Figure 3.2.2).

wealth Fund survey reported being with the same physician for at least 5 years, a lower rate than all comparison countries except Sweden (Schoen et al., 2011). In another Commonwealth Fund survey, U.S. patients were more likely than patients in other countries except Canada to report visiting an emergency department for a condition that could have been treated by their regular physician had one been available (Schoen et al., 2009b).

Access to Health Care Facilities The United States has fewer hospital beds per capita than most other countries, but this measure may be confounded by increasing efforts to deliver care in less expensive outpatient settings. The density of hospital beds decreased in most OECD countries between 2000 and 2009 (OECD, 2011b). In a comparison of eight countries, Wunsch and colleagues (2008) reported that the United States had the third highest concentration of critical care beds (beds in intensive care units per 100,000 population). However, the availability of long-term care beds for U.S. adults ages 65 and older is lower than for those in 10 of the 16 peer countries. Where such care is delivered also differs in the United

States: in most high-income countries, long-term care is usually provided in the patient's home, but in the United States, less than half of adults report receiving long-term care at home (OECD, 2011b).

Timeliness of Care Inadequate insurance, limited access to clinicians and facilities, and other delivery system deficiencies can affect how quickly patients receive the care they need. Responses to the Commonwealth Fund surveys suggest that U.S. patients with complex care needs are more likely than those in many other countries to face delays in seeing a physician or nurse within 1-2 days, especially after normal office hours, making it necessary to rely on an emergency department (Schoen et al., 2011). However, waiting times for nonemergency elective care appear to be shorter in the United States than in most other countries (Davis et al., 2010; Schoen et al., 2010).

Quality of Public Health and Medical Care Systems

Although there is evidence of variance in health protection and other public health services across communities and population groups in the United States (Culyer and Lomas, 2006),[11] there is little direct evidence to determine whether and how this differs across high-income countries. Comparing the quality of public health services in the United States to that of other countries is difficult due to the lack of comparable international data on the delivery of core public health functions.[12]

There are also important differences between countries in what types of programs and services are counted within the broad categories of public health, preventive medicine, and medical care. Thus, the only way to compare the public health services of countries is to examine proxy measures, but proxies often miss other important differences in population-based public health protections. This section discusses several measures of the quality of public health and medical care systems: immunizations, health

---

[11]Examples include variations in motor vehicle safety regulations, illegal blood alcohol concentrations, and requirements to wear safety helmets: they vary greatly across the 50 states and the District of Columbia but are often uniform in many high-income countries (Transportation Research Board, 2011).

[12]In the United States, the 10 essential public health services include monitoring health status, diagnosing and investigating health problems, informing the public, mobilizing community partnerships, developing policies and plans, enforcing laws and regulations, linking people to needed health services, assuring a competent workforce, evaluating quality of health services, and research (Institute of Medicine, 2011e). Similar core public health functions are identified globally by WHO (World Health Organization, 2008b). See Chapter 7 for data on differences in the quality of environment health protections in the United States and other countries.

promotion, screening tests, acute care, chronic illness care, medical errors, and optimizing health care delivery.

Immunizations

Childhood immunization coverage in the United States, although much improved in recent decades, is generally worse than in other high-income countries. For example, according to the OECD, 83.9 percent of U.S. children have been vaccinated against pertussis, the lowest rate of all peer countries but Austria; the U.S. rate is the third lowest among 39 OECD countries and well below the OECD average of 95.3 percent (OECD, 2011b). Conversely, immunization rates for older adults appear to be higher in the United States than in most OECD countries. According to the OECD, 66.7 percent of U.S. adults age 65 and older received influenza vaccination in 2009, a rate below that of France, the United Kingdom, Australia, and the Netherlands but higher than those of the other peer countries.

Health Promotion

Although the prevalence of unhealthy behaviors (e.g., smoking) and other modifiable risk factors (e.g., obesity, environmental exposures) can be compared across countries, data are lacking to accurately compare the quality of public health agencies or programs to address these risk factors, including the extent of health promotion programs aimed at controlling behavioral and environmental risks. This information would have relevance to tobacco and obesity-related diseases that claim excess years of life in the United States or to higher death rates from alcohol, other drugs, and transportation-related injuries (see Chapters 1 and 2). Patient surveys show no evidence that U.S. physicians are less likely to offer behavioral counseling than their counterparts in other countries (Davis et al., 2010), and they appear more likely to prescribe pharmacotherapy (e.g., varenicline) to help with smoking cessation (Fix et al., 2011). (See Chapter 7 for further discussion of how the environmental influences on health behaviors might differ between the United States and other countries.)

Screening Tests

The United States appears to administer more screening tests than do other countries (Gohmann, 2010; Howard et al., 2009). According to the OECD, the United States has the third highest rate of mammography screening among peer countries, surpassed only by the Netherlands and Finland, and it has the highest cervical cancer screening rate among peer countries (and all OECD countries) (OECD, 2011b). In an analysis of survey data from the United States and 10 European countries, Howard and colleagues (2009) reported that the European/U.S. ratio for frequency of screening for adults age 50 and older was 0.22-0.60 for mammography, 0.43-0.49 for colon cancer screening, 0.55-0.88 for cervical cancer screening, and 0.58-0.64 for prostate cancer screening—all indicating that the comparison countries screened less often.

Acute Care

Evidence is limited to compare the quality of acute care services in high-income countries. Some data are available regarding the quality of trauma care in the United States, a form of acute care that is especially relevant to the U.S. health disadvantage because of the country's high death toll from transportation-related injuries and homicide (see Chapters 1 and 2). Although there is evidence that outcomes vary across U.S. trauma centers, even after risk adjustment (Haider et al., 2012), there is little empirical evidence to compare the quality of trauma care in the United States to that in other countries. Such comparisons require a close examination of interrelated determinants of trauma care (e.g., health insurance coverage), socioeconomic and policy contexts (discussed in later chapters), and differences in geography (see Box 4-2).

Chronic Illness Care

As detailed in Part I of this report, deaths and morbidity from non-communicable chronic diseases are higher in the United States than in peer countries, which invites speculation about deficiencies in the quality of medical care for these conditions. Evaluating the quality of chronic illness care is complex because of the multifactorial influences on care management and coordination. The Commonwealth Fund Commission on a High Performance Health System (2008) evaluates the quality of health care on four measures: effectiveness, safety, coordinated care, and patient-centered, timely care. We focus here on the following measures of the quality of chronic illness care: achieving treatment targets, case fatality rates, other clinical outcomes, and proxies for health care quality.

Achieving Treatment Targets The United States is making progress in meeting specified treatment targets, especially those established in practice guidelines, quality performance indicators, and criteria used for pay-for-performance incentives. Establishing higher reimbursements and other incentives has spurred many U.S. providers and hospitals to improve their performance outcomes (Epstein, 2007; Institute of Medicine, 2007c). Treatment goals for controlling hypertension, elevated serum lipids, and diabetes

BOX 4-2
Case Study:
Trauma Care in the United States

Circumstances in the United States could affect the ability of the health care system to render aid to victims of transportation-related injuries and violence, two leading contributors to the U.S. health disadvantage relative to other high-income countries. These circumstances illustrate not only the role of the health care system, as discussed in this chapter, but the important interconnections with socioeconomic factors and public policy as discussed in subsequent chapters. This interdependence is illustrated by two barriers to trauma care services in the United States— lack of health insurance and the geography of the United States—both of which may affect survival and rehabilitation (Greene et al., 2010).

Lack of Health Insurance: Case-fatality rates from the National Trauma Data Bank indicate that injury victims are more likely to die at hospitals with a large percentage of minority patients, and this risk is compounded if they are uninsured (Haider et al., 2012). A separate study reported that risk of death on the first hospital day after injury differs by insurance status, and this disparity becomes more pronounced throughout the hospital stay (Downing et al., 2011). After excluding on-scene deaths, Harris and colleagues (2012) found that U.S. assault victims brought to high-level trauma centers were more likely to die if they were black, even after adjusting for other variables.

These associations raise as many questions as they answer and point to "upstream" factors examined in subsequent chapters of the report. For example, disparities in the outcomes of trauma care do not always appear to relate to the quality of care provided at the institutions themselves (Vettukattil et al., 2011). The trauma literature points to the socioeconomic status of patients and the infrastructure and resources available to trauma centers operating in underserved areas. For example, as Haider and colleagues (2012, p. 68) explain:

> [t]he underinsured population with likely much less resources, which is seen at predominantly minority hospitals, may bring significant residual confounding that could not be controlled for. Issues such as treatment delay, health illiteracy, and differential rates of follow-up and access to rehabilitation services have been implicated as potential reasons for the worse quality of care and worse outcomes among uninsured patients. Additional issues at public hospitals include nurse staffing shortages, constrained budgets, and lack of capital and technical support.

This analysis points directly to the relevance of the socioeconomic conditions of trauma victims (see Chapter 6) and to the role of public policies in shaping conditions that affect health (see Chapter 8).

Geography: The large rural expanses in the United States have relevance to the death toll from transportation-related injuries, because 61 percent of traffic fatalities occur in rural locations (Zwerling et al.,

2005). Trauma outcomes appear to be worse for U.S. patients in nonurban areas (Sihler and Hemmila, 2009; Zwerling et al., 2005). Although factors unrelated to health care could contribute greatly to crash survival (e.g., rural road design and vehicles, speed limits, and the age, alcohol levels, and health status of rural drivers) (Transportation Research Board, 2011; Zwerling et al., 2005), an important factor is how quickly victims can be stabilized and transported to trauma centers by emergency medical personnel. Both time and distance have been shown to inversely affect survival from major trauma in rural areas (Durkin et al., 2005; Grossman et al., 1997; Howell et al., 2010). An analysis in the 1990s reported that emergency response time, scene time, and transportation time to the hospital were longer for rural victims of major trauma than for urban victims. Trauma victims transported by helicopter have lower mortality rates than those conveyed by ground transportation (Sullivent et al., 2011).

Rural physicians and hospitals that are close to crash sites may lack capacity to stabilize patients. One study demonstrated that mortality from motor vehicles crashes was lower in counties with 24-hour availability of a general surgeon, orthopedic surgeon, neurosurgeon, computed tomographic scanner, and operating room and in those with trauma centers (Melton et al., 2003). The absence of onsite specialists can be consequential. For example, survival from traumatic brain injury is improved if there is less than a 4-hour delay between arrival in the emergency department and the performance of a craniotomy or the drainage of a hematoma (Kim, 2011).

Helicopter transportation can therefore be important in saving time and reaching qualified trauma centers, but resources for such services are uneven across rural U.S. counties and are determined by diverse stakeholders. This, too, illustrates the interconnections between health care and public policy. The same applies to the staffing of medical helicopters, which differs in the United States. In a survey of emergency medical services for mountain areas of 14 countries in Europe and North America, Brugger and colleagues (2005) found that 63 percent of European helicopters have a physician on board and 18 percent are staffed with a paramedic, as compared with 32 percent and 60 percent, respectively, of North American helicopters. Policy challenges, among them limited budgets (see Chapter 8), may make it difficult to place more physicians on medical helicopters, especially in rural areas.

Finally, survival from transportation-related injuries or violence cannot be evaluated in isolation from other conditions responsible for the U.S. health disadvantage, such as obesity or diabetes, because comorbidity from chronic illnesses increases the risk of death from injuries (Morris et al., 1990).

rely heavily on the use of prescription drugs, and the United States has higher per capita consumption of pharmaceuticals than peer countries (Morgan and Kennedy, 2010; Squires, 2011). In 2009, per capita spending on pharmaceuticals in the United States was $947, nearly twice the OECD average of $487 (OECD, 2011b). Evidence is available on how the United States compares with other countries in achieving specific cardiovascular and diabetes treatment targets.

Cardiovascular Care U.S. patients appear more likely than those in peer countries to have their blood pressure and serum cholesterol levels checked (Davis et al., 2010; Schoen et al., 2004; Thorpe et al., 2007). The use of preventive drugs for people at risk of cardiovascular disease is more common in the United States than in Europe (Crimmins et al., 2010). In a comparison of medication use in the United States with 10 European countries, Thorpe and colleagues (2007) found that use of antihypertensive agents did not differ significantly but that use of cholesterol-lowering drugs and medication for heart disease was greater in the United States than in Europe. A National Research Council (2011) study also documented that patients with high blood cholesterol and hypertension were more likely to receive medications in the United States than in comparable countries. A 2004 analysis of survey data collected in the 1990s demonstrated that blood pressure was more effectively controlled in the United States than in Canada or Europe (Wolf-Maier et al., 2003), but a more recent patient survey did not reach the same conclusion (Schoen et al., 2011). There is also some evidence that the speed of cardiovascular care for acute coronary syndrome in the United States may match or exceed that of Europe (Goldberg et al., 2009).

Diabetes Care The United States may be less exemplary than other countries in meeting testing and treatment targets for diabetes care. In one survey, patients with diabetes in half the countries were more likely to report a recent hemoglobin A1c test, foot examination, eye examination, and serum cholesterol measurement than patients in the United States (Schoen et al., 2009b). An OECD report found that the United States ranked fourth among 12 countries in the frequency of eye examinations of patients with diabetes (OECD, 2007).

Case-Fatality Rates A measure of the quality of care of life-threatening illnesses is the probability of death following treatment, also known as the case-fatality rate. According to the OECD, U.S. patients admitted for acute myocardial infarction have a relatively low age-adjusted case-fatality rate within 30 days of admission (4.3 per 100 patients) compared with the OECD average (5.4 per 100 patients); however, as shown in Figure 4-2, they have a higher rate than patients in six peer countries. An earlier OECD analysis, based on mortality data from the 1990s, reported that the United States had low case-fatality rates at 30 days, 90 days, and 1 year after acute myocardial infarction (Moise et al., 2003). In a comparison of 5-year mortality rates following acute myocardial infarction among U.S. and Canadian patients, Kaul and colleagues (2004) found that U.S. patients had significantly lower rates, 19.6 percent versus 21.4 percent for Canadians.

The U.S. age-adjusted 30-day case-fatality rate for ischemic stroke is 3.0 per 100 patients, which is below the OECD average of 5.2 per 100 patients, but it is higher than those of four peer countries (Denmark, Finland, Japan, and Norway) (OECD, 2011b). An earlier OECD analysis reported that the U.S. 1-year case-fatality rate from stroke was higher than the OECD average (Moon et al., 2003).

One study calculated the ratio between diabetes mortality for 1994-1998 and incidence at ages 0-39 in 29 industrialized countries. The United States had the 10th highest ratio—higher than all Western European countries,

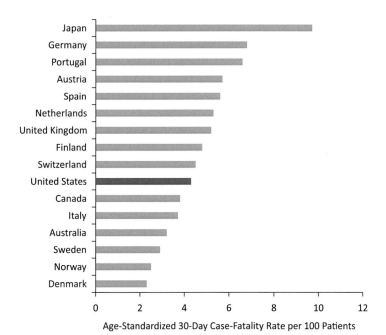

Age-Standardized 30-Day Case-Fatality Rate per 100 Patients

FIGURE 4-2 In-hospital case-fatality rates for acute myocardial infarction in 16 peer countries.
NOTES: Data are for 2009 or nearest year; data apply to deaths within 30 days of admission for acute myocardial infarction.
SOURCE: Data from OECD (2011b, Figure 5.3.1).

Canada, Australia, and New Zealand—but the comparison was subject to a variety of limitations (Nolte et al., 2006).

Other Clinical Outcomes Apart from time-limited case-fatality rates, the panel found no comparable data for comparing the effectiveness of medical care across countries. Data are available for comparing cancer survival rates, which are generally higher in the United States, but cancer survival is confounded by lead-time and length biases introduced by screening (Ciccolallo et al., 2005), a more common practice in the United States than elsewhere. U.S. patients may live longer after their cancer diagnosis simply because the disease is detected at an earlier stage, not because death is delayed. This screening artifact could explain both the higher incidence (Thorpe et al., 2007) and survival rates (Gatta et al., 2000; Verdecchia et al., 2007) for cancer reported by the United States.

Proxies of Health Care Quality There is some evidence that U.S. patients may be more likely to experience postdischarge complications and require readmission to the hospital than do patients in other countries. In one survey, U.S. patients were more likely than those in other surveyed countries to report visiting the emergency department or being readmitted after discharge from the hospital (Schoen et al., 2009). Another study reported that 30-day readmission rates for a common form of myocardial infarction were higher in the United States than in Canada, Australia, New Zealand, and 13 European countries (Kociol et al., 2012).[13]

Little evidence exists to compare the frequency of hospitalization for ambulatory care-sensitive conditions (Institute of Medicine, 2009d)—a proxy for the quality of outpatient care—except for two conditions (asthma and diabetes), and they portray different patterns. Although OECD (2011b) data for peer countries indicate that the United States has the highest asthma hospitalization rate among persons age 15 and older, the U.S. admission rate for uncontrolled diabetes in the same age group is below the OECD average (see Figures 4-3 and 4-4).[14] These proxies are imperfect because countries may differ in their capacity to manage uncontrolled disease complications outside the hospital.

Outcomes after organ transplantation offer an interesting comparative picture of the quality of perioperative care and subsequent chronic care in the United States. Dawwas and colleagues (2007) compared outcomes for

---

[13]In a pattern observed by other health services researchers, Kociol and colleagues (2012) observed that differences in readmission diminished after adjusting for length of stay. Lengths of stay in the United States are shorter than those in other countries and may contribute to higher readmission rates (see Baker et al., 2004).

[14]Earlier OECD data (from 2007) reported that the United States had the highest rate of lower extremity amputations for diabetes among the peer countries (OECD, 2009c).

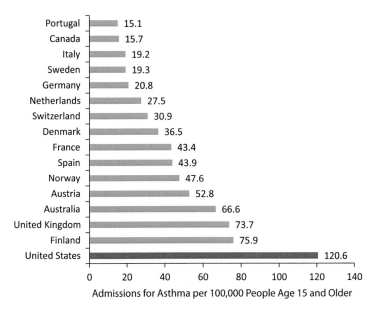

FIGURE 4-3 Hospital admissions for asthma in 16 peer countries.
NOTE: Rates are age-standardized and based on data for 2009 or nearest year.
SOURCE: Data from OECD (2011b, Figure 5.1.1, p. 107).

adults who underwent a first single organ liver transplant between 1994 and 1995 in the United Kingdom or Ireland with those in the United States. Risk-adjusted mortality in both countries was generally higher than in the United States during the first 90 days, equivalent between 90 days and 1 year post-transplantation, and lower than the United States after the first post-transplant year. "Our results are consistent with the notion that the United States has superior acute perioperative care whereas the UK appears to provide better quality chronic care following liver transplantation surgery" (Dawwas et al., 2007, p. 1,606).

Another imperfect measure of the performance of health care systems is to estimate the mortality that is considered amenable to health care (Nolte and McKee, 2003; Rutstein et al., 1976). Relying on the assumption that all deaths from a list of more than 30 causes (and 50 percent of deaths from ischemic heart disease) could be averted by better health care,[15] Nolte and McKee concluded that the United States had the highest amenable mortality

---

[15]This measure, which has been used in a number of studies (Bunker et al., 1994; Commonwealth Fund Commission on a High Performance System, 2011), relies on certain assumptions about attributable mortality and does not adjust for disease prevalence (Gay et al., 2011).

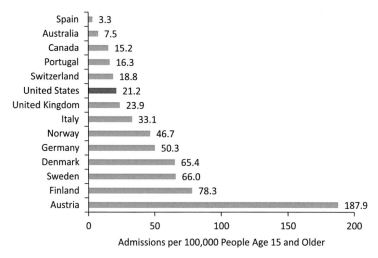

FIGURE 4-4 Hospital admissions for uncontrolled diabetes in 14 peer countries.
NOTE: Rates are age-sex standardized, and they are based on data for 2009 or nearest year.
SOURCE: Data from OECD (2011b, Figure 5.1.1, p. 107).

rate among 16 countries (Nolte and McKee, 2011). Building on this analysis for a larger set of countries, the Commonwealth Fund Commission on a High Performance System concluded (2011, p. 9):

> The U.S. now ranks last out of 19 countries on a measure of mortality amenable to medical care, falling from 15th as other countries raised the bar on performance. Up to 101,000 fewer people would die prematurely if the U.S. could achieve leading, benchmark country rates.

Medical Errors

U.S. patients surveyed by the Commonwealth Fund were more likely to report certain medical errors and delays in receiving abnormal test results than were patients in most other countries (Schoen et al., 2011). U.S. patients with chronic illnesses were more likely than those in all comparable countries included in the survey to recall a medical error (Schoen et al., 2009b).[16] Survey data about perceived errors must be interpreted cautiously, however, because contextual variables may influence perceptions

---

[16]Conversely, another survey found that U.S. patients with chronic illnesses or recent needs for acute care were least likely to report a hospital-related infection (Schoen et al., 2011).

and interpretations of events (Davis et al., 2010). Clinically recorded errors are also imperfect and are only available across countries for a few indicators. According to OECD data, the incidence of postoperative pulmonary embolism or deep vein thrombosis in the United States is 1,019 per 100,000 discharges (the second highest rate among peer countries), three peer countries have higher rates than the United States for postoperative sepsis, and five have higher rates for accidental puncture/laceration and leaving a foreign body in during a procedure (OECD, 2011b).[17]

Optimizing Care Delivery

A factor that could diminish the effectiveness of health care in the United States is disruptions in the care delivery process. For many years, quality improvement programs and health services research have recognized that the fragmented nature of the U.S. health care system, miscommunication, and incompatible information systems foment lapses in care; oversights and errors; and unnecessary repetition of testing, treatment, and associated risks because records of prior services are unavailable (Fineberg, 2012; Institute of Medicine, 2000, 2010).[18] Problems are more pronounced during "handoffs," when patients transition from one care setting to another. Differences in medical error rates between countries have an independent association with breakdowns in care coordination (Lu and Roughead, 2011).

The only detailed data to compare care delivery practices across countries come from surveys conducted each year by the Commonwealth Fund. These data have a variety of limitations. For example, they rely on perceptions (of patients and physicians) rather than independently documented outcomes. Although the surveys have been administered annually since 1998 to thousands of patients and physicians in up to 11 countries, they include dozens of questions about care delivery practices that have varied in wording and administration methods over the years. However, a consistent pattern emerges in the U.S. responses (see Box 4-3). U.S. patients generally give their physicians high marks in the attention they pay to clinical details, to engaging

---

[17]The United States (12.5 percent) and Canada (13.7 percent) have the highest rate of obstetrical trauma among 20 OECD countries (OECD, 2011b).

[18]The question of whether physicians in the U.S. system are less effective in producing health than are physicians in other OECD countries has also been studied. Although specific results varied with the health indicator chosen, Or et al. (2005) found that the productivity of U.S. physicians was typically near the middle of the range. Unusually low physician productivity would not, therefore, appear to contribute to the U.S. health disadvantage.

BOX 4-3
Quality of Care:
Survey Findings from Commonwealth Fund Surveys

Strengths: United States at or Better Than Average
- *Attention to clinical detail*
  Practice knows important information about medical history
  Pharmacist/physician reviews and discusses medications
  Tracking adverse events
  Regular self-assessment of outcomes and patient satisfaction
- *Patient-centered communication*
  Encourages questions
  Discusses goals and priorities
  Explains treatment options
  Involves patient in decision as much as wanted
  Helps make daily treatment plan
  Gives clear instructions about warning symptoms
- *Hospital discharge planning*
  Instructions about symptoms to watch for and when to seek
      further care[a]
  Who to contact for questions about condition or treatment[a]
  Written care plan for care after discharge[a]
  Arrangements made for follow-up visits
  Clear instructions about medications to take[a]

Weaknesses: United States Worse Than Average
- *Coordination of care*
  Wasted time
  Wasted time a "major problem" for primary care physician
  Unnecessary treatment
  Duplicate testing[b]
  No tracking system to ensure results reach clinician
  No system for physician to send reminders
  Not using nonphysician staff to coordinate care
  Not using written guidelines
- *Medical errors*
  Medical mistake made in treatment
  Given wrong medication or wrong dose
  Given incorrect test results on diagnostic test
  Delays in being notified of abnormal results

- **Dissatisfaction with health care system**
  Patient dissatisfaction with the health system[b]
  Primary care provider dissatisfaction with the health system
- **Miscommunication**
  Communication between providers:
    Not sharing medical information
    Not informed about specialist consultation
    Regular doctor not informed about hospitalization or surgery
  Communication between provider and patient:
    Getting answer on day called
    Obtaining advice from help line
    Spends enough time with them
    Explanations easy to understand
- **Inadequate information systems**
  No electronic medical record
  No capacity for electronic ordering of laboratory tests
  No capacity for electronic entry of clinical notes
  No electronic access to test results
  No capacity to electronically prescribe medications
  No electronic alerts/prompts about drug dose/interactions
  No computerized reminders for guideline-based interventions or
    screening tests
  Medical record system cannot generate list of:
  ○ patients due for tests or preventive care
  ○ all medications taken by patient
  ○ patients due for tests or preventive care
  ○ patients by diagnosis or test result

NOTES: Survey findings are based on self-report of survey participants (patients and providers). Data based on surveys conducted in 2004, 2008, 2009, 2010, and 2011. The countries included in the 2011 survey were Australia, Canada, France, Germany, the Netherlands, New Zealand, Norway, Sweden, Switzerland, the United Kingdom, and the United States. Earlier surveys included fewer countries.

[a]More than half of surveyed countries reported a higher prevalence of problems than in the United States.

[b]Half or fewer surveyed countries reported a higher prevalence of problems.

SOURCES: Schoen et al. (2004, 2009a, 2009b, 2010, 2011).

patients in decision-making conversations, and to discharge planning[19] after hospitalization or surgery. However, U.S. respondents are more likely than those in the other surveyed countries to have problems in four key areas that could affect the quality of care outside the hospital, particularly management of chronic illnesses: confusion and poorly coordinated care, inadequate information systems to access needed clinical data, miscommunication between providers and between patients and providers, and medical errors.

- **Poor coordination:** reported problems included unnecessary treatment, duplicate testing, wasted time, not ensuring laboratory results reach the clinician, not sending reminders to patients, not using nonphysician staff to coordinate care, and not spending enough time with the patient;[20]
- **Inadequate information technology:** reported problems include lack of electronic medical records; inability to electronically order laboratory tests, access test results, prescribe medications, enter clinical notes, or receive drug alerts; and inability to generate lists of patients with specific conditions (e.g., diabetes), laboratory abnormalities, overdue tests or vaccines, or medications;[21]
- **Miscommunication:** reported problems include physicians not sharing important medical information with each other; "regular" physicians not being informed about specialist care or hospitalizations; test results, medical records, or reasons for referral not being available in time for appointments; and patients not getting a quick telephone response from their regular provider on the day they call with a medical question or from help lines; and

---

[19]U.S. patients who had been hospitalized were more likely than their counterparts in all other countries to report receiving written care plans, arrangements for follow-up visits, instructions about medications warning symptoms, and information about whom to contact with questions (Davis et al., 2010).

[20]Such problems are compounded when multiple providers are involved. When four or more physicians were involved, 45 percent of U.S. patients reported a medical test or record coordination problem, compared with 21-35 percent in the seven comparison countries (Schoen et al., 2009b).

[21]Some national health systems have centralized databases that are used to identify people in need of public health and preventive services or for outreach for chronic illness management. Long-standing population-based cancer registry systems with national coverage (often regionally organized) and with virtually complete case follow-up exist in all Nordic countries, the United Kingdom, and many Baltic and central European countries (Quinn, 2003). In most European countries, organized breast and cervical cancer screening programs can use these databases to mail periodic screening invitations to all women in the target age group (Howard et al., 2009). Use of such registries in Sweden and other countries has been shown to improve health outcomes, often at lower cost (Larsson et al., 2012).

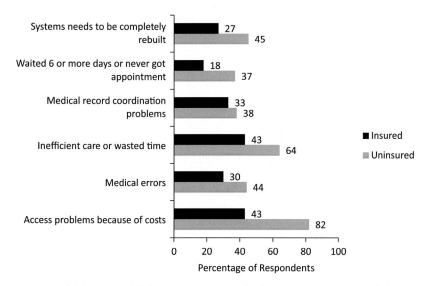

FIGURE 4-5 Frequency of complaints among insured and uninsured U.S. patients with chronic conditions.
NOTE: Based on surveys of patients with chronic illnesses conducted by the Commonwealth Fund.
SOURCE: Adapted from Schoen et al. (2009b, Exhibit 6, p. w12).

- **Medical errors:** reported problems included medical mistakes, incorrect medication or dosage, incorrect results on diagnostic tests, and delays in being notified of abnormal results.

Among surveyed countries, U.S. patients and physicians are most likely to express dissatisfaction with the health system and to recommend rebuilding it (Davis et al., 2010; Schoen et al., 2009a, 2009b, 2011).[22]

Could these coordination problems reflect the large proportion of U.S. patients who lack health insurance coverage? In 2008, the Commonwealth Fund stratified the survey responses of chronically ill patients based on their insurance status. As shown in Figure 4-5, coordination problems were more common among the uninsured, as would be expected, but large proportions of insured patients (up to 43 percent) also reported difficulties getting appointments, inefficient care or wasted time, and medical test or record coordination problems. One in four insured patients was sufficiently dissatisfied to recommend rebuilding the health system (Schoen et al., 2009b).

---

[22]In the 2009 survey, German physicians were more likely than U.S. physicians to recommend completely rebuilding the health care system (Schoen et al., 2009a).

QUESTION 3:
DO U.S. HEALTH SYSTEMS EXPLAIN THE
U.S. HEALTH DISADVANTAGE?

The evidence reviewed above supports the following conclusions: The U.S. public health system is more fragmented than those in other countries, but there are insufficient data to compare core public health functions cross-nationally. More data are available for comparing health care systems across countries. American patients and primary care physicians are more dissatisfied with their health care system and are more likely to want major reforms than are patients and physicians in other countries. A conspicuous problem in the United States is the lack of universal health insurance, something recent reforms have sought to address, but deficiencies in access and quality are pervasive and plague even insured and high-income patients. Notably, U.S. patients with complex care needs—insured and uninsured alike—are more likely than those in other countries to complain of medical costs or defer recommended care as a result.

The United States has fewer practicing physicians per capita than comparable countries. Specialty care is relatively strong and waiting times for elective procedures are relatively short, but Americans have less access to primary care. Continuity of care is weaker in the United States than in other countries: U.S. patients with complex illnesses are less likely to keep the same physician for more than 5 years. Compared to people living in comparable countries, Americans do better than average in being able to see a physician within 1-2 days of a request, but they find it more difficult to obtain medical advice after business hours or to get calls returned promptly by their regular physicians.

There appear to be differences in the quality of hospital and ambulatory care across countries. Compared with most peer countries, U.S. patients who are hospitalized with acute myocardial infarction or ischemic stroke are less likely to die within the first 30 days. And U.S. hospitals also appear to excel in discharge planning. However, quality appears to drop off in the transition to long-term outpatient care. U.S. patients appear more likely than those in other countries to require emergency department visits or readmissions after hospital discharge, perhaps because of premature discharge or problems with ambulatory care.

The U.S. health system shows certain strengths: cancer screening is more common in the United States, enough to create a potential lead-time increase in 5-year survival. Pharmacotherapy and control of blood pressure and serum lipids are above the average for comparable countries. However, systems to manage illnesses with ongoing, complex care needs appear to be weaker. Long-term care for older adults is less common. U.S. primary care physicians are more likely to lack electronic medical records, registry capacities, tracking systems for test results, and nonphysician staff to help with care management. Confusion, poor coordination, and miscommunication are reported more often in the United States than in comparable countries. Moreover, these problems are reported in large numbers by insured and above-average income patients.

Whether poor coordination of complex care needs for chronic conditions—such as asthma, congestive heart failure, depression, and diabetes—is contributing to the U.S. health disadvantage is still unclear. The current evidence is mixed. For example, U.S. hospitalizations for asthma are among the highest of peer countries, but asthma is influenced by factors outside of health care (e.g., air pollution, housing quality) (Etzel, 2003; Lanphear et al., 2001; Sly and Flack, 2008). Testing of patients with diabetes may be less common in the United States than in some other countries, but only five peer countries have a lower rate of hospitalizations for uncontrolled diabetes.

The quality problems with U.S. ambulatory care, though recognized, should not be overstated. The same surveys that describe coordination problems also suggest that U.S. primary care physicians perform as well as those in other countries in some aspects of care coordination, such as being attentive to clinical details, using reminders to monitor test results, and giving patients medication lists and written instructions. U.S. physicians reportedly perform better than their counterparts in providing patient-centered communication.

WHAT U.S. HEALTH SYSTEMS CANNOT EXPLAIN

Problems with health care in the United States are important, but at best, they can explain only part of the U.S. health disadvantage for three reasons. First, some causes of death and morbidity discussed in Part I are only marginally influenced by health care. For example, homicide and suicide together account for 23 percent of the extra years of life lost among U.S. males relative to other countries (see Chapter 1), but victims often die on the scene before the health care system is involved, especially when firearms are involved. Deficiencies in ambulatory care in the United States bear little on the large number of deaths from transportation-related injuries. Access to emergency medical services and skilled surgical facilities could play a role, but there is no evidence that rescue services or trauma care in the United States are inferior to the care available in other countries (see Box 4-2). Other factors, ranging from road safety to drunk driving and socioeconomic conditions, may matter more (Transportation Research Board, 2011).

Second, although poor medical care could be plausibly linked to communicable and noncommunicable diseases, which claim 20-30 percent of the extra years of life lost in the United States (see Chapter 1), the available evidence for two common noncommunicable diseases—myocardial infarction and ischemic stroke—suggests that U.S. outcomes are better than the OECD average. The United States excels in performing screening tests that are known to reduce mortality. However, it is possible that the health disadvantage arises from shortcomings in care outcomes that are not currently measured and from gaps in insurance, access, and coordination. Even the measures that are available for myocardial infarction and stroke are limited to short follow-up periods after the acute event, and outcomes may deteriorate thereafter.

Part I lists nine domains in which the U.S. health disadvantage is documented: (1) adverse birth outcomes (e.g., low birth weight and infant mortality); (2) injuries, accidents, and homicides; (3) adolescent pregnancy and sexually transmitted infections; (4) HIV and AIDS; (5) drug-related mortality; (6) obesity and diabetes; (7) heart disease; (8) chronic lung disease; and (9) disability. Deficiencies in public health systems or in access to quality health care could conceivably play a role in each of these domains. For example, the United States has a high rate of preterm births (see Chapter 2), a large proportion of which appear to be initiated by health care providers (Blencowe et al., 2012). Higher death rates from HIV infection could relate to deficiencies in care. Other U.S. health disadvantages may reflect some degree of inferior medical care, but empirical evidence for any such hypotheses is lacking.

Third, even conditions that are treatable by health care have many origins, and causal factors outside the clinic may matter as much as the benefits or limitations of medical care. For example, smoking and obesity are heavily influenced by the environment and policy decisions (see Chapters 5 and 7). Physicians play an important, but marginal, role in screening for unhealthy behaviors, measuring body weight, prescribing adjunctive pharmacotherapy to support smoking cessation or weight management, performing bariatric surgery for morbid obesity, and referring patients to telephone quit lines and other intensive behavioral counseling programs (Fielding and Teutsch, 2009; Ogden et al., 2012b; Woolf et al., 2005). Physicians can write prescriptions for antihypertensive drugs, statins, oral contraceptives, and antibiotics and antiretroviral agents for sexually transmitted infections and HIV infection. They can encourage healthy behaviors, but other factors exert greater influences on diet, physical activity, sexual habits, alcohol and other drug use, and needle exchange practices (Woolf et al., 2011). Pediatricians can remind parents to secure their children in car seats, but they cannot control motor vehicle crashes. Physicians can screen for and treat depression and be alert for suicidal ideation and signs of family violence but they have limited influence on the prevalence of firearms or the societal conditions that precipitate crime and violence.

CONCLUSIONS

One difficulty in attributing the U.S. health disadvantage to deficiencies in the public health or medical care system is that countries with better health outcomes lack consistent evidence that their systems perform better. In some countries, patients are more likely to report problems. For example, Sweden consistently ranks among the healthiest countries in the OECD, but, in the Commonwealth Fund surveys, its patients were more likely than U.S. patients to report problems with chronic illness care. Sweden has high hospitalization rates for uncontrolled diabetes (Figure 4-4). In 2007, Switzerland had the highest male life expectancy among the 17 peer countries (see Table 1-3, in Chapter 1), but the availability of general practitioners is the second lowest (see Figure 4-1). Australia has the second highest male life expectancy of the peer countries (see Table 1-3, in Chapter 1), but it has the fifth highest case-fatality rate for ischemic stroke (OECD, 2011b). The Netherlands, which ranks highly on many surveys by the Commonwealth Fund, has historically had shorter life expectancy than some other comparable countries.

Various potential explanations could account for these inconsistencies. The simplest is that medical care matters little to health, a thesis that some have advanced as part of a more general argument that health is shaped primarily by the social and physical environment. Indeed, some studies have already questioned whether there is specific evidence to implicate the health care system as the cause of the U.S. mortality disadvantage after age 50 (Ho and Preston, 2010; National Research Council, 2011).[23]

A second possibility is that health care does matter but that only certain aspects affect outcomes. For example, deficiencies in mammography screening or printing medication lists may not matter, and countries with consistently superior health outcomes may excel in the facets of health care that are consequential. Health care may also matter more in certain places or for certain patient populations.

A third explanation—which the panel deems most likely—is that health care exerts a partial influence on health outcomes in concert with other important determinants of health such as lifestyle, socioeconomic status, and public policy. Longer life expectancy and improved health is probably traceable to some combination of health system characteristics and these

---

[23]That study focused on the population age 50 and older, for whom deficiencies in medical care in the United States may be less of an issue because of Medicare, which serves adults age 65 and older and the disabled. The study also examined a smaller set of indicators than are reviewed in this chapter, and based on those indicators, found little evidence to suspect that the quality of health care was responsible for the growing mortality disadvantage among older Americans compared with seniors in other countries.

other individual and community conditions, but the exact contribution of each factor is unknown and may vary over place and time.

A life-course perspective adds additional complexity to the analysis because differences in health outcomes may relate not only to contemporaneous characteristics of health systems, but also to those that existed years earlier when current conditions or diseases were developing. This scenario is especially true for chronic diseases like diabetes and heart failure, which claim lives decades after problems with cardiovascular risk factors and glycemic control first appear. For such conditions, deficiencies in primary care in the 1970s and 1980s may explain current death rates better than the features of today's health systems. The current health system matters more for care conditions that lead directly to health outcomes, such as birth outcomes and survival after a car crash or gunshot wound.

The research comparing health care systems cross-nationally is still evolving and cannot yet support any definitive conclusions about how the U.S. health system might contribute to or ameliorate the U.S. health disadvantage. Comparable international data for meaningful inferences require better data on both dependent (health outcomes) and independent variables (health systems). Although data from the OECD and WHO provide some comparative information on a handful of health system measures, these are much like the keys under the lamppost. A richer and more comprehensive set of data on a variety of carefully selected dimensions of morbidity and mortality and outcomes of care would be needed across countries to make valid comparisons.[24]

Few indicators for assessing the various dimensions of health care have been developed or undergone proper scientific validation. In particular, questions used on surveys such as those conducted by the Commonwealth Fund, which are widely cited in this chapter, have unknown correlations with health outcomes and may have variable meanings across countries. Limitations in statistical power and wide confidence intervals may limit the significance of rankings between one country and another or changes in ranking from year to year. Some questions used by the Commonwealth Fund change from year to year; these changes offer new insights on health systems, but they make it difficult to compare outcomes across time. The Commonwealth Fund gives equal weight to each measure; some weighting is probably warranted, but an empirical basis is lacking to know which characteristics patients value more highly or are more predictive of health outcomes.

Even the proper domains for assessing the performance of health sys-

---

[24]Such data are lacking even within the United States. A recent Institute of Medicine (2011e) report indicated the lack of adequate data to evaluate the health of the American public or the performance of governmental public health agencies and recommended bold transformation of the nation's health statistics enterprise.

tems have yet to be identified. In the first major attempt to rank health care systems, the WHO *World Health Report 2000* introduced a ranking based on health attainment, equity of health outcomes, "patient responsiveness," and "fairness of financial contributions" (World Health Organization, 2000b). The U.S. health system ranked 37th based on this methodology, but the measures, methods, and data were criticized (Jamison and Sandbu, 2001; Navarro, 2002). Another such effort is that of the Commonwealth Fund, which established a Commission on a High Performance Health System in 2005 that regularly issues a "national scorecard" based on five dimensions: quality, access, efficiency, equity, and long, healthy, and productive lives (Commonwealth Fund Commission on a High Performance Health System, 2011). In 2008, WHO identified five shortcomings in health care delivery that are found in systems around the world: inverse care, impoverishing care, fragmented and fragmenting care, unsafe care, and misdirected care (World Health Organization, 2008b). International health experts have not reached consensus on the optimal parameters for measuring and tracking the performance of national health systems.

Statistics for all these dimensions are difficult to capture. The capacity of different countries to collect appropriate data and to do so systematically—using consistent sampling procedures, data collection techniques, coding practices, and measurement intervals (e.g., annually)—is challenging for practical reasons and limited budgets. To cite just one example, patient safety indicators for hospital care are not standardized across countries (Drösler et al., 2012). Access to medical records or administrative data is uneven across countries. International surveys face methodological challenges that introduce sampling biases. One example is survey methodology: some surveys have used a combination of landlines and mobile telephones to conduct interviews, and some countries have low response rates or mobile telephone usage. Adults with complex conditions, low income, or language barriers may be undersampled. Surveys of patients or physicians' perceptions of the quality of care are ultimately perceptions and may not correspond with objective measures. The research challenges and priorities to address these gaps in the science are discussed further in Chapter 9, along with recommendations to remedy the problem.

Despite these limitations, the existing evidence is certainly sufficient for the panel to conclude that public health and medical systems in the United States have important shortcomings, some of which appear to be more pronounced in the United States than in other high-income countries. Subsequent chapters address the factors outside the clinic that may lead to greater illness and injury among Americans, but health problems ultimately lead most people to the health care system, or at least to attempt to obtain clinical assistance. The difficulties Americans experience in accessing these services and receiving high-quality care, as documented in this chapter, cannot be ignored as a potential contributor to the U.S. health disadvantage.

5

Individual Behaviors

Chronic diseases, such as heart disease, stroke, and cancer, often arise as a by-product of how people live and behave. The epidemiologic transition to chronic diseases—in which conditions like atherosclerosis and diabetes have replaced infectious diseases as the leading causes of death—has focused increasing attention on the central role of personal health behaviors in health (Olshansky and Ault, 1986). Two decades ago, in an article titled "Actual Causes of Death," tobacco use, diet, physical inactivity, and other personal behaviors were identified as the leading killers in the modern age (McGinnis and Foege, 1993). That work has since been refined in the United States and replicated throughout the world. It is now widely recognized that health behaviors are the leading contributors to the global burden of disease, especially in high-income countries where the epidemiologic transition has been longstanding (Lopez et al., 2006; Olshansky and Ault, 1986; World Health Organization, 2008a).

Can the U.S. health disadvantage be explained by a prevalence of unhealthy behaviors that is higher among Americans than among people in other high-income countries? Some analyses have questioned whether health behaviors are at the root of the U.S. health disadvantage (see, e.g., Avendano et al., 2009). This chapter examines these behaviors, and it also considers sexual practices, drug use, and injurious behaviors that might also contribute to the U.S. health disadvantage documented in Part I, particularly among younger adults.

A life-course perspective is important in studying health behaviors, many of which are adopted at young ages (especially in adolescence) before becoming life-long habits (see Chapter 3). Smoking is a good example. According to Viner (2012, p. 8):

> [T]he tracking history of smoking is of initiation and increasing use in adolescence, followed by relative stability from the late teens onwards. Compared with other forms of drug use, cigarette smoking shows the least decline in young adulthood. Those who start early in adolescence, e.g., less than 14 years, are more likely to become lifelong smokers.

Comparing health behaviors across high-income countries is difficult due to a scarcity of data and extensive challenges with measurement. In contrast with such health indicators as blood pressure or serum lipid concentrations, which are easy to measure in standardized units across countries, health behaviors are rarely examined in a uniform fashion, even within the same country, and in some countries they are not measured at all. For example, a survey about tobacco use can ask about cigarette smoking only or can also include cigar and pipe smoking or the use of smokeless tobacco. Respondents can be asked how many cigarettes they smoke each day or whether they smoke "currently," have "ever smoked," have ever smoked 100 cigarettes, or have smoked in the past 30 days.

Similarly, questions about physical activity can differentiate between leisure-time "exercise" and physical activity, between regular and episodic physical activity, between levels of exertion, and between duration and frequency of activity per day, per week, or per month. They can measure sedentary behavior, moderate activity, and intense physical activity. They can document types of activity, such as sitting, walking, cycling, running, or gardening. No single measure is the established standard and, unlike smoking, no single measure is the strongest predictor of premature death or morbidity or has the strongest link to obesity.

Questions about diet are perhaps the most varied. They can address food groups (e.g., fruits and vegetables, grains, meats), specific nutrients (e.g., saturated and unsaturated fats, *trans* fats, sodium, calcium), and caloric intake. A proper dietary history involves numerous questions (Paxton et al., 2011) and can produce different results depending on how the data are collected (Erinosho et al., 2011). Nutrition science has yet to identify the most important dietary predictors of longevity or of specific disease outcomes, such as cancer or myocardial infarctions. Few countries include sufficient questions on population surveys to evaluate the national diet, and very few questions are asked in common across countries, confounding efforts to make meaningful or accurate comparisons. Often, the only common data across countries are governmental statistics on national caloric expenditures and food consumption divided by the population size to derive per capita estimates.

With these caveats in mind, this chapter examines six behaviors: (1) tobacco use, (2) diet, (3) physical inactivity, (4) alcohol and other drug use, (5) sexual practices, and (6) injurious behaviors.[1] As in other chapters, the panel poses three questions in reference to each behavior:

- Does the health behavior matter to health?
- Is the behavior more prevalent in the United States than in other high-income countries?
- Does the difference in behavior explain the U.S. health disadvantage?

TOBACCO USE

Question 1: Does Tobacco Use Matter to Health?

The enormous health consequences associated with tobacco use were first recognized in the 1950s. Beginning with the landmark 1964 report by U.S. Surgeon General Luther Terry, the health consequences of tobacco use have since become one of the most extensively documented topics in public health history (U.S. Department of Health and Human Services, 1964, 2000). The tobacco literature is too extensive to recapitulate here, nor is it necessary, other than to list the tobacco-related diseases that contribute to the U.S. health disadvantage (see Part I): noncommunicable diseases, such as coronary artery disease, stroke, cancer, and chronic lung disease, as well as adverse birth outcomes, such as low birth weight and infant mortality.

Question 2: Is Tobacco Use More Prevalent in the United States Than in Other High-Income Countries?

A generation ago, Americans led the Western world in tobacco use (Forey et al., 2002)—reaching peak rates in the 1950s—but decades of tobacco control efforts following the 1964 Luther Terry report led to the marked reductions in smoking rates shown in Figure 5-1, from 42 percent in 1965 to 21 percent in 2009 (Garrett et al., 2011). Cigarette consumption rose and fell earlier and more dramatically in the United States than in many other countries (Cutler and Glaeser, 2006; Pampel, 2010). Figure 5-2 shows that the United States now has the lowest adult smoking rates of all peer countries but Sweden and one of the lowest among OECD countries. In many countries, the mass adoption of smoking occurred earlier among males than among

---

[1]The list of behaviors examined in this chapter is not exhaustive. Other behaviors or health practices, such as getting adequate sleep or reducing stress, are also important to health promotion and injury prevention (Smolensky et al., 2011) but are not examined here.

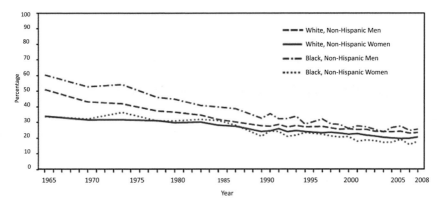

FIGURE 5-1 Percentage of U.S. adults age 18 and older who were current smokers, by sex and race/ethnicity, 1965-2008.
NOTES: Data are based on the National Health Interview Survey (NHIS). For NHIS survey years 1965-1991, current smokers included adults who reported that they had smoked 100 or more cigarettes in their lifetimes and currently smoked. Since 1992, current smokers included adults who reported smoking 100 or more cigarettes during their lifetimes and specified that they currently smoked every day or on some days. Figure depicts trend over time. Data were not available for 1967-1969, 1971-1973, 1975, 1981, 1982, 1984, 1986, 1989, and 1996, because the NHIS did not include questions about smoking.
SOURCE: Garrett et al. (2011, p. 110).

females (Pampel, 2010). In the United States, the highest rates of smoking among males occurred in cohorts born around 1915-1920 and among women in cohorts born around 1940-1944 (Preston and Wang, 2006).

Question 3: Does Tobacco Use Explain the U.S. Health Disadvantage?

Various studies suggest that a large proportion of differences in mortality (notably sex differences) among high-income countries are attributable to tobacco use (Bongaarts, 2006; Janssen et al., 2007; Pampel, 2002; Peto et al., 1992; Retherford, 1975; Valkonen and Van Poppel, 1997; Waldron, 1986). There is a 20-30 year lag between changes in smoking rates and the resulting effects on smoking-related mortality (Lopez et al., 1994): thus, these effects are manifest mainly among older adults. According to Staetsky (2009, p. 892): "The impact of smoking related mortality on old-age mortality in the end of the 20th century is a function of the exposure of middle-aged adults to smoking approximately during the 1970s." Using coefficients derived from lung cancer, she concluded that smoking accounted for the slower decline in mortality among women in Denmark,

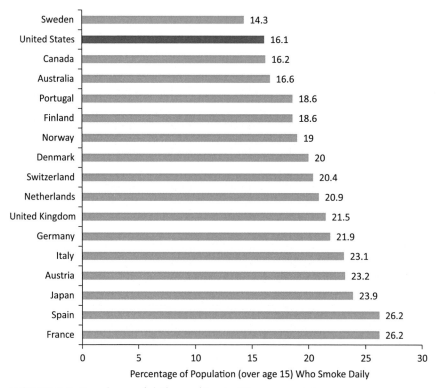

FIGURE 5-2 Prevalence of daily smoking in 17 peer countries.
NOTES: Prevalence rates are for 2005-2009. Prevalence rates for the most current year available are reported.
SOURCE: Adapted from OECD (2011c).

the United States, and the Netherlands when compared with France and Japan (Staetsky, 2009).

Using an innovative macrostatistical method, Preston and colleagues (2010a, 2010b) estimated the attributable fraction of deaths after age 50 from smoking[2] and its effect on life expectancy at age 50 among 10 high-income countries in 1955, 1980, and 2003. The authors calculated that by 2003 smoking accounted for 41 percent of the difference in male life expectancy at age 50 between the United States and 9 comparison countries and for 78 percent of the difference in female life expectancy at age 50 (Preston et al., 2010b).[3] Smoking appeared to have a larger impact

---

[2]The results were modeled on lung cancer mortality, see Preston et al. (2010b).

[3]Attributable fractions produce estimates that are subject to some errors, as reported by the authors. Slightly different results are reported in Preston et al. (2010a).

on women because of the later uptake of smoking by U.S. women (U.S. Department of Health and Human Services, 2000, 2002) (see Figure 5-3). The smoking-attributable fraction of U.S. deaths among males age 50 and older was 23 and 22 percent in 1980 and 2003, respectively, but during the same years increased from 8 to 20 percent among females of the same age (Preston et al., 2010b). Based on the researchers' assumptions, smoking accounted for 67 percent of the shortfall in life expectancy gains that U.S. women experienced relative to 20 other countries between 1950 and 2003.

These findings implicate smoking as a potential cause of the shorter life expectancy of adults age 50 and older, but they do not explain the lower life expectancy observed in younger people. The U.S. health disadvantage before age 50 has worsened over the same time that smoking prevalence rates in this population have decreased. The reduction in smoking rates will produce benefits in years to come. Wang and Preston (2009) predicted that the future will bring a decline in deaths attributable to smoking among men but that improvements for women will occur later.

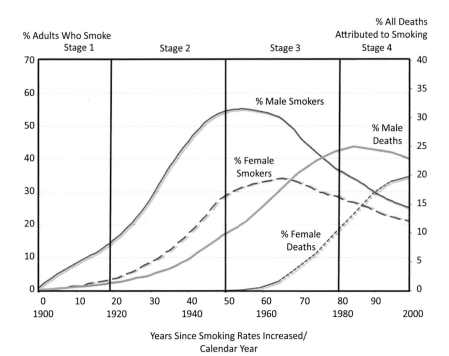

FIGURE 5-3 Four stages of the U.S. tobacco epidemic.
SOURCE: Thun et al. (2012, Figure 1).

DIET

Question 1: Does Diet Matter to Health?

As with tobacco, the evidence linking diet to health is extensive and unnecessary to review here.[4] Caloric intake relative to energy expenditure is a major contributor to obesity and obesity-related diseases such as diabetes. Levels of consumption of specific nutrients (e.g., saturated fats, *trans* fats, sodium, calcium, iron, vitamins) are linked to a variety of diseases (e.g., hypertension, cardiovascular disease, osteoporosis, anemia) (U.S. Department of Agriculture and U.S. Department of Health and Human Services, 2010; Willett, 1998). Consumption of food groups such as fibers and grains are associated with lower risks of hyperlipidemia and some cancers. Low fruit and vegetable intake is associated with increased risk of obesity, coronary heart disease, stroke, diabetes, hypertension, and colorectal cancer (World Health Organization, 2004b). Worldwide in 2000, 5 percent of deaths were attributable to low fruit and vegetable intake (Lock et al., 2004). The World Cancer Research Fund (1997) estimated that up to 30-40 percent of global cancers could be prevented through a change in diet. However, considerable controversy and scientific uncertainty still surround the strength of the evidence linking specific nutrients to disease risks and longevity.

Question 2:
Is an Unhealthy Diet More Prevalent in the United States Than in Other High-Income Countries?

Comparable cross-national data on dietary practices are limited for the reasons noted above, including the challenges that countries face in evaluating the diets of their populations and inconsistencies across countries in food culture, defining indicators, sampling respondents, and administering surveys. Data collected within the United States suggest that the American diet has become less nutritious over time. Between 1971 and 2000, average daily caloric consumption increased from 2,450 kcals to 2,618 kcals among men and from 1,542 kcals to 1,877 kcals among women;[5] similarly, carbohydrate intake increased by 67.7 grams and 62.4 grams, respectively for men and women, and total fat intake increased by 6.5 grams and 5.3

---

[4]We do not review the extensive evidence regarding the dietary benefits of breastfeeding, but we do discuss below the relatively low prevalence of breastfeeding in the United States.

[5]The physical unit "calorie" is the energy required to increase the temperature of 1 gram of water by 1 degree Celsius. The commonly used dietary term "calorie" is shorthand for the scientific term kilocalorie (kcal), which equals 1,000 calories.

grams, respectively, for men and women (Centers for Disease Control and Prevention, 2004).

Between 1950 and 2000, annual per capita food consumption in the United States increased by 20 percent for fruits and vegetables but also for grains (by 44.5 pounds, a 29 percent increase), meats (by 57 pounds, a 41 percent increase), cheese (by 22.1 pounds, a 287 percent increase), and caloric sweeteners (by 42.8 pounds, a 39 percent increase). High-fructose corn syrup consumption per capita rose from zero in 1950 to 85.3 pounds by 2000. Some of these increases may be associated with an increase in dining out, which increased from 18 percent of total food energy consumption in 1977-1978 to 32 percent in 1994-1996 (U.S. Department of Agriculture, 2012).

How do these trends compare with other rich nations? Americans consumed 3,770 kcals per person per day in 2005-2007,[6] more than any other country in the world (see Figure 5-4). This trend is not new: the United States also had the highest caloric consumption in 2003-2005 and ranked fourth in the world in 1999-2001 (behind Austria, Belgium, and Italy). Between 1999-2001 and 2005-2007, the U.S. ranking on fat intake rose from seventh to fourth in the world, with Americans consuming an average of 161 grams per person per day. By comparison, in 2005-2007 the average Swede consumed 17 percent fewer calories and 24 percent less fat (Food and Agriculture Organization, 2010).

Whether other eating habits in the United States differ from those in other countries is less clear due to inadequate data. Only 23 percent of Americans consume fruits or vegetables five times a day (Centers for Disease Control and Prevention, 2012a), as recommended by dietary guidelines (U.S. Department of Agriculture and U.S. Department of Health and Human Services, 2010), but the European Union also reports inadequate intake of fruits and vegetables in its member countries (European Food Information Council, 2012). Some studies have reported that Americans consume fewer fruits and vegetables than people in other countries, such as Canada (Richards and Patterson, 2005) and France (Tamers et al., 2009), but the evidence is inconclusive. Another study reported that Americans consumed more calories per minute than those in Austria, France, Germany, Italy, and the Netherlands; they also spent the least time cooking or eating at home and the most time eating at restaurants (after the French) (Brunello et al., 2008; see also Michaud et al., 2007).[7] Among the 17 peer countries

---

[6]Per capita consumption statistics do not account for wastage and therefore overestimate actual nutrient intake.

[7]A two-country comparison showed a rate of consumption of 53.8 calories per minute for the United States and 28.4 minutes for France (Brunello et al., 2008).

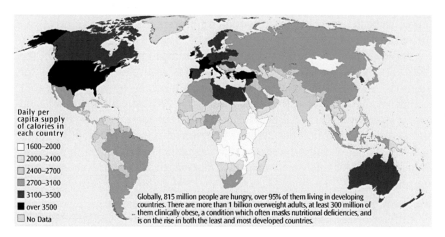

FIGURE 5-4 Global map of per capita caloric intake.
SOURCES: United Nations Development Programme, "Human Development Report 2000," Table 23, http://www.undp.org/hdro/HDR2000.htmlFAO, "The Developing World's New Burden: Obesity," http://www.fao.org/FOCUS/E/Obesity/obes1.htm; World Health Organization, "Global Strategy on Diet, Physical Activity, and Health: Obesity and Overweight," http://www.who.int/dietphysicalactivity/publications/facts/obesity/en/; World Health Organization, "Nutrition for Health and Development," http://www.who.int/nut/publications.htm.
SOURCE: Global Education Project (2004).

examined in Part I, only two countries (Spain and France) report lower rates of breastfeeding of infants (Palloni and Yonker, 2011).

Question 3: Does Diet Explain the U.S. Health Disadvantage?

Being the global leader in caloric intake (see Figure 5-4, above), combined with inadequate physical activity, helps explain escalating obesity rates in the United States. It could plausibly contribute to higher rates of diseases attributable to obesity, such as diabetes and heart disease. High intake of saturated fats and inadequate intake of fresh produce and other healthy foods could explain a variety of diet-related noncommunicable diseases that are more prevalent in the United States than in comparable countries (see Part I). Using the macrostatistical method discussed above for tobacco but with a different set of comparison countries, Preston and Stokes (2011) estimated that obesity accounted for 42 percent of the shortfall in female life expectancy at age 50 years in the United States relative to countries with higher life expectancies, and 67 percent of the shortfall among males.

PHYSICAL INACTIVITY

Question 1: Does Physical Activity Matter to Health?

Although the health consequences of physical inactivity[8] are often difficult to disentangle from the morbidity and mortality associated with obesity and unhealthy diet, decades of research suggest that avoiding sedentary behavior and engaging in regular physical activity (independent of body weight, body mass index, and dietary habits) exerts its own protective effect on the risk of heart disease and stroke and possibly other conditions, including cancer, depression, and dementia (Lee et al., 2012; U.S. Department of Health and Human Services, 2008a). Physical activity among children offers unique age-specific benefits (Janssen and Leblanc, 2010), as does exercise among elderly people. Weight-bearing exercise plays a role in maintaining bone density and preventing osteoporosis and related morbidity (e.g., from hip fractures). The risk of injuries (e.g., falls among the elderly) is related to muscle strength and flexibility and obesity.

Question 2: Is Physical Inactivity More Prevalent in the United States Than in Other High-Income Countries?

As noted above, comparisons of the activity levels of Americans and people in other countries are difficult because of differences in definitions of activity and intensity levels and in survey and sampling methods. There is also inherent imprecision in self-reported activity levels (Sallis and Saelens, 2000). Furthermore, the same questions are rarely asked in each country (Hallal et al., 2012). One study administered the International Physical Activity Questionnaire to adults ages 18-65 in 20 countries. In the United States, 84.1 percent of respondents reported engaging in moderate or vigorous activity in a typical week, higher than all but five countries, only one of which was a high-income country (New Zealand) (Bauman et al., 2009).

In a study of physical activity in adults age 50 and older in the United States and Europe, Steptoe and Wikman (2010) found that the proportion of respondents who were moderately or vigorously active at least once a week ranged from 56 percent in Poland to 83 percent in Sweden, with the U.S. rate (69 percent) in the middle.[9] However, the United States

---

[8]Physical activity is any body movement that activates muscles and requires more energy than resting; exercise (including leisure-time exercise) is a type of physical activity that is planned and structured.

[9]The data were from the Health and Retirement Study (HRS) in the United States, the Survey of Health, Ageing and Retirement in Europe (SHARE), and the English Longitudinal Study of Ageing (ELSA) in England, all of which use similar measures of physical activity.

and Poland had the highest proportion (22 percent) of adults age 50 and older who reported no moderate or vigorous physical activity (Steptoe and Wikman, 2010). Michaud and colleagues (2007, p. 1) concluded that "sedentary lifestyle or a lack of vigorous physical activity may also explain a substantial share of the cross-country differences" between the United States and Europe. However, a recent cross-national comparison of data from 155 surveys in 122 countries reported that the prevalence of physical inactivity in the United States among those age 15 and older (40.5 percent) was lower than in 7 of the 16 peer countries that provided data (Hallal et al., 2012).

In a comparison with Austria, France, Germany, Italy, and the Netherlands, American adults age 20 and older spent 16.8 minutes per day doing sports, more than people in any other country but the Netherlands; however, Americans also spent 157.9 minutes per day watching television or listening to the radio, more than people in any other country (Brunello et al., 2008).

The international Health Behaviour in School-Aged Children (HBSC) survey of children ages 11, 13, and 15 ranked the United States in the bottom third of countries based on the percentage of children who exercised at least two times a week (World Health Organization, 2000a). However, these data are from the 1990s, before the popularity of video games and electronic devices. When the same survey was administered in 2001-2002, the United States had the largest proportion among 34 countries of children ages 10-16 who reported regular physical activity, but 46.7 percent watched 3 or more hours of television per day, more than all high-income countries but Norway (48 percent), Scotland (50.1 percent), and England (51.9 percent) (Janssen et al., 2005b). Other cross-national data on regular physical activity (e.g., walking), rather than leisure-time activity, are limited.

Question 3: Does Physical Inactivity Explain the U.S. Health Disadvantage?

Current evidence is inadequate to compare the physical activity of Americans with people in other countries, let alone to explore its role in explaining the U.S. health disadvantage. The limited data available paint a conflicting picture of whether Americans are more active than others, and many of the data predate the mass popularity of electronic devices that may encourage more sedentary life-styles. Even with ideal physical activity data, strong causal relationships between activity and health outcomes remain inadequately understood. Steptoe and Wikman (2010, p. 208) found that "countries with a higher proportion of individuals who are physically active have a lower prevalence of fair or poor self-rated health" but emphasized that cross-sectional associations do not prove causality.[10] In one study, countries' child obesity rates correlated with reported physical activity levels but not with time spent on computers (Janssen et al., 2005b). Studies based on current life-styles will clarify these relationships and may elucidate whether physical activity levels can ultimately explain some of the U.S. health disadvantages reported in Part I, such as chronic diseases or accidents (e.g., falls among the elderly) that could be attributed to lack of exercise or obesity. The explanation for other conditions lies elsewhere.

ALCOHOL AND OTHER DRUG USE

Question 1: Does Alcohol or Other Drug Use Matter to Health?

Although there is evidence that moderate consumption of alcohol can offer some health benefits, excessive drinking increases the risk of liver disease, various forms of cancer, hypertensive heart disease, stroke, pregnancy complications, and many other conditions (see, e.g., Rehm et al., 2009). Worldwide, alcohol is implicated in 30 percent of violent injuries, 21 percent of road traffic accidents, 19 percent of poisonings, 16 percent of drownings, and 11 percent of falls and self-inflicted injuries (World Health Organization, 2011a). In the United States, alcohol use during adolescence has an important linkage to violence and homicide (Green et al., 2011; Parker et al., 2011b).

Drug use also affects the risk of disease and injury. Among fatally injured drivers in the United States in 2009, 18 percent tested positive for at least one drug (e.g., illicit, prescription, or over-the-counter) (National Highway Traffic Safety Administration, 2010). Use of illicit drugs (e.g., marijuana, cocaine, methamphetamines, heroin), misuse of prescription medication, including medical errors in the clinical setting (Chen and Lin, 2009), and the growing problem of prescription opiate abuse (Centers for Disease Control and Prevention, 2011g; Kuehn, 2012; U.S. Food and Drug Administration, 2011) can affect health in ways that range from mild side effects to permanent complications, including accidental or intentional overdoses. Cocaine was involved in almost 40 percent of U.S. drug deaths in 2008 (United Nations Office on Drugs and Crime, 2010). Injection drug users are at risk of infection with the hepatitis B and C viruses and HIV.

Both alcohol and other drug dependency can increase vulnerability to crime and related violent injuries, such as drug-related homicides. Intoxicated individuals are prone to other risky behaviors, such as unprotected sexual activity. Forensic testing of U.S. victims of violent deaths in 2008

---

[10]It is also important to note that self-rated health does not necessarily correlate with objective indicators of health.

commonly revealed toxicological evidence of alcohol intoxication[11] (62 percent) and ingestion of opiates (26 percent), antidepressants (20 percent), cocaine (11 percent), marijuana (11 percent), amphetamines (3 percent), and other drugs (42 percent) (Karch et al., 2011).

Question 2:
Is Alcohol or Other Drug Use More Prevalent in the United States
Than in Other High-Income Countries?

Alcohol consumption has decreased in recent years in most OECD countries, and U.S. alcohol consumption in 2009 (8.8 liters per capita) was below the consumption levels of 11 peer countries and below the OECD average (OECD, 2011b). The World Health Organization (WHO) also reported that people in the United States consume less alcohol per capita than those in France, Germany, Italy, Sweden, and the United Kingdom, although they consume slightly more than the Japanese (World Health Organization, 2011). Whether *patterns* of alcohol consumption (e.g., binge drinking, alcohol dependence) differ in the United States is less clear. According to the *WHO Global Status Report on Alcohol* (World Health Organization, 2004a),[12] the prevalence of both heavy drinking among U.S. adults (6.4 percent for males, 5.0 percent for females)[13] and of heavy episodic drinking among U.S. youth was lower than in other high-income countries and the prevalence of lifetime abstinence was among the highest. However, another WHO study reported that the prevalence of heavy alcohol use[14] among U.S. and Canadian males ages 15-29 was 23.8 percent, a higher prevalence than in any other region of the world (Chisholm et al., 2004). This could bear on the high rate of adolescent deaths from car crashes in the United States (see Chapter 1). In contrast, the 2007 European School Survey Project on Alcohol and Other Drugs reported higher rates of drinking among young people in Europe than in the United States, including a greater percentage of youth ages 15-16 reporting drinking in the past 30 days, higher intoxication rates, and a greater percentage reporting intoxication before age 13 (Friese and Grube, 2001).

---

[11]In most jurisdictions in the United States, intoxication is defined as a blood alcohol content of 0.08 milligrams per deciliter (mg/dL) or greater.

[12]Also see WHO's Global Information System on Alcohol and Health (GISAH) at http://apps.who.int/ghodata/?theme=GISAH.

[13]Higher rates of binge drinking are reported in U.S. surveys. According to the 2010 Behavioral Risk Factor Surveillance System (BRFSS) survey, the prevalence of binge drinking in the United States was 17.1 percent (Centers for Disease Control and Prevention, 2012d).

[14]Heavy alcohol use was defined as an average rate of consumption of more than 20 grams of pure alcohol daily for women and more than 40 grams daily for men (Chisholm et al., 2004).

United Nations data suggest that drug use may be more prevalent in the United States than in many comparable high-income countries. Fully 12.5 percent of Americans ages 15-64 report using cannabis (marijuana) in the past year, a larger proportion than in all of Europe (except Italy and the Czech Republic) and Australia, although less than in Canada and New Zealand. Although cocaine use has declined significantly in the United States, 2.6 percent of people ages 15-64 report using cocaine each year, a higher rate than in Canada, all of Europe (except Spain and the United Kingdom), Australia, and New Zealand. The United States accounts for 36 percent of the global consumption of cocaine.

The reported use of amphetamine-type stimulants was 1.3 percent in the United States, higher than in Europe (except for Scotland and the Czech Republic) but lower than in Canada, Australia, and New Zealand. Use of opiates and ecstasy is not exceptionally higher in the United States than in other countries, but use of prescription opioids (pain killers) is markedly higher (United Nations Office on Drugs and Crime, 2010). Substance use is common among U.S. adolescents—37 percent of those ages 12-17 report using alcohol, marijuana, analgesic opioids, or other drugs in the past year, and 8 percent meet criteria for a substance-related disorder (Wu et al., 2011)—but comparable data from other countries are lacking.

Question 3: Does Alcohol or Other Drug Use Explain the U.S. Health Disadvantage?

As noted in Chapter 1, alcohol- and other drug-related deaths claim extra years of life in the United States relative to other high-income countries. However, there is little definitive evidence that heavy drinking is more prevalent in the United States than in other high-income countries, except perhaps for young adults. Use of marijuana, cocaine, and amphetamine-type stimulants appears to be more prevalent in the United States than in many high-income countries. In theory, higher rates of substance abuse could explain higher rates of drug-related outcomes—from deaths due to medication errors and poisonings[15] to intoxication-related injuries (e.g., motor vehicle crashes), violence (e.g., homicides), unsafe sex (e.g., unintended pregnancies, sexually transmitted infections), and injection drug use (e.g., hepatitis B virus and HIV infections). However, empirical data are lacking to assert a causal link between substance abuse and the U.S. health disadvantage. Conditions that give rise to substance abuse, such as stressful living conditions (see Chapters 6-7) and mental illness (see Chapter 2), may have a more direct causal link to the health outcomes.

---

[15]In 2008, 9 of 10 poisoning deaths in the United States were caused by medications or illicit drugs (Kuehn, 2012).

SEXUAL PRACTICES

Question 1: Do Sexual Practices Matter to Health?

Failure to use contraception and sexual activity with multiple partners can lead to unintended pregnancies and its consequences and can expose the participants to sexually transmitted infections, such as chlamydia, gonorrhea, syphilis, herpes zoster, HIV, and human papilloma virus (which can cause cervical cancer). Exposure to these risks is highest among sexually active adolescents and young adults and even higher when sexual activity begins at an early age or occurs under conditions (e.g., while intoxicated) when protective measures are often neglected. The factors that influence adolescent pregnancy rates include the timing of first intercourse, the frequency of intercourse, condom and contraceptive use, and the efficacy of the chosen contraceptive method(s) (Hatcher et al., 2007).

Question 2:
Are High-Risk Sexual Practices More Prevalent in the United States Than in Other High-Income Countries?

Oral contraceptives are generally used less regularly in the United States than in other countries (Mosher and Jones, 2010). International data on the prevalence of unsafe sex among adults, such as a history of multiple sexual partners or the prevalence of unprotected intercourse among men engaging in sex with other men, are lacking. Data comparing adolescent sexual behavior in the United States and other countries are somewhat outdated, but they suggest that unsafe sex is more common among U.S. adolescents than their peers in other countries.

A comparison of survey data in the United States and four other countries—Canada, France, Sweden, and the United Kingdom—found few cross-country differences in the proportion of adolescents who were sexually active (Darroch et al., 2001). However, the study found that U.S. adolescents were more likely to report having sex before age 15 (14 percent) than were those in Canada, France, and the United Kingdom (4-9 percent) and more likely to report having multiple sexual partners (two or more in the past year). Female U.S. teenagers were less likely to report using contraceptives at either their first or most recent intercourse (25 and 20 percent, respectively) than were those in France (11 and 12 percent, respectively), the United Kingdom (21 and 4 percent, respectively), and Sweden (22 and 7 percent, respectively). Condom use at the first sexual encounter was lower in the United States than in France, similar to the United Kingdom, and higher than in Sweden, but sole use of condoms at last intercourse (an unreliable contraceptive method) was higher in the United States than in the other study countries. Overall condom use (including dual use with hormonal contraceptives, a more effective method) was lower in the United States than in the United Kingdom and similar to levels in France (Darroch et al., 2001).

In an examination of sexual practices among 15-year-olds, Santelli and colleagues (2008) compared responses to the 2001 and 2003 Youth Risk Behavioral Survey (YRBS) in the United States to responses to similar questions about sexual behavior that adolescents in 24 countries completed on the HBSC survey of 2001-2002, as reported by Godeau and colleagues (2008). In the United States, reported use of condoms after last intercourse was 66 percent among adolescent females and 75 percent among adolescent males compared with 74 percent for both genders in the HBSC survey.[16] Reported use of oral contraceptives was 11 and 6 percent, respectively, among U.S. adolescent females and males but 24 percent for adolescents in other countries. Dual use (condom and oral contraceptive) at last intercourse was 4 and 3 percent, respectively, in adolescent females and males, compared with 16 percent for adolescents in other countries. Use of either the pill or condoms was 72 and 77 percent, respectively, for U.S. adolescent girls and boys and 82 percent in the HBSC survey.

Emergency contraception was not examined in the U.S. survey, but Santelli et al. (2008) noted that the rates in other countries for emergency contraception after last sexual intercourse (9 percent) matched the *lifetime* rate (8 percent) reported for youths aged 15-19 in the U.S. National Survey of Family Growth (Mosher et al., 2004). Santelli and colleagues (2009) examined subsequent data, which showed a similar pattern: in the 2005-2006 HBSC survey (Currie et al., 2008), 72 and 81 percent, respectively, of female and male 15-year-olds in the surveyed countries reported using condoms at last intercourse, compared with 62 and 75 percent for females and males, respectively, in the 2005 and 2007 YRBS. Contraceptive use was also much higher in Europe (Eaton et al., 2006, 2008; Santelli et al., 2009).

Question 3:
Do High-Risk Sexual Practices Explain the U.S. Health Disadvantage?

The apparent tendency of U.S. adolescents to have multiple sexual partners, to not use oral contraceptives as often as their peers in other countries, to rely on less effective barrier methods, such as condoms, and to use them less often than their counterparts in some other countries could explain the higher rates of pregnancies and sexually transmitted infections among U.S. adolescents, the high burden of HIV/AIDS in the United States, and perhaps the excess deaths from some congenital anomalies. However, the

---

[16]In the high-income countries included in the HBSC survey (Austria, Canada, England, Finland, France, the Netherlands, Portugal, Spain, Sweden, Switzerland), condom use ranged from 53 to 86 percent (Godeau et al., 2008).

data available to compare countries are not equivalent due to differences in instrument design and sampling methods.

INJURIOUS BEHAVIORS

Question 1: Do Injurious Behaviors Matter to Health?

Injuries are the leading cause of death among U.S. children and adults from ages 1-45 (National Center for Health Statistics, 2012) and, in the case of nonfatal injuries, are responsible for a heavy burden of lifelong neurologic and other disabilities. As detailed in Chapter 1, transportation-related injuries and violence account for many of the extra years of life lost in the United States relative to other high-income countries. Unintentional injuries include poisonings, motor vehicle crashes, falls, drowning, fires, asphyxiation, and burns (National Center for Health Statistics, 2012). Although such injuries are sometimes true "accidents" that could not have been prevented—and others result from unsafe product designs or weak safety provisions (see Chapter 8)—a large fraction of injuries result from the behavior of individuals. Examples include operating a motor vehicle while intoxicated, not using occupant restraints such as seatbelts and child safety seats, riding motorcycles or bicycles without a helmet, not installing smoke detectors, setting water heaters at scalding temperatures, unsafe boating practices, and failure to secure firearms and medications from children. Intentional injuries (including assault, murder, rape, child abuse and neglect, and intimate partner violence) and self-inflicted injuries (including suicidal behaviors) claim lives but also inflict both physical disabilities and emotional scars (e.g., posttraumatic stress disorder) (Felitti et al., 1998).

Question 2:
Are Injurious Behaviors More Prevalent in the United States Than in Other High-Income Countries?

Data are lacking to determine whether injurious behaviors are more common in the United States than elsewhere. For example, although Chapter 1 reported that poisoning accounts for two-thirds of U.S. nontransportation-related injury deaths before age 50 (and has recently replaced motor vehicles crashes as the leading cause of U.S. injury deaths) (Warner et al., 2011), there are no data to assess whether, for example, U.S. children have easier access to unsecured medications or toxic chemicals than children in peer countries. Nor are there data to know whether Americans are more susceptible to risks from falls, drowning, asphyxiation, or burns.

There is some evidence to compare driving practices. For example, data

TABLE 5-1 Driving Practices in 16 Peer Countries

Country	Drivers Wearing Seatbelts (%)		Motorcyclists Wearing Helmets[a] (estimated %)	Road Traffic Deaths Attributable to Alcohol (%)
	Front	Rear		
Australia	97	92	—	30
Austria	89	49	95	8
Canada	93	87	99	30
Finland	89	80	95	24
France	98	83	95	27
Germany	95-96	88	97	12
Italy	65	10	60	—
Japan	95-96	9-14	—	8
Netherlands	94	73	92	25
Norway	93	85	100	20-30
Spain	89	69	98	—
Sweden	96	90	95	20
Switzerland	86	61	100	16
United Kingdom	91	84-90	98	17
United States	82	76	58	32

[a]Use of bicycle safety helmets in Europe is thought to be less common.
SOURCE: Data from World Health Organization (2009, Table A.3, Table A.4, Table A.6).

in Table 5-1, taken from the WHO *Global Status Report on Road Safety*, suggest that Americans are less likely to fasten front[17] seatbelts than those in most high-income countries (World Health Organization, 2009). And only 58 percent of motorcyclists in the United States wear helmets, far less than the rate reported in most other high-income countries, where more than 95 percent of motorcyclists reportedly wear helmets (World Health Organization, 2009). However, other reports suggest that helmet use in Europe is probably lower, especially among bicyclists. For example, the International Transport Forum (2011) reports that helmets are worn by 49 percent of cyclists in Norway and 31 percent of cyclists in Finland, and these figures may be overestimates of actual use.

Almost one-third (32 percent) of U.S. road traffic deaths are attribut-

---

[17]Use of rear seatbelts in the United States is 76 percent, a higher rate than in many (but not all) high-income countries.

able to alcohol, a higher proportion than in other high-income countries (see Table 5-1), including countries with greater per capita alcohol consumption.[18] (See Chapter 8 for a more detailed discussion of cross-national differences in traffic fatalities.)

There is little evidence that violent acts occur more frequently in the United States than elsewhere.[19] Crime statistics cannot be compared accurately across countries because offences have different legal definitions, but the rate of criminal (police-recorded) nonviolent assaults in the United States is in the middle of the range reported by high-income countries (United Nations Office on Drugs and Crime, 2012). Among the study's 17 peer countries (see Chapter 1), the United States had the sixth highest rate of physical or sexual assaults on partners (intimate partner violence) (OECD, 2012k). Studies that have examined the incidence of violent behavior (e.g., engaging in fights) among U.S. adolescents relative to their peers in other countries have reported mixed results (Smith-Khuri et al., 2004; U.S. Department of Health and Human Services, 2001b).

One behavior that probably explains the excess lethality of violence and unintentional injuries in the United States is the widespread possession

---

[18]Data from the 2010 BRFSS indicate that 1.8 percent of Americans report at least one alcohol-impaired driving episode in the past 30 days (Centers for Disease Control and Prevention, 2011f). Drunk driving appears to have decreased in the United States in recent decades. For example, the percentage of weekend nighttime drivers and underage drivers with a blood alcohol concentration (BAC) of 0.8 g/L or greater decreased from 1973 to 2007 (Transportation Research Board, 2011). The percentage of U.S. crash fatalities involving a driver with a BAC greater than zero decreased from 55 to 38 percent between 1982 and 1995, and it was 37 percent in 2008. However, such statistics are of limited value because BAC tests are performed on only 40 percent of U.S. drivers involved in fatal crashes (Transportation Research Board, 2011). Comparisons with other countries are complicated by differences in measurement methods and legal BAC limits. Nonetheless, despite having higher per capita rates of alcohol consumption, Australia, Germany, Sweden, and the United Kingdom appear to have achieved lower rates of alcohol-related traffic fatalities than the United States, both as a percentage of fatalities and as measured per vehicle kilometer of travel (Transportation Research Board, 2011). This may also reflect differences in management, planning, and enforcement policies (see Chapter 8).

[19]Some surveys hint at greater acceptability of violence in the United States. For example, a survey of students in cities in Estonia, Finland, Romania, Russia, and the United States found that American students were the most likely to justify killing to defend property and were more supportive of war (McAlister et al., 2001).

of firearms and the common practice of storing them (often unlocked) at home. The statistics are dramatic:

- The United States has the highest rate of firearm ownership among peer countries. As shown in Figure 5-5, there are 89 civilian-owned firearms for every 100 Americans.[20]
- The United States is home to approximately 35-50 percent of the world's civilian-owned firearms.
- Fully 48 percent of all violent deaths (66 percent of homicides) in the United States involve firearms (Karch et al., 2011).
- As of 2004, 38 percent of U.S. households and 26 percent of individuals reported owning at least one firearm, and almost one-half of individual owners reported owning four or more firearms (Hepburn et al., 2007).
- Although U.S. youth may be no more violent than those in other countries, they are more likely to carry a firearm (Pickett et al., 2005). In a survey of high school students in Boston, 5 percent reported carrying a firearm (Hemenway et al., 2011).
- U.S. civilians own four times the number of automatic and semi-automatic rifles owned by the U.S. Army (Small Arms Survey, 2007).

International comparisons draw an association between firearm ownership rates and the rate of deaths from homicide and suicide (Hemenway and Miller, 2000; Killias, 1993).[21] Among 23 OECD countries, 80 percent of all firearm deaths occurred in the United States (Richardson and Hemenway, 2011). Consistent with the findings reported in Chapter 1, Richardson and Hemenway (2011) reported that the excess homicide rate in the United States relative to 22 high-income OECD countries was driven by firearm homicide rates that were 20 times higher (43 times higher for those ages 15-24). Firearms are also associated with deaths from causes other than homicide. The presence of a firearm in the home is a risk factor for suicide (Johnson et al., 2010; Miller and Hemenway, 2008): fully 52 percent of all U.S. suicides involve a firearm (Karch et al., 2011). Suicide rates are not higher in the United States than in peer countries, but firearm suicide rates are 5.8 times higher, and unintentional firearm deaths are 5.2 times higher (Richardson and Hemenway, 2011).

---

[20]Because many owners have more than one firearm, the actual proportion of Americans who own firearms is far less than 89 percent.

[21]A notable exception is Switzerland, which has one of the highest rates of gun ownership but relatively low crime rates.

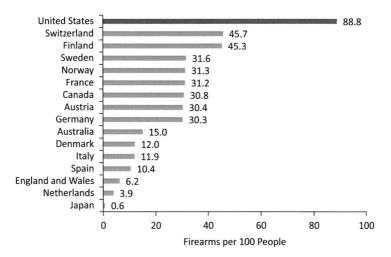

FIGURE 5-5 Civilian firearm ownership in 16 peer countries.
NOTES: The data reflect the number of firearms owned per 100 persons. Because many people own multiple firearms, the proportion of people who own firearms is lower.
SOURCE: Data from Small Arms Survey (2007).

Question 3:
Do Injurious Behaviors Explain the U.S. Health Disadvantage?

There are inadequate data to know whether higher death rates from unintentional injuries in the United States are the result of more injurious behaviors or environmental factors. Countries do not collect similar data on behaviors that affect the risk of falls, poisoning, drowning, or other behaviors, making valid international comparisons of these behaviors impossible. It may be the severity, and not the incidence, of injuries that differs in the United States, a factor influenced not only by personal actions—such as the above evidence that Americans may be less apt to use seatbelts or helmets and are more involved in accidents involving alcohol—but also by deficiencies in product and roadway designs (e.g., crash protection) and resources that protect public safety (e.g., law enforcement). However, the prevalence of firearms in the United States looms large as an explanation for higher death rates from violence, suicidal impulses, and accidental shootings.[22]

---

[22]The validity of the correlation between firearm ownership and homicide is strongly debated by opponents of stricter gun control laws in the United States.

CONCLUSIONS

Individual behaviors contribute to each of the nine domains in which the United States demonstrates a health disadvantage relative to other countries. Smoking contributes to adverse birth outcomes, heart disease, and chronic pulmonary disease, although smoking rates are now lower in the United States than in other countries and would explain little of the U.S. health disadvantage among adults younger than age 50. Unhealthy diet and low physical activity contribute to higher rates of obesity and diabetes. Alcohol consumption, other drug use, and unsafe sexual practices contribute to drug-related mortality, HIV/AIDS, sexually transmitted infections, and adolescent pregnancies. Substance abuse also contributes to injuries (unintentional and intentional), as do injurious practices and the prevalence of firearms in the United States. These conditions are causally interconnected. For example, obesity in early life can give rise to diabetes and, over time, the macrovascular complications of heart disease. Thus, health behaviors may play a pivotal role over the life course in promoting the conditions responsible for the U.S. health disadvantage.

Whether health behaviors in the United States differ significantly from those in other countries and the extent to which they explain the U.S. health disadvantage cannot be determined without better cross-national data. Further research is needed to define the specific behaviors that are predictive of adverse health outcomes, based on rigorous science, and to validate appropriate metrics and sampling methods for measuring those behaviors. Modes of administration, such as using accelerometers and other sensors instead of relying on self-report of physical activity, may also need to evolve. Countries would need to adopt a consistent battery of questions about health behaviors to enable meaningful international comparisons. Crime statistics would have to be more consistent across countries to understand international differences in violence. Historical cohort data on behavior patterns in prior years, or in prior decades, may be important in order to understand current disparities in the prevalence of diseases that result from a lifetime of sedentary behavior, unhealthy diet, or substance abuse.

Although no single behavior can explain the U.S. health disadvantage, the high prevalence of multiple unhealthy behaviors in the United States (Fine et al., 2004) may play a large role. Advocates of "personal responsibility" would note that people choose to engage in all of the behaviors discussed in this chapter, from eating sweets to carrying handguns, and they should be free to make those choices and bear the consequences. But such choices may not always be made "freely": they are made in a societal and environmental context (Brownell et al., 2010). Parents may want to serve healthy meals but may not be able to do so without nearby retailers that sell fresh produce (Larson et al., 2009). They may want their children to play outside but the neighborhood may be unsafe. As Chapter 7 details, Americans do not make product choices in a vacuum, and they face both environmental supports and barriers to healthy behaviors. Advertising and marketing of tobacco, alcohol, and unhealthy foods; sexualized content in television, film, and musical entertainment; and exposure to violence and stress have known consequences to health (Gordon et al., 2010; Harris et al., 2009; Lovato et al., 2003; Mosher, 2011; Nestle, 2007; Robinson et al., 2007; Spano et al., 2012; Vermeiren et al., 2003).

Socioeconomic conditions also limit choices and play an important role in the prevalence of unhealthy behaviors and their effect on mortality (Lantz et al., 2010). Tobacco use and other unhealthy behaviors are significantly more prevalent among adults with limited education and limited incomes (National Center for Health Statistics, 2012; Pampel et al., 2010) (see Chapters 6 and 7). Barriers in access to medical care generally, which is discussed in Chapter 4, and to specific health care services (e.g., pharmacotherapy for smoking cessation, substance abuse counseling, oral contraceptives) also limit the ability of individuals to adopt and sustain behavioral changes. Finally, policy and culture are important factors in understanding why health behaviors might differ between high-income countries: these are discussed in Chapter 8.

6

Social Factors

Chapters 4 and 5 examined the role of health systems and health behaviors in explaining the U.S. health disadvantage, but health is also deeply influenced by "social determinants," such as income and wealth, education, occupation, and experiences based on racial or ethnic identification. These factors have been shown to contribute to large health disparities in the United States and other countries and should be considered in efforts to explain disparities in health among countries. Although the science of the social determinants of health is still evolving, a growing body of biological, epidemiological, and social science research has revealed pervasive and strong links between a range of social factors that shape living and working conditions and a wide array of health outcomes. A rapidly accumulating literature also is elucidating the biological processes that may account for these health effects (Adler and Stewart, 2010; Braveman et al., 2011b; Commission on the Social Determinants of Health, 2008).

Following widespread convention, we use the term "social" to refer to economic as well as psychosocial factors. Access to, and the quality of, medical care are clearly influenced by social policies, such as the legislation creating Medicare and Medicaid in 1965 and the Affordable Care Act in 2010. Generally, however, and in this report, the terms "social factors" and "social determinants of health" refer to factors outside the domain of public health and health care, which are covered in Chapter 4.

As discussed above, the terms "upstream" and "downstream" are often used to denote relative positions of a given health determinant on plausible causal chains. Upstream factors are closer to the fundamental cause and often farther ("distal") from the observed health outcome; downstream ("proximate") factors are closer to the ends of causal chains. Upstream social factors that have repeatedly been linked with important health outcomes in many populations include income, accumulated wealth, educational attainment, and experiences based on racial or ethnic identification. Downstream factors (which may be shaped by upstream factors) include unhealthy diets, lack of exercise, and smoking. Features of neighborhoods and work environments may be thought of as midstream.

This chapter focuses on the social factors that current knowledge suggests may contribute significantly to the U.S. health disadvantage and that can be compared across high-income countries: income and poverty, income inequality, education, employment, social mobility, household composition, and experiences based on racial or ethnic identification (Galea et al., 2011; Link and Phelan, 1995; Marmot, 2005).

The chapter focuses primarily on social characteristics of individuals, families, and populations. The potential role of the social environment—such as features of housing, transportation, and neighborhoods—in contributing to the U.S. health disadvantage is the focus of Chapter 7.

As in other chapters, the panel posed three questions:

- Do social factors matter to health?
- Are adverse social factors more prevalent in the United States than in other high-income countries?
- Do differences in social factors explain the U.S. health disadvantage?

Before turning to these questions, however, we offer an important comment on evidence and the role of social factors on health. Documenting causality and testing the effectiveness of interventions for social factors is inherently challenging (Braveman et al., 2011b; Kawachi et al., 2010). The time intervals between exposures to social factors and a health outcome—such as the psychosocial consequences of poverty—may be quite long (Braveman et al., 2010a; Galobardes et al., 2008; Rychetnik et al., 2002). Exposures may occur during childhood, gestation, or even during the childhood of one's parents (Hertzman, 1999; Kuh et al., 2002; Melchior et al., 2007; Turrell et al., 2007). Furthermore, the causal pathways from fundamental social causes to health outcomes are often complex, with opportunities for effect modification at multiple steps along the way (Braveman et al., 2010a). Relationships are also not unidirectional: cross-sectional associations do not clarify the role of reverse causality, as when poor health limits education or income. Adverse health and socioeconomic circumstances can also negatively affect household stability, family composition, and social support. Because it would be both difficult, if not unethical, to test these hypotheses in randomized controlled trials,[1] researchers use a variety of other methods, including multivariate modeling, instrumental variables, quasi-experimental designs, Bayesian approaches to inference, natural experiments, and "connecting-the-dots" between disparate bodies of knowledge (Braveman et al., 2010a; Dow et al., 2010; Kelly et al., 2006).

Another challenge in analyzing social factors in a cross-national context is that a given factor may have different health implications depending on local circumstances. Standards of living vary across countries, as do social safety nets and other programs designed to alleviate poverty, unemployment, and homelessness. We therefore approach this topic aware of these important caveats and limitations.

QUESTION 1:
DO SOCIAL FACTORS MATTER TO HEALTH?

Recent reviews have documented links between social factors and health, elucidated plausible causal pathways, and discussed the strength of evidence for causality (Adler and Rehkopf, 2008; Adler and Stewart, 2010; Braveman et al., 2010a, 2011b; Commission on the Social Determinants of Health, 2008; Kawachi et al., 2010). As shown in Box 6-1, among the broad types of social factors with strong and pervasive links to a wide array of important health outcomes are income (Mackenbach et al., 2005; Muennig et al., 2010; Woolf et al., 2010), accumulated wealth (Pollack et al., 2007), educational attainment (Elo and Preston, 1996; Jemal et al., 2008a; Woolf et al., 2007), occupational characteristics (An et al., 2011), and social inequality based on racial or ethnic group (Bleich et al., 2012; Marmot, 2005; Williams and Collins, 2001; Williams and Mohammed, 2009). In this section, we briefly summarize this literature.

Income and Wealth

Extensive evidence documents the association between income and mortality. Unhealthy behaviors, such as smoking, tend to be more prevalent among low-income groups. Income or wealth enables one to afford a nutri-

---

[1]The premise that randomized controlled trials are the "gold standard" for establishing causal relationships has put the accumulation of knowledge about the social determinants of health at a distinct disadvantage. It is increasingly recognized that scientifically valid studies of social factors that can answer important questions must draw on a wide variety of well-implemented research designs (Anderson and McQueen, 2009; Black, 1996; Glasgow et al., 2006; McQueen, 2009; Petticrew and Roberts, 2003; Victora et al., 2004).

BOX 6-1
Social Factors That Affect Health Outcomes

Upstream social factors—Laws, policies, and underlying values that shape the following:
- Income and wealth
- Education
- Employment
- Household composition
- Experiences based on race or ethnic group
- Social mobility
- Stressful experiences related to any of the above
- Income inequality

Midstream social factors—Factors that are strongly influenced by upstream factors and that are likely to affect health:
- Housing
- Transportation
- Other conditions in homes, schools, workplaces, neighborhoods, and communities, including conditions that produce stress and family disruption (e.g., parenting skills, parenting stress, leisure time, quality of schooling, physical and psychosocial working conditions)

---

tious diet (Treuhaft and Karpyn, 2010), to buy or rent healthy housing[2] in a healthy neighborhood (Shaw, 2004), and to engage in regular exercise (e.g., through gym membership or living where it is safe and pleasant to exercise outdoors). However, careful analysis of longitudinal data has revealed that the association between adverse economic conditions and mortality persists even after adjusting for unhealthy behaviors (Lantz et al., 2010), suggesting that economic stresses may also affect health through other pathways. Access to employment, educational opportunities, and medical care can be constrained by one's income, particularly in the absence of adequate public transportation (Gordon-Larsen et al., 2006). Exposure to poverty during childhood may have particularly strong and enduring effects on health across the entire life course (Cohen et al., 2010; Pollitt et al., 2005). Material hardship is strongly related to family strife and disruption (Braveman et

---

[2]Healthy housing refers to domiciles that are free of health and safety threats, such as lead, which can affect children's cognitive function, and free of excessive dust, mites, and mold, which can provoke and exacerbate respiratory disease (Lanphear et al., 2001).

al., 2010a; Evans et al., 2012). Accumulated wealth can buffer the adverse effects of temporary periods of lower income.

Income Inequality

Income inequality in a society has repeatedly been shown to be inversely associated with good health, but there is controversy about the health effects of relative income inequality apart from the effects of absolute poverty or economic hardship (Subramanian and Kawachi, 2004). Some experts view relative inequality as a factor with independent effects, which may touch the whole population, perhaps by undermining social cohesion (see Chapter 7) (Daniels et al., 2000; Pickett and Wilkinson, 2009, 2010; Wilkinson and Pickett, 2007, 2009). Other research, however, challenges the premise that relative economic inequality exerts an independent effect apart from its association with absolute levels of material deprivation (Beckfield, 2004; Deaton and Lubotsky, 2009; Lynch et al., 2001, 2004a, 2004b). The apparent association between economic inequality and poor health could reflect other more fundamental factors that shape both economic inequality and health, such as a society's lack of social solidarity. There is, however, consensus about the adverse health implications of absolute material deprivation.[3]

Education

Education and health are strongly interrelated. In 2006, the life expectancy of 25-year-old American men without a high school diploma was 9.3 years shorter than those with a bachelor's degree or higher education; the corresponding disparity for women was 8.6 years (National Center for Health Statistics, 2012).[4] Education is generally a prerequisite for desirable employment and associated income and other resources (e.g., medical insurance, pensions, sick leave). Early childhood experiences and education shape early childhood development, which in turn influences school readiness and, ultimately, educational attainment. Education can confer knowledge, problem-solving skills, and a sense of control over life circumstances.

These psychosocial factors have been strongly tied to healthy behaviors (Dunn, 2010; Pampel et al., 2010; Umberson et al., 2008) and, in some

---

[3]Debates about the role of relative economic inequality isolated from the effects of absolute deprivation have generally been confined to academic settings. In most contexts, economic inequality is assumed to refer to absolute economic adversity for substantial segments of a population alongside extreme wealth for others. High levels of relative inequality and absolute hardship/poverty coexist in the United States and many other nations.

[4]In some research, the association between education and unhealthy behaviors and mortality loses significance after controlling for confounding variables, notably income (Lantz et al., 2010).

cases, more directly to health outcomes (Matthews et al., 2010; Pudrovska et al., 2005). For example, across countries, education and smoking rates are inversely related (Garrett et al., 2011; National Center for Health Statistics, 2012; Pampel and Denney, 2011), and parental education is associated with the health behaviors of children.[5] Other evidence also supports causal connections between education and health outcomes (Fonsenca and Zheng, 2011; Lleras-Muney, 2005), but the two may also have common antecedents. Hopelessness and powerlessness, for example, may contribute both to unhealthy behaviors and to educational and occupational setbacks, and they may link more directly to poor health through plausible physiologic mechanisms (Boehm and Kubzansky, 2012; Downey and Van Willigen, 2005; Goodman et al., 2009; Marmot et al., 1997; Matthews et al., 2010; Pudrovska et al., 2005; Seeman et al., 2010).

Employment

Employment shapes health in diverse ways, in part by determining employment opportunities and income (World Economic Forum, 2011). Low-skilled and low-status employment is more likely to involve exposure to physical hazards, such as toxic chemicals (e.g., pesticides, cleaning solvents), and to occupational injuries. Job loss, unemployment, and economic contraction have been linked with ill health and higher mortality because of psychosocial as well as economic consequences (Bartley and Owen, 1996; McLeod et al., 2012; Paxson and Schady, 2005; Strully, 2009; Sullivan and von Wachter, 2009), although the evidence is not conclusive (Catalano et al., 2011). (See Chapter 7 for additional evidence on the health and injury risks associated with the work environment.)

Social Status

Income, wealth, education, and employment all have implications for prestige and acceptance in society, and hence may affect health through psychosocial pathways involved in perceived position in a social hierarchy. Lower perceived social status has been associated with adverse health outcomes in some studies even after considering objective measures of resources and social status (Singh-Manoux et al., 2003, 2005).

---

[5]In 2007-2010, in U.S. households in which the head of household had less than a high school education, 24 percent of boys and 22 percent of girls were obese. In households where the head had a bachelor's degree or higher, the corresponding figures were 11 percent for boys and 7 percent for girls (National Center for Health Statistics, 2012).

Household Composition

Household composition, which is strongly related to income and education, can influence social factors that in turn influence health.[6] For example, children in low-income single-parent households experience higher rates of poverty, food insecurity, unstable housing, and other adverse living conditions (Center on Human Needs, 2012a). Poverty puts strains on families and creates a greater risk of single-parent households (Center on Human Needs, 2012a; DeNavas-Walt et al., 2011).

Low-income households are often the setting for adolescent childbearing, which is more common in the United States than in other high-income countries (see Chapter 2). Adolescent motherhood affects two generations, children and mothers. Adolescent mothers are less likely than other adolescents to complete their education, and they have more restricted labor market opportunities and more disadvantaged family and household environments (Ashcraft and Lang, 2006; Hoffman and Maynard, 2008). Their children face a greater risk of poor child care, weak maternal attachments, poverty, and other adverse conditions (Baldwin and Cain, 1980; Card, 1981). The female children of adolescent mothers are also at increased risk of becoming adolescent mothers themselves, thus perpetuating adverse conditions over two generations (Kahn and Anderson, 1992).

Racial and Ethnic Factors

In many countries, a variety of health outcomes vary markedly by race and ethnicity (Agency for Healthcare Research and Quality, 2011; Commission on Social Determinants of Health, 2008). These health disparities often mirror large differences in income, wealth, education, occupation, and neighborhood conditions among people of different races and ethnicities, differences that reflect a historical legacy of discrimination (Acevedo-Garcia et al., 2008; Bleich et al., 2012; Cullen et al., 2012; Williams, 1999; Williams and Collins, 1995, 2001).[7] For example, in the United States, blacks with the same level of education as whites have lower incomes, as well as markedly lower levels of accumulated wealth even at the same level of income (Braveman et al., 2005; Kawachi et al., 2005). Living in a society with a high degree of racial inequality may harm the health of society at large—not only of those who experience disadvantage—in the same ways

---

[6]As noted above, the reverse is also true: illness can influence household composition and stability, as well as education and income opportunities.

[7]It is now widely recognized that racial and ethnic groupings are primarily social, not biological, constructs, and that genetic differences probably make a small contribution to racial or ethnic health disparities (American Anthropological Association, 1998; McCann-Mortimer et al., 2004; Winker, 2004).

that some researchers have argued that relative economic inequality may be detrimental to society at large, for example, by undermining social cohesion and trust (Wilkinson and Pickett, 2009) or by affecting individuals' sense of their relative social standing (Marmot, 2006). Unfortunately, as noted below, data are lacking to compare degrees of racial inequality across high-income countries.

In the United States, racial and ethnic groups that have historically experienced discrimination,[8] including blacks, Native Americans, and Hispanics, may suffer ill health effects from these experiences. The health effects may result both from material deprivation and other conditions that directly damage health and from physiologic mechanisms involved in reactions to stress. Such stress, which has been linked with smoking (Purnell et al., 2012) and hypertension (Sims et al., 2012), can result not only from overtly discriminatory experiences but also from a pervasive vigilance about whether harmful incidents will occur to themselves or their families (Krieger et al., 2011; Nuru-Jeter et al., 2009). A relative difference in social standing or a sense of social exclusion for any reason may induce stress and influence one's sense of self-worth or control, which may in turn influence subsequent economic success, health-related behaviors, and health outcomes (Dunn, 2010; Umberson et al., 2008).

Migration

Migration and associated experiences and cultural traditions have been shown to influence health and health behaviors. Almost 14 percent of the U.S. population in 2008 was born outside the United States (OECD, 2011e). Although some immigrants are at higher risk of certain infectious diseases, most recent immigrants to the United States generally have favorable health profiles compared with the native-born population.

Stress

Psychological distress that arises from any of the above social factors, including from social rejection or exclusion associated with racial or ethnic identification, may lead to worse health through physiologic mechanisms involved in stress (Matthews et al., 2010; McEwen and Gianaros, 2010). Those mechanisms include the effects of stress on the hypothalamic-pituitary-adrenal (HPA) axis, the sympathetic nervous system, and immune

---

[8]This legacy has been perpetuated by deeply rooted societal structures, even in the absence of conscious intent to discriminate. This form of unintentional discrimination is often referred to as structural or institutional racism, deeply rooted ways in which opportunity is differentially structured along racial or ethnic lines (Smedley, 2012).

and inflammatory phenomena (Danese et al., 2007; Halfon and Hochstein, 2002; McEwen and Gianaros, 2010; Shonkoff et al., 2009). These effects are thought to induce end organ damage and cardiovascular disease (Barker, 1998; McEwen and Gianaros, 2010). While life-long stress leads to accumulated damage, early exposure to stress can affect sensitive biological processes, such as brain development, and thereby permanently disrupt stress responses later in life (Gluckman and Hanson, 2006; Shonkoff et al., 2009).

A Life-Course Perspective

Research increasingly confirms that health is shaped by social factors that individuals have faced across their entire life courses, not just current or recent experiences. Social disadvantages—and the health consequences associated with them—may accumulate across an individual's lifetime and span multiple generations, making the unfavorable odds increasingly difficult to overcome (Goodman et al., 2011).

Social disadvantage can therefore play an important contributory role to the development of chronic diseases and other conditions that threaten the health and life expectancy of adults age 50 and older, but they can also foster the health problems of early life, including many of the conditions discussed in Chapters 1 and 2. For example, the educational attainment and cognitive skills of today's youth could influence the behaviors that contribute to infant and child mortality due to rates of accidents and homicides; adolescent births and sexually transmitted infection; HIV/AIDS; and drug-related mortality.

Chronic material hardship or stressful events in childhood may also manifest their effects in mid- or even late adulthood (Cohen et al., 2010; Goodman et al., 2011). Chronic social or economic hardship during childhood has been linked with morbidity and mortality due to cardiovascular disease, diabetes, and other chronic diseases in adulthood (Hertzman, 1999; Kuh et al., 2002; Lawlor et al., 2005; Melchior et al., 2007; Turrell et al., 2007). There is evidence of health consequences from experiences during critical or sensitive periods (e.g., in early childhood and puberty), as well as from the cumulative effects of experiences over an individual's life course (Murray et al., 2011; National Research Council and Institute of Medicine, 2000; Viner, 2012; World Bank, 2007). Over and above the influence of any particular event, the number of such events and the number of domains affected by social disadvantage can determine the health damage associated with poverty (Evans and Kim, 2010; Sexton and Linder, 2011).

Inheritance is a major route of transmission for wealth and its associated advantages. Low social mobility—that is, the low likelihood that a person born to low-income or poorly educated parents will achieve higher income or education levels as an adult—could exacerbate the health effects of adverse social conditions by leading to the accumulation of social disadvantage across generations, thereby producing greater poverty or other consequences that compromise health (Case et al., 2005). Downward social mobility has repeatedly been linked with adverse health outcomes (Case and Paxson, 2010, 2011; Currie and Widom, 2010; Delaney and Smith, 2012). Lack of upward mobility in a society could exacerbate economic and social inequality and could plausibly affect health through a range of pathways, including by shaping optimism (Boehm and Kubzansky, 2012) and health-related behaviors (Dehlendorf et al., 2010; McDade et al., 2011) and possibly by undermining feelings of social solidarity (Pickett and Wilkinson, 2010; Wilkinson and Pickett, 2009).

Epigenetic Effects

Social factors—or their consequences in social and physical environments—may also influence health by interacting with a person's genotype in ways that can trigger or suppress the phenotypic expression of deleterious (or favorable) genes that may be related to obesity, heart and lung disease, diabetes, and cancer. A deleterious gene in one's DNA may not be harmful in the absence of certain triggers that "turn on" gene expression and cause cancers to develop. These modifications in gene expression, which are thought to occur through molecular processes (such as histone modification and DNA methylation) can be inherited and affect the health of offspring. "Epigenetics" refers to the transfer, from one generation to the next, of gene expression patterns that do not rely explicitly on differences in the DNA code (Gluckman and Hanson, 2006; Institute of Medicine, 2006b; Sandoval and Esteller, 2012). Social and environmental factors may therefore influence biological outcomes through their effects on gene expression.

QUESTION 2:
ARE ADVERSE SOCIAL FACTORS MORE PREVALENT IN THE UNITED STATES THAN IN OTHER HIGH-INCOME COUNTRIES?

Cross-country comparisons of social factors can be difficult because of differences in measurement, as well as the meaning of a given factor in different settings. Readily comparable cross-national data are not available on all relevant factors. For example, racial and ethnic disparities are important to health, but data are lacking to compare the United States with peer countries in terms of the magnitudes of racial and ethnic health disparities. Data are available, however, to examine health disparities by income, education, and other socioeconomic determinants. The comparative data produced by the OECD are widely considered to be the best available and are the principal source of cross-national comparisons presented here.

In aggregate, socioeconomic conditions—income and wealth—in the United States are at or above average for high-income countries. Both the size of the U.S. economy and median household income in the United States are among the highest in the world. As of 2007, the United States ranked second in the OECD (after Luxembourg) in annual household income[9] (OECD, 2011e) and seventh in gross domestic product per capita (World Bank, 2012a). However, the United States ranks poorly on the *equitable distribution* of economic resources, with relatively high levels of poverty and income inequality.

Poverty

The relative poverty rate,[10] defined as the proportion of the population with low incomes relative to the median income, has been higher in the United States than in other high-income countries since at least 1980 (Luxembourg Income Study, 2012). Historically, the U.S. poverty rate declined from very high levels in the 1940s to low levels in the late 1970s (Danziger and Gottschalk, 1986): the rate (based on total household income) fell from 40.5 percent in 1949 to 22.1 percent in 1959, 14.4 percent in 1969, and 13.1 percent in 1979 (Ross et al., 1987). During these same decades, many European countries instituted social welfare reforms that were designed to promote social equity and alleviate economic distress (see Chapter 8), lowering the rates of poverty in many of these countries (Brady, 2005). The gap between the levels of income inequality in the United States and other rich democracies began to widen in the 1970s-1980s, possibly because of the adoption of more conservative economic policies in the United States and a retrenchment in public assistance programs (Card and Freedman, 1993; Danziger and Gottschalk, 1995; Hanratty and Blank, 1990).

Absolute poverty is a basis for comparing incomes across countries against a common benchmark (such as a given level of income in U.S. dollars). Analyses that have used a common data set to compare countries in terms of absolute poverty find that other countries seem to have higher rates

---

[9]Household income is defined by the OECD as annual median equivalized household disposable income: gross household income after deduction of direct taxes and payment of social security contributions and excluding in-kind services provided to households by governments and private entities, consumption taxes, and imputed income flows due to home ownership (OECD, 2011e).

[10]Relative poverty is defined by the OECD as the percentage of people living with less than 50 percent of median equivalized household income. "People are classified as poor when their equivalized household income is less than half of the median prevailing in each country. The use of a relative income-threshold means that richer countries have the higher poverty thresholds. Higher poverty thresholds in richer countries capture the notion that avoiding poverty means an ability to access the goods and services that are regarded as customary or the norm in any given county" (OECD, 2011e, p. 68). See above discussion of absolute poverty as an alternate measure.

than the United States (Kenworth, 1998; Sharpe, 2011; Smeeding, 2006). This finding reflects the higher overall standard of living in the United States (Smeeding, 2006). For example, in one analysis, the U.S. absolute poverty rate was lower than 8 of 10 high-income countries (Gornick and Jäntti, 2010).

Beginning in the 1980s, *relative* poverty rates in the United States have consistently exceeded those of other high-income countries (Smeeding, 2006) (see Figure 6-1). This difference has increased over time. By the late 2000s, the relative poverty rate in the United States exceeded that of all 16 peer countries. It also exceeded rates in 31 OECD countries, including Australia, Chile, the Czech Republic, Estonia, Hungary, Japan, Korea, Mexico, New Zealand, Poland, the Slovak Republic, Slovenia, and Turkey (OECD, 2011e).

Measured in terms of relative poverty, the United States also has the highest rate of child poverty of the 17 peer countries (Gornick and Jäntti, 2010; OECD, 2012e). As of 2008, more than one in five (21.6 percent) U.S. children lived in poverty, the fifth highest rate among 34 OECD countries (OECD, 2012e).[11] Similarly, a UNICEF study found the United States to have the highest child poverty rate of the 24 rich countries it examined (UNICEF, 2007).[12] As with poverty overall, the trend first became noticeable in the 1980s, a time of economic transformation in the United States, and the effect on child poverty rates was dramatic: within the short span of the mid-1980s, child poverty increased by almost one-third in the United States (Jäntti and Danziger, 1994). Since then, the country has consistently had the highest relative child poverty rates among all rich nations (OECD, 2012e; Whiteford and Adema, 2007) (see Figure 6-2).

Income Inequality

According to the OECD, income inequality in the United States in the late 2000s was higher than the average of all OECD countries. One common measure of income inequality is the Gini coefficient, which ranges from 0 to 1 with larger values indicating greater inequality: the OECD average was 0.31, and it was 0.38 in the United States. The U.S. Gini coefficient exceeded that of all 16 peer countries, as well as all other OECD countries except Chile, Mexico, and Turkey (OECD, 2011e).[13]

---

[11]The U.S. child poverty rate also exceeds that of many emerging economies in the former Soviet satellite countries of Eastern Europe.

[12]Measured in terms of absolute poverty, a few high-income countries appear to have higher child poverty rates than the United States (Gornick and Jäntti, 2010).

[13]The OECD values for the Gini coefficient in the United States are lower than those reported by the U.S. Census Bureau, which reported a Gini coefficient of 0.47 in 2010 (DeNavas-Walt et al., 2011).

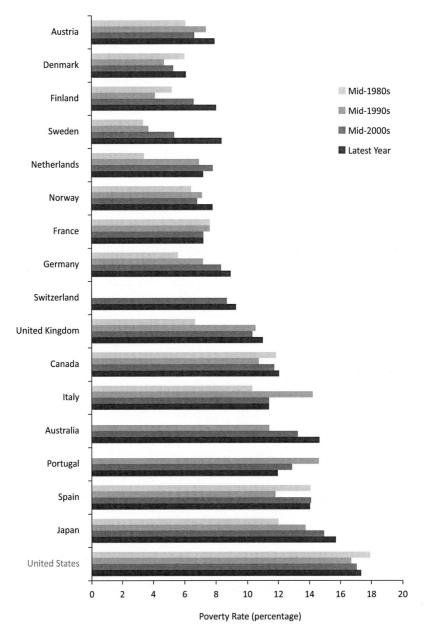

FIGURE 6-1 Poverty rates in 17 peer countries.
NOTES: Poverty rates are based on relative poverty, defined here as incomes below
50 percent of the median income of the country. Poverty rates for Switzerland were
not available for certain years.
SOURCE: Data from OECD (2012l), StatExtracts: Income distribution—Poverty.

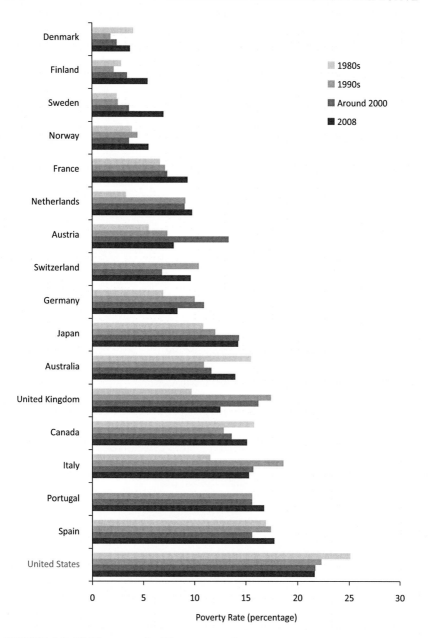

FIGURE 6-2 Child poverty in 17 peer countries.
NOTES: Poverty rates are based on relative poverty. Data for Portugal and Switzerland were not available for certain years.
SOURCES: Data from Whiteford and Adema (2007, Table 1), OECD (2012e), and Table CO2.2A.

Income inequality has been rising in the United States since 1968 (DeNavas-Walt et al., 2011; Taylor et al., 2011). Income inequality has also risen throughout most high-income countries, but not as dramatically.[14] Since the 1980s, levels of income inequality in the United States have been higher than in all other high-income countries except Portugal and Spain.

Education

Educational Attainment

The U.S. ranking on educational attainment when examined in the aggregate appears similar to that of many other high-income countries. However, a closer examination reveals that the nation has been losing ground for decades. Although educational indicators for the United States have generally not worsened over time in their absolute levels, other rich countries have been gradually but substantially increasing their populations' educational attainment and performance over time. This improvement is true not only among the 17 high-income peer countries examined in Part I, but also among other countries (e.g., Korea), whose rapidly improving educational performance is surpassing the United States. According to the World Economic Forum (2011), the education system of the United States ranks 26th on how well its education system meets the needs of a competitive economy. The United States does rank highly among older age groups, but the rankings are mediocre or below average among younger adults or cohorts who have been educated in more recent decades (OECD, 2011a).

Years of Schooling

As of 2003, U.S. adults aged 25-64 had completed an average of 13.3 years of formal education, ranking fourth among the 17 peer countries examined in Part I. The ranking for older adults (ages 45-64) was even higher: U.S. women in that age group had more years of schooling than older women in any peer country. In contrast, the United States ranked seventh in the years of schooling of young adult males (ages 25-34); the country also ranked below some countries that were outside the peer comparison group, such as Korea (OECD, 2012f).

---

[14]Income inequality has been rising in the United States over the past three to four decades but has also risen in Finland, Italy, the Netherlands, New Zealand, Norway, Portugal, Sweden, and the United Kingdom (Förster and d'Ercole, 2005).

Enrollment in Child Care and Preschool

According to OECD data from the 17 peer countries, enrollment rates of children under age 3 in formal child care are also lower in the United States than in eight peer countries (see Figure 6-3). Figure 6-4 shows that enrollment rates in educational preschool (ages 3-5) are also lower in the United States than in all but two peer countries (OECD, 2012j). Comparisons with other countries indicate that Korea, as well as Mexico and most Eastern European countries, have higher rates of participation in educational preschool than the United States (OECD, 2012j).

Completion of Secondary Education

Across all age groups in the United States, approximately 88 percent of adults have completed a secondary education (the U.S. equivalent of high school), which remains the highest ranking among the 17 peer countries. However, other countries (including countries outside the peer group, such as those in Eastern Europe and Korea) are gaining ground, and thus the U.S. ranking among young adults is less favorable. Among adults aged 25-34, the United States ranks fourth among peer countries in the completion of secondary schooling and is now outranked by the Czech Republic, Korea (OECD, 2011a), Poland, and Slovenia.

The United States also has a significant high school dropout problem, although it has declined since the 1970s (Snyder and Dillow, 2011). In 2006, 11 peer countries had higher percentages of 17-year-olds enrolled in secondary education than did the United States (National Center for Education Statistics, 2012b).[15] As shown in Figure 6-5, among 13 peer countries that provided data for 2009, 10 countries[16] exceeded the United States in the percentage of the population who had graduated from upper secondary education. As of 2009, the U.S. upper secondary graduation rate (76.4 percent) had declined below the OECD average and was lower than that of several former Soviet countries, including Hungary, Poland, Slovenia, the Slovak Republic, and the Czech Republic (OECD, 2011a). As the OECD (2011a, p. 44) explains:

> Now, at least 80% of young adults in all OECD countries complete an upper secondary education. Within this general pattern, the United States has seen only a small improvement, having started out from the highest high-school completion rate, while Finland and Korea transformed them-

---

[15]Other countries, including Chile, the Czech Republic, Hungary, Korea, Poland, and the Slovak Republic, also had higher secondary education enrollment rates than the United States.

[16]When the analysis was restricted to young persons of typical graduation age, nine countries had higher rates than the United States (National Center for Education Statistics, 2012b).

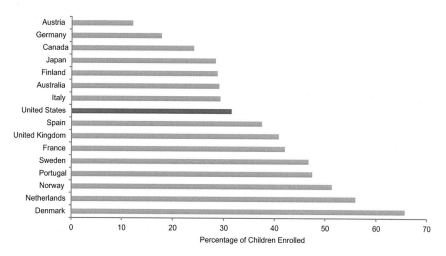

FIGURE 6-3 Enrollment of children aged 0-2 in formal child care in 16 peer countries, 2008.
SOURCE: Data from OECD (2012j), OECD Family Database, Table PF3.2.

selves from countries where only a minority of students graduated from secondary school to those where virtually all students do.

Completion of Tertiary Education

For decades, the United States has led the world in college education, and it still maintains that status: among peer countries, only Canada has a higher rate for completion of tertiary education. Other countries, however, have been rapidly gaining ground, which is reflected in a clear cohort effect across age groups. Among the 17 peer countries examined in Part I, the United States ranks second in tertiary education among people aged 55-64, third for those aged 45-54, fourth for those aged 35-44, and ninth for those aged 25-34.[17] Figure 6-6 shows that college completion among adults aged 25-34 is higher in eight peer countries than in the United States. It is also higher in a number of countries outside the peer group, such as Korea and Russia (OECD, 2011a). According to the OECD (2011a, p. 16):

Half a century ago, employers in the United States and Canada recruited their workforce from a pool of young adults, most of whom had high school diplomas and one in four of whom had degrees—far more than in

---

[17]In 2006, 16 of the 29 OECD countries reporting outperformed the United States on the percentage of the population of the typical graduation age who had received bachelor's degrees (National Center for Education Statistics, 2012c).

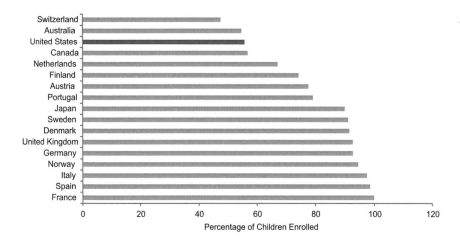

FIGURE 6-4 Enrollment of children aged 3-5 in preschool in 17 peer countries, 2008.
SOURCE: Data from OECD (2012j), OECD Family Database, Table PF3.2.

most European and Asian countries. Today, while North American graduation rates have increased, those of some other countries have done so much faster, to the extent that the United States now shows just over the average proportion of tertiary-level graduates at ages 25-34.

College graduation rates in the United States increased from 33 to 37 percent between 1995 and 2009. The OECD average at the start of the same time period (1995) was 20 percent, but by 2009 it had reached 39 percent, exceeding the U.S. graduation rate (National Center for Education Statistics, 2012b).[18] As of 2006, 10 of the peer countries[19] that reported data outperformed the United States on the percentage of the population of the typical graduation age who had received bachelor's degrees (National Center for Education Statistics, 2012c).

---

[18]Americans who do complete college are more likely to have a degree that the OECD classifies as "less than 3 years" (e.g., associate degrees). The proportion of degrees in this category is 7 percent for the OECD but 35 percent in the United States. Only Turkey has a larger proportion of students who are so classified (OECD, 2011a).

[19]Other countries outside the peer group also outperformed the United States, including Hungary, Korea, Poland, and Russia.

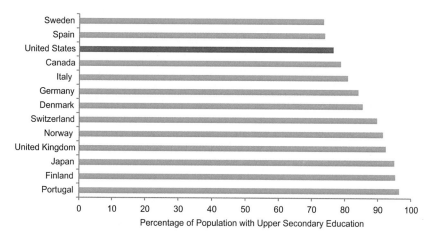

FIGURE 6-5 Upper secondary education rates in 13 peer countries, 2009.
NOTE: Upper secondary education corresponds to International Standard Classification of Education (ISCED) classifications 3 and 4, equivalent to high school in the United States, which prepares students for university-level education, vocationally oriented tertiary education, or workforce or postsecondary nontertiary education.
SOURCE: Data from OECD (2011a, Table A2.1).

Educational Achievement

Educational achievement on cognitive tests may be a more meaningful measure of education than the number of years of schooling. U.S. test scores are about average among high-income countries (Hanushek et al., 2008), but they are above average for U.S. grade school students. International comparisons of the reading literacy of fourth grade students are conducted periodically in the Progress in International Reading Literacy Study (PIRLS).[20] As of the 2001 PIRLS, the United States ranked eighth, fifth, and thirteenth, respectively, of 27 countries, on combined reading literacy, average literary subscale, and average informational subscale scores. The countries participating in the 2001 PIRLS included 9 of the 17 peer countries examined in Part I and, among these, the United States ranked fifth for combined reading literacy of fourth graders. The 2006 PIRLS included 12 peer countries, among which U.S. fourth graders ranked seventh for combined reading score (Baer and McGrath, 2007; Ogle, 2003).

---

[20]The International Association for the Evaluation of Educational Achievement (IEA) is the international coordinating body for PIRLS studies; in the United States, PIRLS is overseen by the National Center for Education Statistics (NCES) in the U.S. Department of Education.

The mathematics and science aptitudes of students in fourth and eighth grade are tested in the Trends in International Mathematics and Science Study (TIMSS), which is administered in the United States by the U.S. Department of Education. The U.S. scores in 1995 and 2007 ranked highly among the peer countries examined in Part I,[21] although the highest mathematics scores in 2007 were reported by China, Hong Kong, Japan, Kazakhstan, the Russian Federation, and Singapore (Gonzales, 2008).

Less encouraging results for the United States have emerged from the OECD Programme for Student International Assessment (PISA), which assesses the cognitive performance of high school students. Since 2000, the PISA has been administered every 3 years to 15-year-old students to assess cognitive skills in reading, mathematics, and science and general problem-solving skills. A National Academy of Sciences et al. (2007, p. 31) report, *Rising Above the Gathering Storm*, reacted with concern to the 2003 PISA scores:

> Students in the United States are not keeping up with their counterparts in other countries. In 2003 [PISA] measured the performance of 15-year-olds in 49 industrialized countries. It found that U.S. students scored in the middle or in the bottom half of the group in three important ways: our students placed 16th in reading, 19th in science literacy, and 24th in mathematics. In 1996 (the most recent data available), U.S. 12th graders performed below the international average of 21 countries on a test of general knowledge in mathematics and science.

In our examination of the 2003 PISA data for the 17 peer countries that are the focus of Part I, we found that U.S. 15-year-olds ranked seventh lowest in reading (just above the OECD average) and science (just below the OECD average), and third lowest in mathematics. By the time of the 2006 PISA, the U.S. rankings had deteriorated further (see Table 6-1). Among the 17 peer countries, U.S. 15-year-olds ranked fifth lowest in science, third lowest in mathematics, and lowest in reading.[22] The U.S. science score was exceeded by many nonpeer countries, led by China and followed by Estonia, New Zealand, Liechtenstein, Korea, Slovenia, the Czech Republic, Belgium, Ireland, Poland, Croatia, and Latvia (OECD, 2012n). The U.S. scores and ranking improved somewhat in the 2009 PISA, but they remained at or below the OECD average. By 2009, the U.S. science score was exceeded by 14 countries, led by China, Singapore, and Korea (OECD, 2012n).

Because those affected by the U.S. health disadvantage include people

---

[21]Several peer countries (e.g., Austria, Denmark, Finland, France, Germany, the Netherlands) were not included in one or both years.

[22]The 2009 PISA also showed that U.S. students were among those least likely in OECD countries to read for enjoyment (OECD, 2011a).

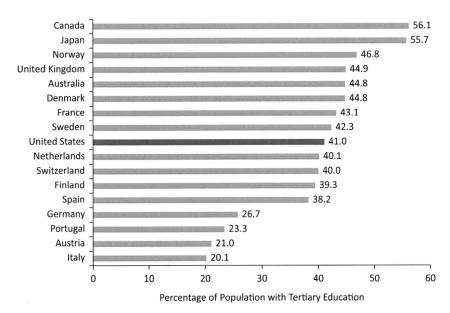

FIGURE 6-6 Percentage of adults aged 25-34 with a tertiary education in 17 peer countries, 2009.
NOTE: Tertiary education corresponds to International Standard Classification of Education (ISCED) classifications 5A and 5B, equivalent to a 2-year or 4-year college education in the United States.
SOURCE: Data from OECD (2011a, Table A1.3a).

who were educated many years ago, educational performance in the past decade, as reported above, may be less relevant to current health outcomes than the quality of education in prior decades. Hanushek and colleagues (2008) compiled data for 50 countries from 1964 to 2003 by standardizing the results of 12 PISA tests and other international mathematics and science assessments. In the 1960s and 1970s, U.S. students had lower scores than those in 13 countries (including 9 of the peer countries examined in Part I). The United States was outranked by 18 countries in the 1980s (including 10 peer countries), by 20 countries in the 1990s (including 9 peer countries), and by 23 countries in the 2000s (including 10 peer countries). By the 2000s, Finland and nonpeer countries such as Korea, Taiwan, and Japan were the top performers (Hanushek et al., 2008).

TABLE 6-1 Comparative Ranking of 15-Year-Old Students in High-Income Countries, 2006

Rank	Reading	Science	Mathematics
1	Finland	Finland	Finland
2	Canada	Canada	Netherlands
3	Australia	Japan	Switzerland
4	Sweden	Australia	Canada
5	Netherlands	Netherlands	Japan
6	Switzerland	Germany	Australia
7	Japan	United Kingdom	Denmark
8	United Kingdom	Switzerland	Austria
9	Germany	Austria	Germany
10	Denmark	Sweden	Sweden
11	Austria	Denmark	France
12	France	France	United Kingdom
13	Norway	United States	Norway
14	Portugal	Spain	Spain
15	Italy	Norway	United States
16	Spain	Italy	Portugal
17	United States	Portugal	Italy

NOTES: Actual scores for United States vs. OECD average were as follows: Reading (est. 460 vs. 495), Science (489 vs. 500), and Mathematics (474 vs. 498). According to the OECD's website, the U.S. reading data that were originally displayed in bar charts were "subsequently removed from the PISA publications for technical reasons." In the 2009 PISA, the U.S. reading score was 487, lower than all but three peer countries.
SOURCE: Data from OECD (2012n).

Employment

The recession-related decline in employment between 2007 and 2009 was more abrupt in the United States than in many countries (OECD, 2011e), but unemployment rates in the United States have traditionally not exceeded those of other high-income countries since the 1980s (Nickell et al., 2005; U.S. Census Bureau, 2012).[23] The United States had notably high unemployment rates in the 1960s and the mid-1970s, rates that were

---

[23]The United States has one of the lowest levels of employment for college graduates (OECD, 2011a).

higher than those in Australia, Canada, France, Italy, Japan, Sweden, the United Kingdom, and West Germany (U.S. Census Bureau, 1970, 1980).[24]

Social Mobility

Multiple studies have demonstrated that the United States has less social mobility than other countries (Blanden et al., 2005; Corak, 2004; Isaacs et al., 2008; Solon, 2002). A report by the Brookings Institution (Isaacs et al., 2008, p. 40) noted:

> Men born into the poorest fifth of families in the United States in 1958 had a higher likelihood of ending up in the bottom fifth of the earnings distribution than did males similarly positioned in five Northern European countries—42 percent in the United States, compared to 25 to 30 percent in the other countries.

In contrast, in the United States now, "only 8 percent make the 'rags to riches' climb from bottom to top rung in one generation, while 11 to 14 percent do so in other countries" (Isaacs et al., 2008, p. 40). A 2011 follow-up study reported that one-third of Americans who grew up in the middle class fall below that category as adults (Acs, 2011).

A 2010 OECD report found that nine other OECD countries out-ranked the United States on the link between individuals' and their parents' earnings, an accepted measure of economic mobility (OECD, 2010a). A 2005 report from Warwick University concluded that the United States has a particularly "high likelihood [compared with Nordic countries, and a higher likelihood than the United Kingdom] that sons of the poorest fathers will remain in the lowest earnings quintile . . . [and a] . . . very low likelihood that sons of the highest earners will show [long-term] downward . . . mobility" (Jäntti et al., 2005, p. 27). Another study concluded that "Intergenerational [economic] mobility in the United States is lower than in France, Germany, Sweden, Canada, Finland, Norway, and Denmark; only the United Kingdom had a lower rate of mobility than the United States" (Hertz, 2006, p. i).[25]

Homelessness

Data are limited to compare housing instability in the United States and other high-income nations. A telephone survey in Belgium, Germany, Italy,

---

[24]Data are lacking to compare the occupational health and safety of workers in the United States and other high-income countries.

[25]Blacks in the United States have less economic mobility than whites (Blanden et al., 2005; Hertz, 2006).

the United Kingdom, and the United States found that the United Kingdom (7.7 percent) and the United States (6.2 percent) had higher lifetime rates of literal homelessness than did the other countries (Toro et al., 2007).

Incarceration

The United States has the highest incarceration rate among affluent countries—approximately 750 of every 100,000 U.S. citizens are in prison—and the rate has been increasing over time (Glaze, 2011; Pew Center on the States, 2008). Between 1987 and 2007, year-end prison counts in the United States nearly tripled from 585,084 to 1,596,127 (Pew Center on the States, 2008). And within the United States, the rate of ever having gone to prison among males was more than six times higher among blacks than whites (Bonczar, 2003). The U.S. Department of Justice estimates that if incarceration rates remain unchanged, 6.6 percent of U.S. residents (and 32.2 percent of black males) born in 2001 will go to prison at some point in their lifetime (Bonczar, 2003).

Household Composition

Between 1940 and 2000, the percentage of U.S. children born to unmarried women increased from approximately 4 percent to almost 35 percent (U.S. Census Bureau, 2003). By 2008, 30 percent of U.S. households with children were headed by a single parent. The corresponding percentages in comparable countries—including Canada, Japan, Denmark, France, Germany, Ireland, Netherlands, Sweden, and the United Kingdom—also increased during these decades, but none had the high percentage of the United States. In 2007, 39.7 percent of live births in the United States were to unmarried women, although some other countries (including Denmark, France, and Sweden) reported even higher rates (U.S. Census Bureau, 2011). A study that compared the United States with 16 other countries (including 9 of the peer countries examined in Part I) found that children born in the United States can expect to live more years with parents apart, with a single mother, without a mother, or in a maternal stepfamily than children in other countries (Heuveline et al., 2003). Single-parent households have a greater need for some social services, such as day care, which are less available in the United States.

Summary

The United States differs from other high-income countries in several social domains that relate to health outcomes[26]:

- The United States has the second highest median household income in the OECD but the fourth highest (among all OECD countries and the highest among peer countries) level of income inequality, the latter having increased in the United States since 1968.
- Since the 1980s, the United States has had among the highest rates of overall poverty and child poverty of all rich nations and many less affluent countries.
- The United States ranks high in average years of schooling and educational attainment, but other countries (including many emerging economies) have been improving educational performance more rapidly, and U.S. adults aged 25-34 now have mediocre rates for completing secondary and tertiary education.
- U.S. preschool enrollment is lower than in most high-income countries.
- Although U.S. grade school students score well relative to children in other countries, by age 15 U.S. students have average or below-average scores on mathematics, science, and reading.
- The United States has low levels of social mobility relative to other high-income countries.
- The United States has the highest rate of incarceration among high-income countries.
- The United States has the highest rate of households with children headed by a single parent.

QUESTION 3:
DO DIFFERENCES IN SOCIAL FACTORS EXPLAIN THE U.S. HEALTH DISADVANTAGE?

It is highly plausible—although not proven—that the social conditions discussed above have contributed to the U.S. health disadvantage relative

---

[26]Racial and ethnic disparities are an important domain that affects health outcomes (Agency for Healthcare Research and Quality, 2011). However, for reasons noted above, notably the lack of comparable evidence, this summary does not include differences among the United States and other countries in the magnitude of racial or ethnic disparities. It focuses on socioeconomic differences, for which data are more readily available, but the panel is acutely aware of the important additional role of race and ethnicity in the health profile of all high-income countries. We also note that the implications of the listed conditions can differ substantially across countries because of differences in social programs and the strength of safety nets.

to other high-income countries. In particular, overall poverty and child poverty are especially plausible explanations for the pervasive U.S. health disadvantage across multiple causes of illness, unhealthy behaviors, and mortality during the first three or possibly four decades of life. Both have been markedly higher in the United States than in almost all other high-income countries during the years when people now in their 20s, 30s, or early 40s were born, growing up, or entering young adulthood. Conditions for children and young people during that period could have shaped risks for low birth weight and infant mortality among the babies born to that cohort, as well as risks for virtually all of the health conditions on which the United States has a disadvantage during the first three or four decades of life. Adult poverty during the past three to four decades may even explain some of the health disadvantage observed among older U.S. adults, but these effects might be less dramatic than those for younger age groups.

Chapters 1 and 2 documented that the U.S. health disadvantage is not confined to minorities or those with low incomes or low educational levels but exists at all socioeconomic levels and for non-Hispanic whites. The high rates of poverty and child poverty in the United States would not explain the persistence of the problem in these advantaged groups, although it is possible that social inequality itself contributes to or reflects conditions that affect the entire population.

Similarly, it is plausible that racial inequality in the United States compounds the societywide effects of economic inequality with which it is intertwined, but empirical cross-national data on this dimension of social inequality are unavailable. The large incarcerated population in the United States suggests a profound degree of multidimensional social disadvantage that affects many people—not only prisoners themselves but also their families and communities (Wildeman and Western, 2010). Racially dispro-portionate incarceration rates in the United States are probably reflections of multiple societal problems and are a likely contributor to the health risks associated with poverty and to social immobility.

Although the United States has a proportionally larger foreign-born population than the OECD average (OECD, 2011e), the large immigrant population does not explain the U.S. health disadvantage because of the "immigrant paradox," the tendency of first-generation immigrants to have better health than the native-born population. This phenomenon is chiefly manifested in the United States as the so-called Hispanic paradox, in which Hispanic Americans tend to have better health outcomes than people born in the United States, especially within 10 years of immigration (Markides and Eschbach, 2005). The phenomenon has also been documented among black immigrants from Africa and the Caribbean (Collins et al., 2002).[27]

Differences in child care and preschool education in the United States could have broad effects that reach beyond disadvantaged groups. Strong evidence demonstrates that high-quality child care and early childhood development programs (from infancy through age 5) lead to higher educational attainment, income, and employment rates in adulthood, and lower rates of criminal behavior. Although the benefits of preschool education are greater for poor children than others, they have been demonstrated among children of diverse socioeconomic backgrounds; and the benefits are themselves strong predictors of subsequent health (Karoly et al., 2005; National Research Council and Institute of Medicine, 2000; Rolnick and Grunewald, 2003).

The mixed but overall mediocre—and in some cases very low—comparative standing of the United States on educational attainment and measures of educational achievement in secondary and tertiary education may contribute to the U.S. health disadvantage. The educational attainment and cognitive skills of today's young adults could, for example, influence the behaviors that can contribute to infant and child mortality due to accidents and homicides; adolescent births and sexually transmitted infections; HIV/AIDS incidence and mortality; and drug-related mortality. That the United States once led the world on educational attainment and has a highly educated cohort of older adults, however, makes education a less likely contributor to the health disadvantages currently observed among older Americans. They were educated many decades ago, when their counterparts in other high-income countries did not, as a group, hold an educational advantage.

Inequalities in life expectancy and all-cause death rates among Americans with different levels of education and income have been increasing for decades (Dow and Rehkopf, 2010; Jemal et al., 2008b; Meara et al., 2008; Pappas et al., 1993), and this gradient may be steeper in the United States than in other high-income countries. In an examination of mortality rates by educational attainment among adults aged 30-74, mortality rates among white men were higher in the United States than in England and Wales, France, Norway, Sweden, and Switzerland, but the excess mortality was more pronounced in lower than in higher educational groups (Avendano et al., 2010). For example, although mortality among U.S. men was 30 percent higher than among Swedish men, it was 24 percent higher among men with a tertiary or higher education, but 55 percent higher among men

---

[27]The panel did not examine evidence regarding whether or not some of the specific health disadvantages observed in the United States (e.g., certain communicable diseases and injuries) are more common among immigrants.

with a lower secondary education or less. The authors concluded that part of the excess mortality among U.S. men is due to larger inequalities in mortality by educational levels in the United States than those in some European countries.

A variety of hypotheses could explain this pattern, among them that prosperity in the United States may be more deeply tied to higher education (Cutler et al., 2011). The consequences of inadequate educational attainment may differ by country, depending on the educational credentials required for desirable jobs, economic security, and other material and psychosocial benefits gained through employment. In comparison with the United States, only Poland, Portugal, Spain, and Turkey have similarly low rates (approximately 20 percent or less) of skilled employment among employed persons who lack tertiary schooling, indicating the larger socioeconomic consequences in the United States of not having a tertiary education (OECD, 2009b).

The health benefits associated with education might be stronger in the United States than in other countries. A recent study reported that life expectancy among people with less than 12 years of education *decreased* in the United States between 1990 and 2008, at the same time that it *increased* among people with 12 or more years of education (Olshansky et al., 2012). Underscoring the importance of education, the study also found that blacks and Hispanic Americans with 16 or more years of education lived 7.5 years and 13.6 years longer, respectively, than whites with less than 12 years of education, although racial and ethnic disparities persisted at all levels of education. The study noted that "[t]he same highly educated black men and women who live longer than less educated whites still live about 4.2 years less than comparably educated whites" (p. 1,806).

More so than in other countries, there appears to be a stronger link in the United States between parental education and children's economic, educational, and socioemotional outcomes (Ermisch et al., 2012). Mediocre performance on education could be a result of disadvantaged home environments, such as a lack of parental stimulation of children's cognitive development, or other disadvantages that could accompany child poverty (Ermisch et al., 2012). Compared with disadvantaged students in many high-income countries, those in the United States appear to have less "resilience" and to show greater deficiencies in reading (OECD, 2011a). Their disadvantage also could be exacerbated by deficiencies in health during childhood and young adulthood (Fletcher and Richards, 2012).

Another factor that could compound the effects of low income and education is the comparatively weak social safety net (i.e., fewer publicly funded transfers and services) in the United States (Avendano and Kawachi, 2011). A weaker safety net may exacerbate the detrimental health effects of poverty, unemployment, and economic insecurity (Bartley et al., 1997; Dow and Rehkopf, 2010). The health effects of unemployment may be buffered, for example, by programs that provide job training and counseling, medical care, and income and housing supports for the unemployed (Bartley and Owen, 1996). In one recent study, prospective data from the United States and Germany covering 1984 to 2005 showed that unemployment was associated with higher mortality in the United States but not in Germany. This relationship was only evident among low- and medium-skilled workers, prompting this analysis by the authors (McLeod et al., 2012, pp. 1,544-1,555):

> In the American cohort there was no relationship between unemployment and mortality for the high-skilled. It appears that individuals with a high level of education may be best suited to take advantage of the more flexible labor markets within the United States. The high-skilled were also more likely to receive benefits, when unemployed, than were those of lower skill levels. These individuals may also have other resources (e.g., savings, familial resources, and social or business contacts from educational or professional organizations) to draw upon that would buffer the effect of unemployment on health. . . . In Germany, the unemployed medium-skilled had the lowest relative risk of dying. This is the strongest evidence that institutional environment can affect the relationship between unemployment and health as institutional protection is targeted toward medium (and vocationally) skilled workers in Germany.

The health consequences of low income may be mitigated by other resources that help individuals and families meet their basic needs (Anand and Ravallion, 1993), such as free or subsidized food, medical care, child care, elder care, education, housing, public transportation, recreational services (e.g., parks, supervised activities for children), and other social protections. A study of 18 countries documented that social security transfers and public health spending significantly reduce poverty levels (Brady, 2005).

As with other social factors, child poverty could potentially have more severe adverse health consequences in the United States than in other affluent nations. Greater public investments in child and family supports—including child care, early childhood development, and preschool programs (see Chapter 8)—appear to help alleviate the effects of child poverty in other countries. For example, in a study that compared Sweden and the United States in the 1980s, the authors found that child poverty rates did not differ substantially when measured by household income *before* social transfers and taxes (Jäntti and Danziger, 1994, p. 50):

> After counting income from both the market and the welfare state, 12.8% of children in the United States and 2.1% of children in Sweden were disposable income-poor. The difference in disposable income poverty rates

between children living in two-parent and single-mother families was 32.5 percentage points in the United States, but virtually nil in Sweden. Over the mid-1980s, child poverty increased by almost one-third in the United States (from 9.6% to 12.8%), while it fell slightly in Sweden (from 2.6% to 2.1%).

The relatively weak social safety net in the United States is a potential explanation not only of a health disadvantage among low-income children and a contributor to low social mobility, but also of the health disadvantages observed among children in all income groups. Stronger safety nets could, at least in theory, lessen the stress and anxiety associated with a potential loss of income or the high costs of medical care, child care, and education (Bartley et al, 1997).[28]

Safety net programs, the quality of schools, and economic conditions in the United States are shaped in part by public policies, such as the relatively low public spending on services for families and young children compared with other high-income countries. The role of public policies in shaping social conditions is discussed further in Chapter 8.

CONCLUSIONS

Part I of this report documents that life expectancy and other health outcomes (e.g., infant mortality) in the United States began to lose pace with other high-income countries in the late 1970s, a trend that has continued to the present. During this same time, as this chapter notes, there has also been a potentially important co-occurrence of worsening social conditions in the United States, notably a rise in income inequality, poverty, child poverty, single-parent households, divorce, and incarceration—all more pronounced than in other rich nations—and the loss of the U.S. leadership position in education. Like the U.S. health disadvantage, many of these social problems began to differentiate the United States from other rich nations in the late 1970s and 1980s.

Whether these co-occurring social trends, individually or in combination, were causally related to the increasing U.S. health disadvantage is still unclear. Answering this question requires a careful examination of historical data to make relevant cross-national comparisons on a range of social conditions over several decades. The cross-country rankings on social indicators discussed in this chapter reflect relatively recent data (since the

---

[28]The role of safety net programs and their interplay with other societal factors is undoubtedly complex, however. As discussed further in Chapter 8, European studies have shown that although stronger safety nets are consistently associated with better aggregate health in countries, they do not necessarily correlate with the size of health inequalities within European countries (Kunst et al., 1998; Mackenbach et al., 1997, 2008).

1980s), but as noted throughout this report, current health outcomes may have been influenced significantly by social conditions experienced much earlier, particularly for children in the post–World War II era. To understand the current U.S. health disadvantage, it would be important to examine cross-nationally the trajectories of social factors, including programs, services, and spending, that were in place four to six decades before the relevant health outcomes appeared.

Whether the worsening social conditions in the United States and its growing health disadvantage are causally interrelated, their co-occurrence during the same time span in recent U.S. history certainly warrants further scrutiny. As documented in this and the next chapter, there have been dramatic changes in the social fabric of the United States; see in particular the discussion of social capital in Chapter 7. These unsettling trends present a potentially important explanation for the U.S. health disadvantage and are shaped by a range of more deeply rooted societal and structural factors. An examination of these underlying causes can shed light on why the United States appears to be losing ground in so many domains: not only health, but also education, economic equality, and child well-being.

Chapter 7 explores the role of the physical and social environment as an explanatory factor. Chapter 8 explores the important role of lifestyles, cultural attributes, public policies, spending priorities, and values as contributors to the patterns observed in this report. These societal factors cannot be ignored when trying to understand either the U.S. health disadvantage or the unfavorable social and economic circumstances reported in this chapter.

7

Physical and Social
Environmental Factors

The previous chapters of this report focused on health systems and individual and household-level risks that might explain the U.S. health disadvantage, but it has been increasingly recognized that these health determinants cannot be fully understood (or influenced) in isolation from the environmental contexts that shape and sustain them. In contrast with traditional environmental health approaches that focus primarily on toxic substances in air, water, and soil, this more recent approach conceptualizes the environment more broadly to encompass a range of human-made physical and social features that are affected by public policy (Frumkin, 2005). These economic, social, urban or rural, transportation, and other policies that affect the environment were not traditionally thought of as relevant to health policy but are now attracting greater attention because decision makers are beginning to recognize their health implications (Cole and Fielding, 2007).

By definition, environmental factors affect large groups that share common living or working spaces. Thus, they are key candidates as explanatory factors for health differences across geographic areas, such as countries. Indeed, a major motivation for the research on environmental determinants of health has been the repeated observation that many health outcomes are spatially patterned. These patterns are present across countries and across regions within countries, as well as at smaller scales, such as across urban neighborhoods (Center on Human Needs, 2012b; Kawachi and Subramanian, 2007). Strong spatial variation is present for a large range of health outcomes, including many of the outcomes for which there are cross-national health differences, such as noncommunicable diseases, associated risk factors, injuries, and violence.

Understanding the reasons for the spatial patterns of health within countries may shed light on environmental factors that may contribute to differences across countries. Several factors may explain the strong spatial patterns that are observed within countries. A key contender is the spatial sorting of people based on their socioeconomic position, race, or ethnicity. However, evidence suggests that regional and neighborhood differences in health persist even after adjusting for these socioeconomic and demographic factors (Diez Roux and Mair, 2010; Mair et al., 2008; Paczkowski and Galea, 2010; Pickett and Pearl, 2001). This evidence suggests that broad environmental factors may play an important role in health. Moreover, environmental factors linked to space and place may in turn contribute to and reinforce socioeconomic and racial or ethnic health disparities (Bleich et al., 2012; Laveist et al., 2011). Thus, individual and environmental factors may be part of a reinforcing cycle that creates and perpetuates health differences. These reinforcing processes by which environmental factors and individual-, family-, and community-level factors reinforce each other over time may also play an important role in generating cross-national differences in health.

This chapter focuses on both the physical and social environment in the United States as potential contributors to its health disadvantage relative to other high-income countries. This chapter, like others before it, focuses on three questions:

- Do environmental factors matter to health?
- Are environmental factors worse in the United States than in other high-income countries?
- Do environmental factors explain the U.S. health disadvantage?

QUESTION 1: DO ENVIRONMENTAL FACTORS MATTER TO HEALTH?

Many aspects of the physical and social environment can affect people's health.[1] Spatial contexts linked to regions or neighborhoods are among

---

[1]Although analytically distinct, physical and social environments may also influence and reinforce each other: for example, physical features related to walkability may contribute to social norms regarding walking, which may in turn promote more walkable urban designs and community planning.

the most frequently studied,[2] but other contexts may also be important for certain segments of the population.[3]

Physical Environmental Factors

The factors in the physical environment that are important to health include harmful substances, such as air pollution or proximity to toxic sites (the focus of classic environmental epidemiology); access to various health-related resources (e.g., healthy or unhealthy foods, recreational resources, medical care); and community design and the "built environment" (e.g., land use mix, street connectivity, transportation systems).

The environment can affect health through physical exposures, such as air pollution (OECD, 2012b). A large body of work has documented the effects of exposure to particulate matter (solid particles and liquid droplets found in the air) on cardiovascular and respiratory mortality and morbidity (Brook et al., 2010; Laumbach and Kipen, 2012; Mustafić et al., 2012; Tzivian, 2011). Research has identified specific physiologic mechanisms by which these exposures affect inflammatory, autonomic, and vascular processes (Brook et al., 2010; Tzivian, 2011).

The effects of particulate matter on mortality appear to be consistent across countries. For example, a recent review of studies from the late 1990s to mid-2000s found a consistent inverse relationship between airborne particulate matter and birth weight in Australia, Brazil, Canada, France, Italy, the Netherlands, South Korea, the United Kingdom, and the United States (Parker et al., 2011a). Another notable example is the evidence linking lead exposures to cognitive development in children (Bellinger, 2008; Levin et al., 2008). The evidence of environmental effects of air pollution and lead has been reflected in legislation in many countries directed at reducing levels of these pollutants in the environment.

Increasing attention has focused on the implications for health behaviors and social interactions that are created by the built environment. The

---

[2]Much early work on the spatial patterns of health used variables such as aggregate summaries of area socioeconomic or race/ethnic composition or measures of residential segregation by various attributes as proxies for a range of broadly defined environmental factors that may be relevant to health (see, e.g., Diez Roux and Mair, 2010). The identification of causal effects using these aggregate summaries raises a number of methodological challenges and does not allow one to identify the specific environmental attributes that may be relevant. More recent work has attempted to identify the specific environmental factors that may be important to specific health outcomes, as well as the pathways through which these factors may operate.

[3]The environment can also be considered on a larger geographic scale, especially in seeking explanations for cross-national health differences. For example, the health of some nations is affected by their geography or climate.

built environment refers to the presence of (and proximity to) health-relevant resources as well as to aspects of the ways in which neighborhoods are designed and built (including land use patterns, transportation systems, and urban planning and design features). An important example is evidence that links proximity to healthy or unhealthy food stores with dietary behaviors and related chronic disease outcomes (Babey et al., 2008; Larson et al., 2009; Moore et al., 2008; Morland et al., 2006).[4] Food availability and food advertising influence energy intake and the nutritional value of foods consumed (Grier and Kumanyika, 2008; Harris et al., 2009; Institute of Medicine, 2006a).

Another large body of work has documented how walking and physical activity levels are affected by access to recreational facilities, land use mix, transportation systems, and urban planning and design (Auchinloss et al., 2008; Diez Roux et al., 2001; Ding et al., 2011; Durand et al., 2011; Gordon-Larsen et al., 2006; Heath, 2009; Kaczynski and Henderson, 2008; McCormack and Shiell, 2011; Transportation Research Board, 2005). Studies conducted in the United States and other high-income countries have found that "walkability" (which is measured by such proxies as building density, land use mix, and street connectivity) predicts walking patterns (Durand et al., 2011; Inoue et al., 2009; Sundquist et al., 2011; Van Dyck et al., 2010). Across countries, studies have also shown that physical activity by children is associated with features of the built environment, including walking-related features, and physical activity resources (Bringolf-Isler et al., 2010; Davison and Lawson, 2006; Galvez et al., 2010; Sallis and Glanz, 2006).[5]

Although more definitive evidence is needed (see Feng et al., 2010), it has been hypothesized that these environmental features may contribute to the obesity epidemic (Galvez et al., 2010; Papas et al., 2007; Sallis and Glanz, 2009). The importance of residential environments to obesity and related conditions, such as diabetes, was recently highlighted by a randomized housing intervention: low-income participants who were randomly assigned to move into low-poverty areas experienced significant improve-

---

[4]Although in the U.S. context a number of studies have reported associations of local access to healthy foods with diet, some studies have not detected such associations (Cummins et al., 2005; Pearce et al., 2008). An important difficulty in comparing results across countries is that the proxy measure for the local food environment is often the type of food stores or restaurants available (such as supermarkets or fast food outlets), but the extent to which these typologies reflect relevant differences in the foods actually available to consumers may differ significantly across countries.

[5]Studies that compare the effects of built environment features across countries are limited and inconclusive. One recent review found that access to open space (parks and other green spaces) in neighborhoods was associated with physical activity levels in both the United States and Australia (Pearce and Maddison, 2011).

ments in weight and diabetes indicators (Ludwig et al., 2011). Unfortunately, the study was not designed to identify the specific environmental features responsible for the observed effect.

A range of other physical environmental features have been linked to other health outcomes. For example, the density of alcohol retail outlets has been linked to alcohol-related health complications (Campbell et al., 2009; Popova et al., 2009), including injury and violence (Cunradi et al., 2012; Toomey et al., 2012). Transportation systems and other aspects of physical environments that influence driving behaviors are also related to injury morbidity and mortality (Douglas et al., 2011). Living in socioeconomically disadvantaged neighborhoods (as a proxy for a range of environmental exposures) has been linked to higher rates of injury in both adults and children (Cubbin et al., 2000; Durkin et al., 1994).

Social Environmental Factors

Factors in the social environment that are important to health include those related to safety, violence, and social disorder in general, and more specific factors related to the type, quality, and stability of social connections, including social participation, social cohesion, social capital, and the collective efficacy of the neighborhood (or work) environment (Ahern and Galea, 2011).[6] Social participation and integration in the immediate social environment (e.g., school, work, neighborhood) appear to be important to both mental and physical health (DeSilva et al., 2005). What also seems important is the stability of social connections, such as the composition and stability of households[7] and the existence of stable and supportive local social environments or neighborhoods in which to live and work.

A network of social relationships is an important source of support and appears to be an important influence on health behaviors. Work on the "transmission" of obesity through social networks has highlighted the possible importance of social norms in shaping many health-related behaviors (Christakis and Fowler, 2007; Hruschka et al., 2011; Kawachi and Berkman, 2000).[8] A long tradition of sociological research links these social features not only to illness, but also to risks of violence (Morenoff et al., 2001; Sampson et al., 1997). Social environments may also operate

---

[6]Other factors that are also frequently discussed, such as social norms, have been more difficult to study because of a variety of methodological and data challenges.

[7]As noted in Chapter 6, divorces and single-parent households have become more prevalent in the United States over time than in other high-income countries.

[8]Analytical complexities make the isolation of these effects difficult in observational studies.

through effects on drug use, which also has consequences for violence and mental-health-related outcomes.[9]

Neighborhood conditions can create stress (Cutrona et al., 2006; Do et al., 2011; Merkin et al., 2009), which have biological consequences (see Chapter 6). Features of social environments that may operate as stressors (including perceptions of safety and social disorder) have been linked to mental health, as have factors that could buffer the adverse effects of stress (e.g., social cohesion, social capital) (DeSilva et al., 2005; Mair et al., 2008).

One mechanism through which the social environment can enhance health is through social support. Social support has appeared in many (but not all) studies to buffer the effects of stress (Cohen and Wills, 1985; Matthews and Gallo, 2011; Ozbay et al., 2007, 2008). Resilience to the adverse health effects of stress has also been tied to factors that could influence how one perceives a situation (threat versus challenge) and how one responds to stressors (Harrell et al., 2011; Hennessy et al., 2009; Matthews and Gallo, 2011; Ziersch et al., 2011). One theory for the tendency of some immigrant groups to have better health outcomes than might be expected on the basis of their incomes and education (see Chapter 6) is the social support immigrants often provide one another (Matthews et al., 2010).

Social capital refers to "features of social organization, such as trust, norms, and networks, that can improve the efficiency of society by facilitating coordinated actions" (Putnam, 1993, p. 167). Studies have shown consistent relationships between social capital and self-reported health status, as well as to some measures of mortality (Barefoot et al., 1998; Blakeley et al., 2001; Kawachi, 1999; Kawachi et al., 1997; OECD, 2010c; Schultz et al., 2008; Subramanian et al., 2002). Social capital depends on the ability of people to form and maintain relationships and networks with their neighbors. Characteristics of communities that foster distrust among neighbors, such as neglected properties and criminal activity, can affect both the cohesiveness of neighbors as well as the frequency of poor health outcomes (Center on Human Needs, 2012b).

Spatial Distribution of Environmental Factors

In addition to considering differences between the United States and other countries in the absolute levels of environmental factors, it is also important to consider how these factors are distributed within countries. Levels of residential segregation shape environmental differences across neighborhoods (Reardon and Bischoff, 2011; Subramanian et al., 2005).

---

[9]Although findings have not always been consistent, levels of safety, violence, and other social environmental features have also been found to be associated with walking and physical activity (Foster and Giles-Corti, 2008).

Neighborhoods with residents who are mostly low-income or minorities may be less able to advocate for resources and services. Perceptions and stereotypes about area reputation, local demand for products and services, and the purchasing power of residents may also influence the location of health-relevant resources. Physical environmental threats (such as proximity to hazardous sites) may be more prevalent in low-income or minority neighborhoods, a concern of the environmental justice movement (Brulle and Pellow, 2006; Evans and Kantrowitz, 2002; Mohai et al., 2009; Morello-Frosch et al., 2011). These neighborhoods may also lack the social connections and political power that can help remedy adverse conditions.

Other Environmental Considerations

The panel focused its attention on the role of local physical and social environments as potential contributors to the U.S. health disadvantage and did not systematically examine whether other contexts, such as school or work environments, differ substantially across high-income countries. Nor did the panel examine whether neighborhood conditions exert a greater influence on access to health care in the United States than in peer countries. However, these conditions are important to health. For example, the school environments of children, adolescents, and college students can affect diet, physical activity, and the use of alcohol, tobacco, and other drugs (Katz, 2009; Wechsler and Nelson, 2008). Dietary options on cafeteria menus and in vending machines, opportunities for physical activity, and health education curricula are all important to children's health.

Workplaces have also long been recognized as important determinants of health and health inequalities, occupational safety, and access to preventive services (Anderson et al., 2009; Schulte et al., 2011). Physical working conditions (e.g., exposure to dangerous substances, such as lead, asbestos, mercury), as well as physical demands (e.g., carrying heavy loads), human factors, and ergonomic problems can affect the health and safety of employees. Stressful psychosocial work environments and "job strain"—which refers to high external demands on a worker with low levels of control or rewards—have become recognized as prominent determinants of health and have been linked to self-reported ill health (Stansfeld et al., 1998), adverse mental health outcomes (Clougherty et al., 2010; Low et al., 2010; Stansfeld and Candy, 2006), and markers of chronic disease (Fujishiro et al., 2011). Exposure to job strain exhibits a strong social gradient, which influences inequalities in the health of workers (Bambra, 2011).[10]

---

[10]Findings on job strain have not been consistent, raising the question of whether these are primarily markers of socioeconomic position, which can influence health through other plausible material or psychosocial pathways (Eaker et al., 2004; Greenlund et al., 2010).

Although the panel did not undertake a systematic comparison of workplace conditions in the United States and other countries, it did note that U.S. employees work substantially longer hours than their counterparts in many other high-income countries. In 2005, annual hours worked in the United States were 15 percent higher than the European Union average (OECD, 2008a). Other working conditions and work-related policies for U.S. employees often differ from those of workers in peer countries. For example, U.S. workers have a larger gender gap in earnings, which could potentially affect the health of women, and U.S. workers spend more time commuting to work (OECD, 2012g), which decreases cardiorespiratory fitness (Hoehner et al., 2012). Other important differences in work-related policies include employment protection and unemployment benefits, as well as family and sickness leave (see Chapter 8). However, cross-national comparisons of workplace safety, other occupational health characteristics, labor market patterns, and work-related policies were beyond the scope of the panel's review.

QUESTION 2: ARE ENVIRONMENTAL FACTORS WORSE IN THE UNITED STATES THAN IN OTHER HIGH-INCOME COUNTRIES?

There is scant literature comparing social and physical environmental features across countries. Here we provide selected examples of the ways in which levels or distributions of physical and social environments relevant to health might differ between the United States and other high-income countries.

Physical Exposures

Few data are available to make cross-national comparisons of exposure to harmful physical or chemical environmental hazards. There is, for example, little evidence that air pollution is a more severe problem in the United States than in other high-income countries (Baldasano et al., 2003; OECD, 2012a; Parker et al., 2011a).[11] Although cross-national comparisons of the volume of emissions and carbon production per gross domestic product show that the United States is a major emitter, this finding does not provide a basis for comparing the cleanliness or healthfulness of air, water, or other resources. The heavy reliance on automobile transportation in the United States is linked to traffic levels, which contribute to air pollution and its health consequences (Brook et al., 2010; Laumbach and Kipen,

---

[11]Averages could mask important spatial heterogeneity in air pollution, and this heterogeneity could have important implications for differences in aggregate health if some populations are systematically exposed to high levels of pollution.

2012). Data on population exposures to air pollution across countries are relatively scarce (OECD, 2008b). One available measure is the concentration of particulate matter less than 10 micrometers in diameter (PM-10):[12] in the United States, the concentration of PM-10 levels is 19.4 micrograms per cubic meter, lower than the OECD average of 22 micrograms per cubic meter (OECD, 2012a).

An important factor that influences a range of environmental features relates to patterns of land use and transportation. In general, U.S. residential environments are highly dominated by Americans' reliance on private automobile transportation. This characteristic has promoted dispersed automobile-dependent development patterns (Transportation Research Board, 2009) with consequences for population density, land use mix, and walkability (Richardson, 2004), all of which may have health implications. In 2008, the United States had 800 motor vehicles per 1,000 people compared with 526 in the United Kingdom, 521 in Sweden, 598 in France, and 554 in Germany (World Bank, 2012b). Cities in the United States tend to be less compact and have fewer public transportation and nonmotorized travel options and longer commuting distances than cities in other high-income countries (Richardson and Bae, 2004). Many European countries have strong antisprawl and pro-urban centralization policies that may contribute to environments that encourage walking and physical activity as part of daily life (Richardson and Bae, 2004).[13]

Social Factors

International comparisons of the social environment are complicated by difficulties in obtaining comparable measures of social environments. For example, aside from their direct links to injury mortality (see Chapter 1), violence and drug use may be indirect markers of social environmental features that affect other health outcomes. As noted in Chapters 1 and 2, homicide rates in the United States are markedly higher than in other rich nations. There are fewer data to compare rates of other crimes across countries. As noted in Chapter 5, certain forms of drug use (which is often linked to other social environmental features) also appear to be more prevalent in the United States than in other high-income countries.

Although Chapter 6 documented a long-standing trend of greater poverty and other social problems in the United States than in peer countries,

---

[12]Particulate matter less than 10 micrometers in diameter (PM-10) poses a health concern because it can accumulate in the respiratory system. In particular, particles that are less than 2.5 micrometers in diameter ("fine" particles) are thought to pose the largest health risks (U.S. Environmental Protection Agency, 2007).

[13]Even in these countries, however, automobile use is rising quickly.

evidence is more limited to compare these countries in terms of social cohesion, social capital, or social participation. For example, OECD data indicate that the United States has the highest prevalence of "pro-social behavior," defined as volunteering time, donating to charities, and helping strangers (OECD, 2011e). At least one study of cross-national differences in social capital found that the United States ranked at an intermediate level compared with other high-income countries in measures of interpersonal trust; the study also found that the United States ranked higher than many other countries on indicators of membership in organizations (Schyns and Koop, 2010). A previous National Research Council (2011) report and a paper prepared for that study (Banks et al., 2010) did not find much evidence that the United States had unique social networks, social support, or social integration. However, the focus of that paper was on the social isolation of individuals rather than on social cohesion or social capital measured as a group-level construct. Other data indicate that nearly 3 percent of people in the United States report "rarely" or "never" spending time with friends, colleagues, or others in social settings. This figure is one of the lowest in the OECD (2012a).

On another measure, OECD data suggest that levels of trust[14] are lower in the United States than the OECD average and than in all peer countries but Portugal, with Nordic countries showing the highest levels (OECD, 2011e). According to the World Gallup Poll, people in the United States are less likely than people in other high-income countries to express confidence in social institutions, and Americans also have the lowest voting participation rates of OECD countries.

In an interesting link between physical and social environments, Putnam (2000) has argued that increasing sprawl could contribute to declining social capital in the United States because suburban commutes leave less time for social interactions. However, it remains unclear whether sprawl helps explain differences in levels of social capital, or health, across countries.

Spatial Distribution of Environmental Factors

Research in the 1990s demonstrated that people of low socioeconomic status were more likely to experience residential segregation in the United States than in some European countries (Sellers, 1999). More

---

[14]Trust data are based on the question: "Generally speaking would you say that most people can be trusted or that you need to be very careful in dealing with people?" Data come from two different surveys: the European Social Survey (2008 wave 4) for OECD European countries and the International Social Survey Programme (2007 wave) for non-OECD Europe (OECD, 2011e).

recent evidence also suggests that residential segregation by income and neighborhood disadvantage has been increasing over time in the United States (Reardon and Bischoff, 2011). Given the established correlation between neighborhood, race, and socioeconomic composition and various health-related neighborhood resources in the United States, this greater segregation could also result in greater exposure of some population sectors to harmful environments (Lovasi et al., 2009). Although studies of residential segregation do not directly assess environmental factors, to the extent that segregation is related to differences in exposure to environmental factors, countries with greater segregation may also experience greater spatial inequities in the distribution of environmental factors, resulting in greater health inequalities and possible consequences for overall health status. Studies that use measures of area socioeconomic characteristics as proxies for environmental features have generally reported similar associations of area features with health in both the United States and other countries (van Lenthe et al., 2005), but there is some evidence that area effects may be greater in countries, like the United States, which have relatively greater residential segregation (Moore et al., 2008; Stafford et al., 2004).

At least two studies have suggested that spatial variation in health-related resources may have very different distributions in the United States than in other countries. A review of spatial variability in access to healthy foods found that food deserts—areas with limited proximity to stores that sell healthy foods—were more prevalent in the United States than in other high-income countries (Beaulac et al., 2009). A New Zealand study found that area deprivation was not always consistently associated with lack of community resources (including recreational amenities, shopping, educational and health facilities) (Pearce et al., 2007). This finding is in sharp contrast to studies of the United States, which have found associations between neighborhood socioeconomic disadvantage and the absence of resources that are important to public health (Diez Roux and Mair, 2010).

Large geographic disparities in toxic exposures to environmental hazards and in healthy food access have been repeatedly noted in U.S. communities (Diez Roux and Mair, 2010; Mohai et al., 2009; Pastor et al., 2005). Similar geographic disparities may exist for other environmental features. For example, the distribution of walkable environments may be more variable in the United States than in other countries, creating "unwalkable" islands, where walking is not a viable transportation alternative to driving. These barriers may inhibit physical activity for parts of the population, resulting in worse overall health. Levels of safety and violence may also be more strongly spatially segregated in the United States than in other countries, resulting in areas with greater exposure to violence and its harmful health consequences.

QUESTION 3: DO ENVIRONMENTAL FACTORS EXPLAIN THE U.S. HEALTH DISADVANTAGE?

Although no studies have collected the necessary data to determine directly the contribution of the environment to the U.S. health disadvantage, existing evidence on the health effects of environmental factors and on differences in levels and distributions of environmental factors between the United States and other high-income countries suggest that environmental factors could be important contributors to the U.S. health disadvantage. Below we review the possible contributions of the environment to major conditions for which U.S. health disadvantages have been documented.

Obesity, Diabetes, and Cardiovascular Disease

Environmental factors that affect physical activity (primarily through their effect on active life-styles, including walking) and access to healthy foods (rather than calorie-dense foods) may help explain differences in obesity and related conditions between the United States and other high-income countries. As noted above, land use patterns and transportation systems differ starkly between the United States and other high-income countries (Richardson and Bae, 2004; Transportation Research Board, 2009). Transportation behavior also differs between the United States and other high-income countries, with U.S. residents walking and cycling substantially less than Europeans (Bassett et al., 2008; Buehler et al., 2011; Hallal et al., 2012). For example, analyses of comparable travel surveys show that between 2001-2002 and 2008-2009, the proportion of "any walking" was stable in the United States, at 18.5 percent, while it increased in Germany from 36.5 to 42.3 percent. The proportion of "any cycling" was extremely low and stable in the United States, at 1.8 percent, while it increased in Germany from 12.1 to 14.1 percent. There was also less variation in active travel among socioeconomic groups in Germany than in the United States (Buehler et al., 2011). Although the precise effects of these transportation differences on people's energy expenditure is difficult to quantify, it seems reasonable to expect that different transportation patterns would have important implications for U.S. levels of obesity (Pucher et al., 2010a).

The food intake of the U.S. population is influenced by both supply and demand, particularly food availability, advertising, and other aspects of the way in which meals are socially produced, distributed, and consumed (including mass production and marketing of cheap calorie-dense foods and large portion sizes) (Institute of Medicine, 2006a; Nestle, 2002; Story et al., 2008).[15] In addition, there is evidence that food access is more inequitably distributed in the United States than in other high-income countries (Beaulac et al., 2009; Franco et al., 2008; Moore and Diez Roux, 2006), which may create problems of food access for vulnerable populations.

Importantly, these various features of the physical environment may act synergistically, reinforcing their effects and creating an "obesogenic" environment that affects all U.S. residents, at least to some extent. In addition, these environmental effects may contribute to the development of social norms regarding behaviors and weight (Christakis and Fowler, 2007), which then reinforce certain features of the physical environment, making them increasingly difficult to modify. This reinforcement creates a vicious cycle in which the environment contributes to the development of social norms (such as reliance of automobile transportation) and the behavior resulting from the norm reinforces the environmental features (such as absence of bicycle lanes or public transportation) that sustain it.

Injuries

The dominant land use and development pattern espoused in the United States for decades (Richardson and Bae, 2004) has created dependence on private automobile transportation, with important implications for traffic volume and associated traffic injuries and fatalities (Transportation Research Board, 2009). Once established, the land use patterns and transportation systems are self-reinforcing and may in turn hinder the development of efficient and inexpensive public transportation alternatives. A physical environment that promotes and incentivizes automobile transportation also reinforces social norms regarding travel, which complicates efforts to modify the patterns. The existing land use patterns and reliance on private automobile transportation not only contribute to traffic volume and injury fatalities, but probably also contribute to physical inactivity, air pollution, and carbon emissions. In this way, a common physical environmental feature may explain the coexistence of the U.S. health disadvantage on apparently unrelated health domains (obesity and injuries).

Homicides, Violence, Drug-Related Deaths, and HIV Risk

Environmental factors, broadly defined, may also contribute to at least part of the U.S. health disadvantage in homicide, violence, and drug-related deaths. As noted above, residential segregation by income in the United States is associated with violence and related outcomes (Sampson et al.,

---

[15]Advertising also plays an important role in promoting alcohol and tobacco use (Chuang et al., 2005; Kwate and Meyer, 2009; Mosher, 2011; Primack et al., 2007).

1997; U.S. Department of Justice, 2007). Residential segregation by income and race have also been linked to drug use (Cooper et al., 2007) and HIV/AIDS risk (Poundstone et al., 2004), other contributors to the U.S. health disadvantage. Neighborhood violent crime has in turn been linked to low birth weight (Morenoff, 2003) and childhood asthma (Wright, 2006), two other health conditions that appear to be more common in the United States than in other high-income countries. Residential segregation (and its many social and physical correlates) may be another environmental factor that affects multiple, seemingly unrelated health domains in which the United States has a health disadvantage.

Another important environmental influence on homicide and suicide rates is the ease of access to firearms, which has a strong association with homicide rates (Hepburn and Hemenway, 2004). Legislative policies in other countries limit circulation and ownership of firearms by civilians. As stated in a thorough review by Hepburn and Hemenway (2004, p. 429):

> High-income countries outside the United States have much lower rates of handgun ownership than the United States, and the licensing, registration, and safe storage regulations they have make it much harder for known criminals to obtain firearms. Thus, relatively few of the homicides in these countries are firearm homicides.

CONCLUSIONS

There is some evidence that environmental factors that could affect the U.S. health disadvantage are worse or are more inequitably distributed in the United States than in other high-income countries. It is plausible to hypothesize that factors in the built environment related to low-density land development and high reliance on automobile transportation; environmental factors related to the wide availability, distribution, and marketing of unhealthy foods; and residential segregation by income and race (with its social and economic correlates) may be important contributors to the U.S. health disadvantage in many domains.

It is noteworthy that these environmental factors may interact with other factors at both "higher" levels of broad social policy and "lower" levels that operate at the individual level. For example, high levels of residential segregation may create large social inequalities across neighborhoods that, in the presence of easy access to guns, may result in high gun violence and homicide rates. Easy access to unhealthy foods may interact with personal sources of stress (e.g., from work) in promoting the consumption of calorie-dense foods. Environments that discourage physical activity may also limit social interactions, with potential implications for violence and drug use.

Environments also help to create and reinforce social norms (Hruschka et al., 2011) that influence health outcomes. In this way, environmental factors are undoubtedly part of a self-perpetuating cycle that operates across multiple domains, but delineating exactly how this occurs—and how this may differ across place and time—will require further research.

Many of the environmental factors relevant to health are directly amenable to policy. Therefore, identifying which of these factors are important contributors to the U.S. health disadvantage could point to policy interventions that might reduce the disadvantage. For example, cross-national comparisons show that levels of active transportation, such as walking or cycling, can be effectively modified by specific land use and transportation policies (Pucher and Dijkstra, 2003; Pucher et al., 2010b). Although many of the data reviewed in this chapter are highly suggestive of an important role for environmental factors, more empirical evidence is needed to draw definitive conclusions. Important areas for future cross-national research on environmental factors and health include (1) characterizing levels and distributions of environmental risk factors using comparable measures across countries; (2) documenting inequalities in the distribution of these environmental factors; (3) identifying the extent to which these environmental factors affect health and the extent to which their effects are modulated by individual-, community-, or country-level factors; (4) examining directly the contribution of environmental factors to health differences between the United States and other high-income countries; and (5) studying national, regional, and local country policies that may curb levels of adverse environmental exposures, reduce the extent to which they are inequitably distributed, or buffer their effects.

The contribution of environmental factors to the U.S. health disadvantage is likely to result from dynamic and reinforcing relationships between environmental and individual-level factors. Environmental factors also operate over a person's life course, so that the environments one experiences early in life may influence health trajectories over time. Environmental factors are in turn linked to upstream social and policy determinants. In many ways, the environment can be thought of as the mid- or "meso-" level of influence linking macrolevel factors (e.g., economic and social policy) and microlevel processes (e.g., individual behavior). A comprehensive understanding of the causes of the U.S. health disadvantage will require recognizing how the environment interacts with these other factors and helps perpetuate or mitigate the disadvantage across a broad set of health domains.

8

Policies and Social Values

Chapters 4-7 identified intriguing differences between the United States and other high-income countries that might plausibly contribute to the health gap:

- The U.S. health system suffers from a large uninsured population, financial barriers to care, a shortage of primary care providers, and potentially important gaps in the quality of care (Chapter 4).
- Americans have a higher prevalence of certain unhealthy behaviors involving caloric intake, sedentary behavior, drug use, unprotected sex, driving without seatbelts, and the use of firearms (Chapter 5).
- The United States lags in educational achievement, and it has high income inequality and poverty rates and lower social mobility than most other high-income countries (Chapter 6).
- Americans live in an obesogenic built environment that discourages physical activity, and they live in more racially segregated communities (see Chapter 7).

Although each of these unfavorable patterns could be examined in isolation, the panel was struck by a recurring theme: data compiled from unrelated sources show that the United States is losing ground to other high-income countries on multiple measures of health and socioeconomic well-being. This finding is true for the young and old and perhaps even for affluent and well-educated Americans. Other rich nations outperform the

United States not only on health status but also on protecting children from poverty, educating youth, and promoting social mobility.

It is highly likely that the U.S. health disadvantage has multiple causes and involves some combination of unhealthy behaviors, harmful environmental factors, adverse economic and social conditions, and limited access to health care.[1] Although there are a number of explanations for the U.S. health disadvantage, the panel began to consider the possibility that this confluence of problems reflects more upstream, root causes. Is there a "common denominator" that helps explain why the United States is losing ground in multiple domains at once? This pattern began decades ago. As long ago as the 1970s and 1980s, the United States began losing pace with other high-income countries in preventing premature death, infant mortality, and transportation-related fatalities; in alleviating income inequality and poverty; and in promoting education.

More research is needed to determine if there is a common underlying cause, but the panel did discuss possibilities, such as characteristics of life in America that create material interests in certain behaviors or business models. For example, those characteristics include the typically pressured work and child care schedules of the modern American family, the strong reliance on automobile transportation, and delays created by traffic congestion often leave little time for physical activity or shopping for nutritious meals. Busy schedules create a market demand for convenient fast food restaurants.[2] It is plausible, but as yet unproven, that societal changes in the United States in the post–World War II period set the stage for many of the deteriorating conditions that appeared in the 1970s and continue to this day.[3]

Certain character attributes of the quintessential American (e.g., dynamism, rugged individualism) are often invoked to explain the nation's great achievements and perseverance. Might these same characteristics also be associated with risk-taking and potentially unhealthy behaviors? Are there health implications to Americans' dislike of outside (e.g., government)

---

[1]Similarly, there are also probably multiple explanations for the health *advantages* the United States experiences relative to other countries, such as the potential dietary, medical, and policy explanations for the country's below-average rate of stroke mortality.

[2]The panel notes the "chicken and egg" question of whether U.S. preferences—for fast foods, traveling in large automobiles, etc.—originated historically from consumer demand or from efforts by companies to create a market for these products and build an infrastructure for them (e.g., highways, drive-in restaurants) that is less prevalent in other rich nations. The currently strong market demand for these products in a society that has grown accustomed to a life-style that depends on these conveniences provides less incentive for businesses to change and strengthens the argument that they are providing products and services that consumers want.

[3]Some of these trends are increasingly observed in other countries as well.

interference in personal lives and in business and marketing practices? Few quantitative data exist to answer these questions or to assert that these characteristics actually occur more commonly among Americans than among people in other countries.[4] Nor is it reasonable to apply a stereotype to an entire society, especially one with the demographic, geographic, and cultural diversity of the United States. Still, for a variety of social or historical reasons, these values have salience for a large segment of U.S. society and may be important in understanding the pervasiveness of the U.S. health disadvantage.

The nature of the interaction between the free market economy and consumer preferences may also be somewhat distinctive in the United States. Manufacturers and other businesses cater to consumer demand for products and services that may not optimize health (e.g., soft drinks and large portion sizes) or, as in the case of cigarettes, are dangerous (Brownell and Warner, 2009). The tobacco industry's long success in manufacturing and marketing products that have been known for five decades to cause cancer and other major diseases (Kessler, 2001; Lovato et al., 2003) reflects, in part, a symbiotic interdependence between producers and consumers who want (or are addicted to) the products.

Another systemic explanation considered by the panel is whether there is something unique in how decisions are made in the United States, in contrast with other countries, which might produce different policy choices that affect health. Not all of the problems identified in this report are affected by policy decisions—many relate to individual choices or perhaps the inherent nature of life in America—but decisions by government and the private sector may play a role in shaping many of the health determinants discussed throughout this report.

THE ROLE OF PUBLIC- AND PRIVATE-SECTOR POLICIES

The relevance of public policy to health is perhaps most conspicuous in relation to recognized problems in the U.S. health care system—limited access, especially for people who are poor or uninsured; fragmentation, gaps, and duplication of care; inaccessibility of medical records; and misalignment of physician and patient incentives (Institute of Medicine, 2001, 2010)—and the policies that are designed to address them. But the potential causes of the U.S. health disadvantage go beyond health care practice and policy. People are responsible for their individual behaviors, but individual life-styles are also influenced by the policies adopted by communities, states, and national leaders (Brownell et al., 2010). Ciga-

---

[4]However, there is qualitative evidence regarding these characteristics from research in political science, anthropology, and other social science disciplines.

rette smoking, second-hand smoke inhalation, and societal norms about smoking are influenced by the price of cigarettes, bans on indoor smoking, and advertising regulations (Brownson et al., 2006; Garrett et al., 2011). The obesogenic environment reflects decisions by the food industry and restaurants about the content and sizes of their offerings; business strategies about where to locate supermarket chains and fast food outlets; ballot decisions on parks, playgrounds, and pedestrian walkways; school board policies on high-calorie cafeteria menus and vending machine contracts; and the marketing of electronic devices to children (Brownell and Warner, 2009; Institute of Medicine, 2006, 2009b, 2009c, 2011c; Nestle, 2002).

Public- and private-sector policies affect drinking and driving, binge drinking, prescription and illicit drug abuse, and the use of contaminated needles by injection drug users. Policies can also influence access to contraceptives and firearms. Both the incidence and lethality of injuries are affected not only by personal choices, but also by decisions made by manufacturers, builders, lawmakers, and regulatory agencies that control product safety, road design, building codes, traffic congestion, law enforcement of safety regulations (e.g., use of seatbelts, blood alcohol testing), fire hazards, and the availability of firearms.

Policies also affect the social and economic conditions in which people live, and the quality of education—from preschool through college and professional schools (Bambra et al., 2010). Political and economic institutions, which help drive the economic success of nations, are subject to a range of public policies (Acemoglu and Robinson, 2012). Tax policy and decisions by employers, business leaders, government, and voters affect job growth, household income, social mobility, savings, and income inequality. They determine the strength of safety net and assistance programs and the quality of the environment, from its physical characteristics (e.g., pollution, housing quality) to social surroundings (e.g., crime, stress, social cohesion). The relevance of macroeconomic government policies on health was exhibited in a natural experiment when East and West Germany unified in 1989-1990: after unification, the mortality rates for the elderly in the eastern part of the country declined to those of the western part (Scholz and Maier, 2003; Vaupel et al., 2003).[5]

_____

[5]The German experience also provides a useful reminder that interventions to improve health outcomes (and address the U.S. health disadvantage) can be effective among older adults. Notwithstanding the importance of addressing the causes of the U.S. health disadvantage among young people (e.g., violence, transportation-related accidents) and the influence of early life conditions on future health trajectories (see Chapter 3), policies to improve the health of middle-aged and older adults are also vitally important.

THE ROLE OF INSTITUTIONAL ARRANGEMENTS
ON POLICIES AND PROGRAMS

Policies that affect public health, education, and the economy are themselves shaped by the institutional arrangements in a society—the governmental and nongovernmental arrangements that organize social relations, rank people into social hierarchies, assign worth, structure employment and the labor market, and address working conditions (Bambra and Beckfield, 2012). As illustrated in Table 8-1, some studies of what has been described as the political economy of health (Muntaner et al., 2011) have demonstrated a positive association between styles of governance and health outcomes. Institutional arrangements in a society determine the population's entitlement and access to housing, health care, education, pensions, unemployment insurance, collective bargaining, political incorporation, incarceration, and culture (Hall and Lamont, 2009; Krieger et al., 2008; Pinto and Beckfield, 2011). These influences are multilayered and complex. Figure 8-1 presents a model by Hurrelmann and colleagues (2011), which illustrates the multitude of social and political factors that contribute to population health and, by extension, to cross-national differences in health.

The U.S. approach to policies that relate to health and social programs is what sociologists classify as an Anglo-Saxon or *liberal* model

TABLE 8-1 The Association Between Political Themes and Health Outcomes: Findings of 73 Empirical Studies

Political Theme of Countries	Positive Association with Health[a] N (%)	Inverse Association with Health[b] N (%)	Mixed Results[c] N (%)	Total N
Democracy	21 (81)	3 (12)	2 (8)	26
Globalization	1 (17)	4 (67)	1 (17)	6
Egalitarian political tradition	9 (90)	1 (10)	0	10
Welfare state generosity	19 (61)	1 (3)	11 (36)	31
Total N (%)	50 (69)	9 (14)	14 (19)	73 (100)

[a]Political variable demonstrates a positive, direct or indirect, association with a population-related health outcome.

[b]Political variable demonstrates a negative, direct or indirect, association with a population-related health outcome.

[c]Political variable is either unrelated or inconsistently related to a population-related health outcome.

SOURCE: Adapted from Muntaner et al. (2011, Table 2).

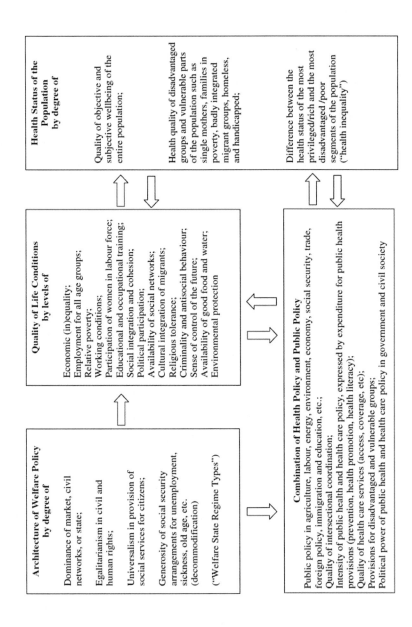

FIGURE 8-1 A model of structural and political influences on population health.
SOURCE: Hurrelmann et al. (2011, Figure 3).

(Esping-Andersen, 1990).[6] In this terminology, "liberal" refers to the many English-speaking countries with economies that are more oriented to the free market (with relatively low levels of regulation, taxes, and government services) than other capitalist economies. Sociologists distinguish the Anglo-Saxon/liberal model of the United States and the United Kingdom from countries like Sweden, which operate under a *social democratic* model in which the state makes generous commitments to full employment, income protection, housing, education, health, and social insurance. Most European welfare programs came into existence after World War II with the goal of providing more universal access to assistance (Bambra and Beckfield, 2012). The social democratic model promotes social equality through wage compression, organized through strong collective bargaining by unions, and tax policies that direct resources to the social security system (Bambra and Beckfield, 2012).[7]

As detailed in Part I of the report, the Scandinavian (social democratic) countries generally have higher health rankings than the United States, along with more favorable measures of social and economic well-being. As a group, these social democratic countries report longer life expectancies, lower infant mortality rates, and better self-rated health than do liberal countries, including both the United States and the United Kingdom (Bambra, 2005, 2006; Chung and Muntaner, 2007; Coburn, 2004; Eikemo et al., 2008b; Lundberg et al., 2008; Navarro et al., 2003).[8] Figure 8-2 shows the high infant mortality rates that exist in liberal countries, especially the United States. Figure 8-3 shows that this pattern has existed for decades (Conley and Springer, 2001).

Sociological research is beginning to suggest that the style of governance in a country may exert its own influence on health outcomes, independent of individual-level variables. One study found that whether a country had a social democratic, Anglo-Saxon/liberal, or other sociopolitical model explained 47 percent of the variation in life expectancy between countries (Karim et al., 2010). Another study concluded that the model type predicted approximately 20 percent of the difference in infant mortality

---

[6] As distinct from the meaning of "liberal" as commonly used in the United States to describe left-leaning or progressive social or political ideology.

[7] A number of other typologies have been proposed: see, for example, Bonoli (1997); Castles and Mitchell (1993); Eikemo and Bambra (2008); Ferrera (1996); Korpi and Palme (1998); Leibfried (1992); and Navarro and Shi (2001).

[8] There is substantial between-country variation within Scandinavia (Christensen et al., 2010), and health outcomes in Scandinavian countries are not always the best. For example, mortality rates in Denmark approach those of the United States, and Finland has high mortality rates for some conditions. Similarly, there is substantial between-country variations in Anglo-Saxon/liberal countries, such as the marked differences between the United States and England discussed in previous chapters.

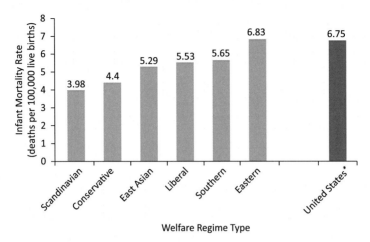

FIGURE 8-2 Infant mortality rate for the United States and 30 other countries, classified by welfare regime type.
*The United States is included in the group of countries classified as having "Liberal" regimes, but it is also presented here in isolation for comparison.
NOTE: *Scandinavian* countries: Denmark, Finland, Norway, Sweden; *Conservative* countries: Austria, Belgium, France, Germany, Luxembourg, the Netherlands, Switzerland; *East Asian* countries: Hong Kong, Japan, Korea, Singapore, Taiwan; *Liberal* countries: Australia, Canada, Ireland, New Zealand, United Kingdom, United States; *Southern* countries: Greece, Italy, Portugal, Spain; *Eastern* countries: the Czech Republic, Hungary, Poland, Slovenia.
SOURCE: Adapted from Karim et al. (2010, Table 5).

rates among countries and 10 percent of the difference in low birth weight (Chung and Muntaner, 2007).

However, the panel notes the limitations of current evidence on this topic, which relies heavily on cross-sectional associations. Such associations often provide only circumstantial evidence; they do not prove a causal effect, and population trends may not apply to individuals (the "ecological fallacy"). Controlled trials to produce more definitive evidence would be untenable, and all studies on this subject must cope with a variety of methodological challenges, such as the potential endogeneity of the political and social environments, as well as issues relating to aggregate efficiency, intertemporal dynamics, and macroeconomic effects. Typologies for regimes, such as welfare states, can be blunt measures that require further refinement to properly differentiate policy nuances across and within countries and to track changes that affect countries over time.[9] For these reasons, among others, research on the effect of welfare states on population health has often produced mixed results and has not fully explained cross-national health patterns. For example, social democratic countries like Sweden had low infant mortality rates early in the 20th century (Regidor et al., 2011), even before the introduction of their social welfare benefits, probably because of improved sanitation and other public health interventions (Burström et al., 2005).

There is little question that the European welfare model is effective in redistributing income and reducing poverty. More universal and generous welfare systems achieve greater income equality than other systems through more generous income transfers through taxes and services (Esping-Andersen and Myles, 2009). These entitlement benefits may buffer the health effects of material deprivation and thereby improve health outcomes but they may have other consequences that are not economically or politically viable in the United States.

Related characteristics of Scandinavian society, such as greater gender equality (Stanistreet et al., 2005) and social cohesion (Putnam, 2000), are also cited as potential explanations for the region's relatively good health outcomes. Political empowerment of minority groups and women appears especially important to health (Beckfield and Krieger, 2009). As noted in Chapter 7, citizen engagement in the United States, such as voting in elections, is lower than in most other OECD countries (2011e), and the United States has one of the lowest rates of female participation in the national legislature (Congress) (Armingeon et al., 2012).

Scandinavian society is also known for having less income inequality than in the United States (see Chapter 6), a likely product of the welfare state. The Luxembourg Income Study provides evidence that social democratic policies have, over time, substantially reduced income inequality (Alderson and Nielsen, 2002). The Scandinavian welfare programs (universalism, generous wage replacement rates, extensive welfare services) may also narrow income inequalities and provide low-income individuals with greater access to services (Coburn, 2004). However, as discussed in Chapter 6, it remains unclear whether income inequality itself, or the policies that affect income inequality, bear more on the U.S. health disadvantage (Beckfield, 2004).

There is some evidence to suggest that aggregate spending on social programs is associated with better health. One study examined spending

---

[9]The categories assume that all the policies in a particular regime reflect a similar approach and that each category reflects a coherent set of principles, neither of which may be true (Kasza, 2002). No single country adheres to all aspects, and there is internal policy variation within individual welfare states and among the countries of each welfare state regime (Bambra, 2007).

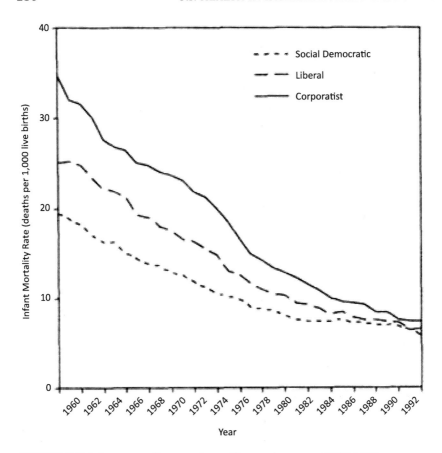

FIGURE 8-3 Infant mortality rates by welfare regime type, 1960-1992.
NOTE: In this study, corporatist countries included Austria, Belgium, France, Germany, and Italy.
SOURCE: Conley and Springer (2001, Figure 3).

on health care and social services in 30 OECD countries and found that U.S. spending on social services (13.3 percent of gross domestic product [GDP]) was less than the OECD average (16.9 percent) and less than that of all countries except Ireland, Korea, Mexico, New Zealand, and the Slovak Republic (Bradley et al., 2011).[10] The ratio between spending on

---

[10]Social services expenditures included public and private spending on old-age pensions and support services for older adults, survivors benefits, disability and sickness cash benefits, family support, employment programs (e.g., public employment services and employment training), unemployment benefits, housing support (e.g., rent subsidies), and other social policy areas excluding health expenditures.

social services and health care was 0.91 in the United States and 2.00 in the OECD. More importantly, the study found a significant association between social spending and life expectancy, infant mortality, and potential years of life lost (Bradley et al., 2011).[11] Another study also found an association between social spending[12] and mortality in an analysis of 15 European countries (Stuckler et al., 2010). According to that study, each additional $100 per capita in social spending was associated with a 1.19 percent decrease in all-cause mortality (Stuckler et al., 2010).

In a commentary about the U.S. health disadvantage, Avendano and Kawachi (2011) noted a number of potentially important differences between the United States and Europe that may affect health: European tax systems are more progressive, child benefits are traditionally available for parents in many countries regardless of income, social programs are generally not restricted to the poor, employment protection is substantially higher, unemployment benefits are more generous, and labor standards for working parents are more extensive. Authors of another study also noted that the United States ranks poorly on measures of full-time employment, public child care, union representation, and parental leave (Pettit and Hook, 2009) (see Table 8-2). Many of these may be less acceptable in the United States because of related tax burdens and other implications.

In seeking a systemic cause for the U.S. health disadvantage, Avendano and Kawachi (2011, p. 4) noted the following:

> We have suggested a potentially promising line of inquiry based upon differences in social policy contexts. However, the challenge is obviously to identify the particular social and labor policies that have a causal impact on health and that may contribute to cross-national health differences. For example, do the more generous parental leave policies in Europe contribute to their comparative health advantage? Have employment protection policies contributed to the better health of European workers compared with their U.S. counterparts? The great variation in policy reform during the last 50 years across Europe and the United States provides us with a potentially fruitful set of natural experiments to consider. Broadening the scope of our inquiry to include the social and policy context of nations might help to solve the puzzle of the U.S. health disadvantage.

---

[11]Social spending was also associated with low birth weight, a finding the authors speculated might reflect genetic factors or sociocultural features of the population that were not controlled for in the analysis.

[12]This study defined social spending as spending related to family support programs (such as preschool education, child care, and maternity or paternity leave), old-age pensions and survivors benefits, health care, housing (such as rent subsidies), unemployment benefits, active labor market programs (to maintain employment or help the unemployed obtain jobs), and support for people with disabilities.

TABLE 8-2 Macro-Level Conditions That Affect Work-Family Policy, by Country, Mid-1990s

Country	Part-Time Workers* (%)	Children Aged 0-2 in Publicly Funded Child Care (%)	Union Membership of Workforce (%)	Parental Leave, Maximum Weeks*
Australia	21	2	43	0
Austria	13	3	37	112
Belgium	14	30	60	67
Canada	19	5	33	25
Denmark	23	48	76	28
Finland	7	32	75	160
France	15	23	9	162
Germany	16	11	27	162
Italy	6	6	32	48
Luxembourg	8	3	50	16
Netherlands	36	8	23	68
Norway	27	20	53	64
Spain	5	5	9	162
Sweden	24	33	88	85
United Kingdom	22	2	34	18
United States	19	5	14	12

*Part-time employment was calculated from self-reports of usual hours worked. Employment was classified as part time when a respondent reported 1-30 hours of work per week. Parental leave represents the maximum number of weeks (paid or unpaid) available.
SOURCE: Data from Pettit and Hook (2009, Table A.2).

Research to date "has far too many black boxes," note Bambra and Beckfield (2012, p. 29). "Surprisingly, despite vast cross-national variation in population health, and vast cross-national variation in institutional arrangements, very little work connects the two." It is doubtful that any single aspect of the welfare model could be responsible for the better health outcomes observed in certain countries. Even the reduction in income inequality achieved by this form of capitalism probably results from the interaction and combination of multiple policies (e.g., universal access to welfare services) (Bambra and Beckfield, 2012; Chung and Muntaner, 2007; Navarro et al., 2006).

Nor is the social democratic model a panacea for public health. In what Hurrelmann and colleagues (2011) describe as the "Scandinavian welfare paradox of health," social democratic countries that have favorable *aggregate* health statistics—e.g., average disease rates or life expectancy for the population as a whole—sometimes have steeper within-population health gradients (i.e., larger *health inequalities*) than do countries governed by other models (Bambra, 2007; Bambra and Eikemo, 2009; Dahl et al., 2006; Eikemo et al., 2008a; Huijts and Eikemo, 2009; Kunst et al., 1998; Lahelma and Lundberg, 2009; Mackenbach et al., 1997, 2000, 2008; Stirbu et al., 2010). Compared with Scandinavian countries, health gradients tend to be lower in Bismarkian countries (e.g., Austria, France, Germany) and Southern Europe (e.g., Italy, Spain) and highest in Eastern European and Baltic countries (Eikemo et al., 2008a). As discussed in Box 8-1, a variety of explanations for the paradox have been proposed (Bambra, 2011; Dahl et al., 2006; Huijts and Eikemo, 2009; Hurrelmann et al., 2011; Mackenbach, 2012).

The United States can take little comfort in debates about why some European countries do better than others in reducing health gradients, because it is still the case that the United States and the United Kingdom generally fare worse than all of them—on both aggregate health status and the steepness of the health gradient (Avendano et al., 2010; Eikemo et al., 2008a). As noted in Chapter 6, at least one study has reported that the health gradient by education is steeper in the United States than in Western European countries (Avendano et al., 2010). These cross-national comparisons certainly shed light on the U.S. health disadvantage, but other factors unique to the United States may also be important in understanding the relatively poor health of Americans, as discussed in the next section.

SOCIETAL VALUES

Social, economic, and public health policies are often an expression of societal values, set against the backdrop of other exigencies (e.g., economic turmoil). For example, the emergence of social democratic regimes in post–World War II Scandinavia was driven in part by their shared values, including a belief in the obligation of society to promote equity and guarantee universal access to assistance as a right of citizenship, regardless of one's economic means (Esping-Andersen, 1987, p. 86). Social rights were extended at minimal cost to the entire population in a social contract that sought to eliminate status privilege (Bambra and Beckfield, 2012).[13]

---

[13]Even now, the ministers of the G20 (the group of 19 countries with major economies and the European Union) have been discussing plans to extend "social protection floors" to ensure their populations expanded social protection systems amid current fiscal constraints (International Labour Office, 2011).

BOX 8-1
Explanations for the Scandinavian Welfare Paradox

One proposed explanation for the paradox of greater health inequalities in many Scandinavian welfare states is that lower social strata may have a higher relative concentration of individuals at increased risk of disease. Decades of upward intergenerational social mobility may have increased opportunities for social selection and created more homogenous disadvantaged social groups with such characteristics as low cognitive ability and less favorable personality profiles. The increase of intergenerational social mobility is due primarily to changes in the economy that have led to an expansion of higher education, but to the extent that welfare policies have contributed to making the education system more merit based, they may paradoxically have contributed to a widening of health inequalities (Mackenbach, 2012).

Another proposed explanation is that some European welfare states happen to be further in their epidemiological development, and have now reached the fourth stage of the epidemiological transition in which health improvement depends largely on behavior change (Olshansky and Ault, 1986). This explanation increases the importance of nonmaterial factors—including cultural capital and such personal characteristics as cognitive ability in relation to health—that have become more socially differentiated because they have largely been untouched by the welfare state. To the extent that welfare policies have contributed to making an affluent life-style widely affordable, they may have paradoxically contributed to a widening of health inequalities (Mackenbach, 2012).

Hurrelmann and colleagues raised the following hypothesis in their call for further research on the paradox (2011, p. 16):

> If public social expenditures are accompanied by decreased responsibility and influence of private actors and informal corporate institutions, then the social system disincentivises and devaluates the social activities of closely knit social networks and diminishes individual's perception that they can help themselves. In economic terms, this is tan-

---

Conversely, the limited state welfare assistance that exists in East Asian countries (Hong Kong, Singapore, South Korea, Taiwan, and sometimes Japan)—which rely instead on the family and voluntary sector for the social safety net—reflect Confucian social ethics, such as obligation for immediate family members, thrift, diligence, and a strong education and work ethic (Aspalter, 2006; Bambra and Beckfield, 2012; Croissant, 2004; Walker and Wong, 2005).

As in all countries, institutional arrangements in the United States are

tamount to a "crowding out" of informal health institutions. As a consequence, health-promoting strategies within the family, leisure and work settings may be neglected or deemphasized in the social democratic countries. In this respect, conservative countries* with their somewhat stronger reliance on informal social networks may have an advantage vis-à-vis the Scandinavian countries.

Some of these international patterns can probably be explained by between-country variations in the social patterning of health-related behaviors, such as smoking and alcohol consumption. In the Scandinavian countries, inequalities in mortality from smoking-related causes (such as lung cancer and chronic lung disease) and from alcohol-related causes tend to be larger than in many other Western European countries (Mackenbach et al., 2008; Van der Heyden et al., 2009). Survey data show that inequalities in smoking are larger in the north and west of Western Europe than in the south (Cavelaars et al., 2000; Huisman et al., 2005). These geographic patterns reflect differences between countries in the progression of the smoking epidemic: countries in Southern Europe tend to be at an earlier stage in the progression, in which smoking is not yet as strongly socially patterned as in later stages (Lopez et al., 1994).

There is also some evidence to suggest that cross-national variations in inequalities in access or quality of health care may play a role. Socioeconomic inequalities in mortality from conditions that are amenable to medical intervention are particularly large in Eastern Europe (Mackenbach et al., 2008; Stirbu et al., 2010), where inequalities in utilization of health care services also tend to be large (Plug et al., 2012). This proposed explanation would also need further confirmation.

*By "conservative," the authors refer to European political systems such as those in Austria, Belgium, France, Germany, and Italy.

affected by social values and the nation's historical legacy. On issues that pertain directly to the U.S. health disadvantage—ranging from government regulation to attitudes toward contentious issues, such as firearms or birth control—the policy of the U.S. government and of states and localities often reflect societal priorities and beliefs.

The United States ranks poorly on a number of factors that could explain its health disadvantages. As detailed in Chapters 4-7:

- The public health and medical care systems are more fragmented, and greater barriers exist in access, affordability, and some measures of quality than those in other high-income countries.
- Certain important unhealthy or injurious behaviors are more common in the United States than peer countries, including high-caloric intake, drug misuse, unsafe driving practices, high-risk sex, and the use of firearms.
- Poverty, unemployment, and income inequality are more prevalent than in comparable countries, education has not kept pace with other countries, and social mobility is more limited.
- Land use, the built environment, and the transportation model are less conducive to physical activity; food availability, distribution, and marketing discourage healthy diets; and communities are more heavily segregated by income and race than in other high-income countries.

These conditions reflect multiple factors, including history, governance models, societal values, and priorities that cannot be ignored in trying to understand the U.S. health disadvantage. For example, as discussed in Chapter 4, the lack of universal health insurance coverage sets the United States apart from most high-income nations, but there are reasons this situation has existed for generations. At the same time that countries in Europe were establishing universal access to health services, attempts to do so in the United States, beginning with the Truman administration in the 1940s, met with political resistance, as it still does today (Altman and Shachtman, 2011; Starr, 2011). The resistance has been shaped not only by interest groups, such as medical organizations and health insurance companies, but also by societal beliefs about the proper role of government and the private sector in health care (Freeman and Marmor, 2003). Similarly, to make sense of why Americans are more likely to engage in certain unhealthy behaviors or injurious practices (see Chapter 5), the role of societal values in enacting or resisting countermeasures cannot be ignored.

Yet there is little empirical evidence to prove that values in the United States differ substantially from those in other high-income countries.[14] Some evidence on the subject has been gathered from the World Values Survey (World Values Survey Association, 2012) and polling organizations (e.g., World Gallup Poll). The internal and external validity of the indicators and sampling techniques used in such surveys is less than ideal.

---

[14]It is also true that cultural values are not uniform, either within the United States or within other countries, and that such values are dynamic and shift over time (Byrne, 2004; McKee, 2002; Staley, 2001).

However, the panel believes that the individual behaviors and policies of Americans in relation to public health and socioeconomic issues are, to at least some extent, influenced by prevailing values and priorities (Goldberg, 2011). For example, five iconic American beliefs seem especially relevant: individual freedom, free enterprise, self-reliance, the role of religion, and federalism.

Individual Freedom   Strong beliefs in individual freedom, as expressed in the Declaration of Independence and Bill of Rights, remain powerful drivers in modern America (Fairchild et al., 2010). As in other countries, Americans struggle with the natural tension between the state's responsibility to safeguard public health (Institute of Medicine, 2011d) and the rights of people to freely make their own decisions about eating habits, tobacco or alcohol use, and other health-related risky behaviors (Nathanson, 2009). Some personal freedoms carry special significance in the United States, such as the right to bear arms, a constitutional protection that does not exist in most other countries (Glantz and Annas, 2009).

Free Enterprise   American society is committed to free-market capitalism and generally eschews restrictions on industries, especially when they impede economic activity or involve an expansion of governmental regulatory authorities. Many aspects of the political process, including the campaign finance system in the United States, give large donors and special interests a degree of influence over the formulation of policy than may exist in other countries (Mann and Ornstein, 2008). Whether regulations are meant to protect public health, assist vulnerable populations, or meet other needs, a popular refrain in the United States is that effective solutions for social and economic problems are best achieved through the free market and more directly by families and their communities (see discussion on self-reliance, below). It is also true that a vibrant and growing economy is good for public health and for the health of the population, that anything that impedes economic growth and flexibility can have detrimental population health effects, and that strategies to boost employment and raise levels of income and wealth can yield important health benefits.

Self-Reliance   In a nation founded by pioneers, many Americans believe in the responsibility of individuals, not the state, to solve personal problems: dependency on government welfare programs or "handouts" is discouraged. Thus, raising taxes for state-financed social or health programs is often unpopular with a large proportion of American voters.

In contrast, there is a consensus in many other high-income countries around shared responsibility, solidarity, and the principle that a certain standard of living is a right of citizenship (Bambra and Beckfield, 2012; Esping-Andersen, 1990); this consensus that may not be as pervasive in the United States.

Role of Religion   Although the separation of church and state is a core principle in the United States, the United States is less secular than most other high-income countries (Taylor, 2003; World Values Survey Association, 2012), and religious beliefs are often raised in public discourse. Sensitive public health policies, such as those related to contraception or adolescent sexuality, may not be as contentious in other countries (Darroch et al., 2001; Hofstede, 1998; Inglehart and Welzel, 2005; cited by Santelli and Schalet, 2009).[15]

Federalism   The United States originated with a revolution against an overbearing government, and Americans continue to seek limits on the size and budget of government, including agencies responsible for social services, safety, and even health itself. The federalist principles adopted by the nation's founders reserved limited authorities for the federal government and divided the remainder across the states, which in turn have delegated many authorities to counties and municipalities. Although this decentralized model is an ingenious strategy for separation of powers and is of growing appeal in other countries (Charbit, 2011), the resulting fragmentation complicates attempts to set national policy priorities that many smaller countries with more centralized governments can pursue more easily. It also creates an uneven distribution of resources that might not exist in countries with more centralized models and that often affect the neediest. For example, because authority for so many services rests with the states, the poorest Americans often live in states (e.g., Louisiana, Mississippi) with low tax revenue and small budgets for Medicaid, public schools, and social services.

The potential relevance of these societal values cannot be ignored in attempting to explain the findings documented in this report. Values and priorities create configurations of policies that can act as an upstream underlying cause, or as a systemic explanation for downstream health consequences, as the following examples illustrate:

---

[15]However, sexual content is increasingly prominent in U.S. entertainment media, such as film, television, music, and advertising.

- **Constitutional law:** The Second Amendment provides an essential context for understanding why civilian firearm ownership is so common in the United States.
- **Deregulation:** The dramatic rise in the consumption of high-fructose corn syrup in the United States—climbing from zero in 1950 to 63.8 pounds per capita in 2000 (U.S. Department of Agriculture, 2003)—did not occur solely because of consumer demand. The placement of such products on store shelves and restaurant menus was also influenced by business decisions by the food industry, and the reluctance of government to impose regulatory limits on commercial marketing practices (Brunello et al., 2008; Nestle, 2002).
- **Taxation:** Failed attempts in some U.S. jurisdictions to discourage caloric consumption by applying a tax on carbonated beverages reflects, at least in part, the political influence of strong industry lobbies and consumer resistance to government taxation (Wang et al., 2012).
- **Free markets:** For years, tobacco companies have defended their right to market and profit from a product that customers purchase and that supports tobacco farmers, despite the major health risks associated with tobacco.

As Box 8-2 discusses in more detail, the policies that may be responsible for the high rate of traffic fatalities in the United States offer a case study of this phenomenon and of the combined impact of social-ecological influences (from individual behavior to public policy) in producing unfavorable health patterns in the United States.

POLICIES FOR CHILDREN AND FAMILIES

Just as the high rate of traffic fatalities could arise from multiple causes, other areas of health disadvantage in the United States are equally complex, both in origin and policy solutions. Many of the problems, such as obesity and diabetes, can be addressed by policies directed at middle-aged or older adults, but a life-course perspective becomes important to fully analyze underlying causes.

Consider the example of childhood obesity. Figure 8-4 shows that energy imbalances that cause weight gain and obesity-related health outcomes originate as early as the prenatal period. While scientists study the responsible physiological mechanisms, such as effects on mood, metabolism, appetite, genes, and the hypothalamic-pituitary-adrenal (HPA) axis

BOX 8-2
The Role of Public Policies on U.S. Traffic Fatalities

In 2011, the National Academies' Transportation Research Board (TRB) issued a special report, *Achieving Traffic Safety Goals in the United States: Lessons from Other Nations.* The report was of great interest to this panel both because it examined a major cause of death that sets the United States apart from other high-income countries and because it provides a powerful example of how an in-depth cross-national study can shed light on potential strategies to address an important aspect of the U.S. health disadvantage. As noted in Chapter 1, transportation-related accidents account for 18 and 16 percent for men and women, respectively, of the excess years of life lost before age 50 in the United States.

The TRB study clearly documents how motor vehicle injury is influenced by many different factors:

- economic conditions (traffic fatalities increase with economic growth and higher employment levels);
- traffic characteristics (such as traffic congestion and the mix of pedestrians, cyclists, and vehicle types);
- demographic characteristics (younger populations have higher crash rates);
- alcohol abuse;
- land use (such as miles of road and urban-rural balance) and geographic context, which can influence the success of regulatory measures on traffic accidents;
- vehicle characteristics (such as fleet age and passenger restraints);
- road design and maintenance standards;
- driver behaviors (such as the prevalence of drunk driving, seatbelt use, speeding, and compliance with speed limits and other traffic laws);
- the timeliness and quality of medical care; and
- government safety policies (including those that affect vehicle and road design standards, traffic regulations, enforcement practices, and education and communication activities).

It is worth noting how this list spans multiple health determinants covered in this part of the report—including medical care (Chapter 4), personal/driver behaviors (Chapter 5), environmental factors (Chapter 7), and governmental policies (Chapter 8). Despite this complex array of causal factors, the panel was struck by how closely the TRB committee's overall findings parallel those documented in Part I of this report. Indeed, the TRB report's opening paragraph could have been used (with only minor changes) for this report (Transportation Research Board, 2011, p. 7):

The United States is missing significant opportunities to reduce traffic fatalities and injuries. The experiences of other high-income nations and of the U.S. states with the best improvement records indicate the benefits from more rigorous safety programs. Most high-income countries are reducing traffic fatalities and fatality rates (per kilometer of travel) faster than is the United States, and several countries that experienced higher fatality rates 20 years ago now are below the U.S. rate. From 1995 to 2009, annual traffic fatalities declined by 52 percent in France, 39 percent in the United Kingdom, 25 percent in Australia, and 50 percent in total in 15 high-income countries (excluding the United States) for which long-term fatality and traffic data are available, but by only 19 percent in the United States. Some U.S. states have fatality rates comparable to those of the countries with the safest roads; however, no state matches the typical speed of improvement in safety in other countries.

The TRB report's findings relating to alcohol, seatbelts, and speeding are presented below, followed by the report's findings on cross-national differences in road design, and finally, the report's observations about differences in policy making and enforcement.

Alcohol-Related Fatalities

According to the TRB report (p. 150):

[I]n the past decade almost no reduction has been achieved in the annual numbers of fatalities in alcohol-related crashes in the United States. . . . Although differences in measurement methods complicate comparisons, Germany, Great Britain, Sweden, and Australia all appear to have attained lower rates of alcohol-involved traffic fatalities, per vehicle kilometer of travel and as a fraction of all fatalities, than the United States.

The report went on to identify public policy as the way forward in addressing the problem (pp. 150-151):

Getting progress started again in the United States apparently will require more widespread and systematic application of the proven countermeasures and greater coordination of strategy among law enforcement agencies, the court system, and public health programs aimed at alcohol abuse. [. . .] In countries that have introduced sustained, high-frequency programs of random sobriety testing, including Australia, Finland, and France, reductions of 13 to 36 percent in the frequency of alcohol-involved fatal injury crashes have been achieved. Evaluations

continued

BOX 8-2 Continued

of intensive campaigns of selective testing at sobriety checkpoints in U.S. jurisdictions (following procedures now legal in most states) have reported reductions of 20 to 26 percent in alcohol fatal injury crashes (Shults et al., 2001, 76; Fell et al., 2004, 226). In the United States in 2008, 12,000 persons were killed in crashes involving a driver who was alcohol-impaired (National Highway Traffic Safety Administration, 2009). Therefore, widespread implementation of sustained, high-frequency sobriety testing programs in the United States could be expected to save 1,500 to 3,000 lives annually.

Seatbelts

The TRB report noted that almost every high-income country requires the use of seatbelts, but the share of front seat occupants who use seatbelts is lower in the United States than in many of these countries (see Table 5-1, in Chapter 5). The report noted the effects of decentralized safety regulation (a major theme of the report) and political opposition (p. 181):

> The cases of seat belts and of motorcycle helmets . . . provide clear illustrations of how public and political attitudes can restrain risk-reducing measures despite the availability of effective and well-managed countermeasure programs in many states. The effectiveness of seat belts in reducing casualties and of specific interventions (primary laws and high-visibility enforcement) in increasing usage are well established by research and by the experience of many states. The interventions are not complex or expensive compared with the efforts required for speed control or impaired-driving control. Nonetheless, some jurisdictions have chosen not to apply these measures.

Speed Control

Speeding may contribute to as many as one-third of fatal accidents (Aarts and van Schagen 2006, pp. 220, 223), and speed is an aggravating factor in the severity of all accidents. In light of these findings, the results of a survey by the Governors Highway Safety Association (2005, p. 5) are especially troubling: "[S]tates are becoming increasingly concerned that gains made in the areas of safety restraint usage and impaired driving have been offset by increased fatalities and injuries due to higher speeds." In addition, the TRB report noted (p. 151):

> [I]n contrast, in several of the countries that are making the greatest progress in highway safety, speed control is one of the interventions receiving the greatest attention and resources. If speed control is weakening in the United States, this trend may explain part of the safety performance gap between the United States and other countries.

The TRB report documented how U.S. failures in addressing this problem span research, planning, practice, and policy: see table below. In comparing U.S. policy with efforts in benchmark countries, the report concluded (p. 233):

> Successful speed management initiatives in other countries are of high visibility (through publicity and endorsement of elected officials), are long term (sustained for periods of years), target major portions of the road system, use intensive enforcement (e.g., automated enforcement and high penalties), sometimes use traffic-calming road features (such as narrow lanes and traffic circles that cause drivers to reduce speed), and monitor progress toward publicly declared speed and crash reduction objectives. No U.S. speed management program today is comparable in scale, visibility, and political commitment to the most ambitious programs in other countries.

Driving Speed Management in Selected Countries

Speed Management Strategy	France, United Kingdom, and Australia	United States[a]
Management and Planning	Focused program with goals, strategy, and budget Timely monitoring and publication of relevant speed and crash data Long-term, multiyear, or permanent perspective	Routine, low-level activity; reactive management; no long-term plan No speed data; no meaningful crash data Episodic attention; occasional enforcement crackdowns
Technical Implementation of Countermeasures	Major portions of national or state road network targeted Automated plus traditional enforcement Penalties designed as part of the integrated program	Haphazard or spot enforcement Automated enforcement not authorized or rarely used Little attention to effectiveness of penalties
Political and Public Support	Active support and leadership of elected officials; management held accountable for results	Politically invisible except when speed limits altered or automated enforcement proposed

[a]Does not necessarily include all states.
SOURCE: Transportation Research Board (2011, Table 4-3).

continued

BOX 8-2 Continued

The role of societal values is central in a striking observation by the AAA Foundation for Traffic Safety: "[C]urrent methods for controlling speed are virtually powerless in the face of this [U.S.] speeding culture" (Harsha and Hedlund, 2007, p. 1). This report notes that a successful nationwide program to reduce speeding will require political leadership at the federal, state, and local levels, starting with congressional action, as well as a staged approach to speed control campaigns that includes efforts to increase public awareness and support for these efforts.

Safe Road Design and Highway Network Screening

The TRB report found that definitive studies and data linking highway screening to safety improvements are still missing, but it also found that all countries have design standards for new construction and reconstruction that are intended to improve safety.* The TRB report noted a shift in some benchmark countries' road programs, which emphasize research on the relationship of design to crash and casualty risk, give higher priority and earlier attention to risk reduction in the design of projects and in project programming, and are more willing to trade a degree of traveler convenience for the sake of safety. Road designers in these countries are expected to quantify the predicted crash frequency and to justify the level of risk in the design.

Political Factors and Public Attitudes

Interestingly, just as the TRB report's opening paragraph parallels the cross-national mortality patterns observed by this panel, the committee that wrote that report also looked upstream in search of explanations

---

*Unlike laws that proscribe risky individual behaviors (such as speeding), highway screening and safe road design aim to make roads inherently safer. Highway screening programs use data to identify places with frequent crashes and then modify these locations to reduce accident risk. The changes can include adjusting alignment, widening shoulders, removing roadside obstacles, improving signage and pavement markings, changing intersections, installing barriers, and increasing traffic law enforcement. When new roads are built or old ones rehabilitated, various design standards can be used to introduce safer road characteristics, including alignment; lane, shoulder, and median widths; sight distance; superelevation (i.e., banking on curves); pavement surface; number of lanes; intersection design; and the roadside environment.

for the U.S. poor performance on traffic safety. The committee noted the following obstacles (p. 14):

- Decentralization: in most benchmark countries, regulation and enforcement are highly centralized, often the responsibility of a single national authority, whereas in the United States, 50 states and thousands of local jurisdictions are responsible for traffic safety and the operation of the highway system;
- Public attitudes that oppose measures common elsewhere: for example, in the United States, motorcycle helmet laws and speed enforcement using automated cameras often encounter active public opposition;
- Weak support for or opposition to rigorous enforcement in legislatures and among the judiciary;
- The constitutional prohibition of unreasonable searches, which prevents police from conducting driver sobriety testing without probable cause, a common practice in some other countries; and
- Resource limitations that prevent enforcement of the intensity common in other countries.

The obstacles are, to an extent, the product of differences in political systems and in the physical characteristics of transportation systems, and possibly of other social and cultural factors.

Many of these underlying explanations are not only applicable to traffic fatalities but also may contribute to other health and injury risks that are more prevalent in the United States than elsewhere (as detailed in this and other chapters). For example, decentralization contributes to lapses in traffic safety, to fragmented public health, and medical care systems in the United States (Institute of Medicine, 2011d). Opposition to rigorous enforcement applies to speed control, life-style choices, and restrictions on industry. Constitutional prohibitions restrict not only unreasonable searches but also proscribe interventions on gun possession. Resource limitations apply not only to law enforcement but also explain deficiencies in public health programs (Institute of Medicine, 2012), the foods chosen for school lunch menus (Institute of Medicine, 2010b), and weakness in social and safety net services.

(portrayed in the bottom of the diagram), policy solutions occupy the diverse domains at the top: macro issues, such as the built environment that enables children to engage in outdoor physical activity and farm subsidies for corn-based food products, as well as other obesogenic influences, such as cultural norms about body image, commercial messaging, local food environments, and the effects of material deprivation and psychological stresses.

A key finding of this report is the alarming scale of health disadvantage among children and adolescents in the United States compared with their peers in other high-income countries. This finding has major implications not only for public health (especially when today's children become tomorrow's older adults), but also for the economy and national security (World Economic Forum, 2011). The spectrum of problems that disproportionately affect youth in the United States relative to other countries covers virtually every aspect of their lives: the risk of infant mortality and low birth

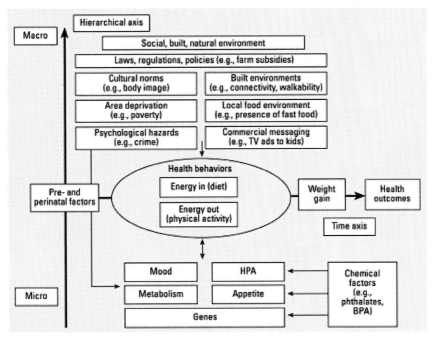

FIGURE 8-4 A life-course perspective on childhood obesity.
NOTES: BPA: bisphenol A; HPA: hypothalamic-pituitary-adrenal axis. The life span is depicted horizontally; factors are depicted hierarchically, from the individual level at the bottom of the figure to the community level at the top of the figure.
SOURCE: Trasande et al. (2009, Figure 1).

weight; injuries and homicide; behavioral health problems involving drug use, high-risk sex, and depression; high rates of childhood disease (e.g., obesity, diabetes, asthma); high rates of child poverty; lower educational achievement; and lower social mobility. This list is a powerful signal for greater attention and investment in policies and programs for children and families (National Research Council and Institute of Medicine, 2008) but, historically and even now, the United States has made greater investments in assisting the elderly than the nation's youth. Some analysts have concluded that the underinvestment in children and adolescents may be the product of their limited political power compared with older voters (Isaacs et al., 2012; Preston, 1984). Those investments in older adults have produced important social and public health benefits for older Americans and offer an important avenue for addressing the U.S. health disadvantage, but the problems that affect the nation's youth deserve greater investment.

Maternal and child well-being are clearly important to any nation's health, and a comprehensive review of this component of population health in the United States is beyond the scope of this panel. However, the areas of disadvantage among U.S. children and adolescents relative to other rich nations that we document point to a number of important areas that should be considered. These include environmental factors—at home, school, and elsewhere—that promote obesity and limit physical activity; the need for child care and early childhood education; reducing barriers that children and mothers face in obtaining essential preventive services and health care; providing a range of supports for youth, especially around sexual health and preventing tobacco, alcohol, and other drug use; and interventions to prevent car crashes and fatalities that involve children or young drivers. Child protection policies would also be important to reduce children's exposure to family violence, crime, and the risk of violent deaths (especially from firearms), to unhealthy air and housing, to the material deprivations of poverty, and to schools and home environments that compromise learning, educational opportunities, and social mobility.

The life-course perspective is a reminder that adverse exposures during childhood—from fetal life through other critical periods of children's physical, sexual, and emotional development—have profound implications in shaping health outcomes later in life and, increasingly, the chances of even surviving to old age. Investing in today's youth is thus an investment in all age groups.

SPENDING PRIORITIES

The familiar adage to "follow the money" is a reminder that a society's policy priorities are often reflected in budget decisions. The panel's review of data on the U.S. health disadvantage and its potential causes shows that the United States often spends less per capita in many of the areas in which its performance is lagging, with the obvious exception of health care. Levels of spending should be interpreted with caution because they say little about the efficiency or effectiveness of programs, but the spending patterns of the United States stand in contrast to those of other high-income countries with better health outcomes. Examples include early childhood education, family and children's services, education, and public health.

- **Early childhood education:** In 2007, the United States spent only 0.3 percent of its GDP on formal preschool programs (for children aged 3-5 years), less than that of seven peer countries and even some emerging economies in Eastern Europe (OECD, 2012i).
- **Family and children's services:** Total public spending by the United States on services for families and young children places the United States last among the 13 peer countries studied. In 2004, the most recent year reported by the OECD, the United States devoted only 0.78 percent of GDP to public services for families and young children, whereas Nordic countries spent approximately 4 percent (OECD, 2006). Only Korea ranked lower than the United States on the proportion of its economy devoted to public services for families and young children.
- **Public health:** According to many analyses, public health is systematically underfunded in the United States (Institute of Medicine, 2012; Mays and Smith, 2011), for a variety of reasons (Hemenway, 2010), but valid data for international comparisons are lacking. The OECD does measure the proportion of public expenditures devoted to health and to public health, but classification schemes are too variable by country to draw meaningful inferences.
- **Social services:** Compared with other countries, the United States spends less on social programs, subsidies, and income transfers than do other countries (see Figure 8-5). As noted above, U.S. spending on social services (13.3 percent of GDP) was less than the OECD average (16.9 percent) and that of all 30 countries except Ireland, Korea, Mexico, New Zealand, and the Slovak Republic (Bradley et al., 2011). A recent report found that the United States spent less on public social protection (as a percentage of GDP) than any peer country but Australia and less than some emerging economies, including Russia and Brazil (International Labour Office, 2011).

In contrast, however, the United States ranks high on public spending on education. In 2008, U.S. spending per student on public education (primary through tertiary levels) was equaled only by Switzerland. Among all

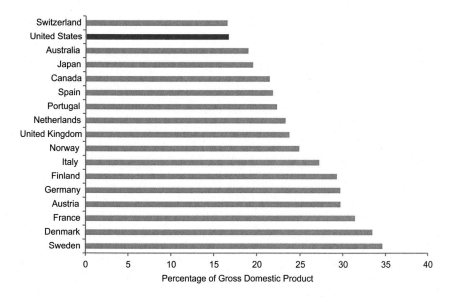

FIGURE 8-5 Social benefits and transfers, 17 peer countries, 2000.
NOTES: Social benefits reflect current transfers to households in cash or in kind to provide for the needs that arise from certain events or circumstances (e.g., sickness, unemployment, retirement, housing, education, family circumstances) that may adversely affect the well-being of households either by imposing additional demands on resources or by reducing incomes. Transfers are typically made by governments. SOURCE: Data from National Accounts at a Glance: 5. General Government, OECD (2012l).

OECD countries, the United States had the fifth highest public expenditure per student on primary education, the fourth highest for secondary education, and the highest for tertiary education (OECD, 2011a). Measured as a percentage of GDP, U.S. public expenditures on education ranked eighth (tied with France, Ireland, Israel, the Netherlands, Switzerland, and the United Kingdom) (OECD, 2012h).

Many of the programs discussed above are financed in other countries by taxes, an approach with limited political support in the United States. Of the 17 peer countries that are the focus of Part I of this report, 11 report a higher tax burden than the United States (U.S. Census Bureau, 2008).[16] Since the 1980s, no country in this peer group except Switzerland has spent less than the United States (as a percentage of employee-employer payroll

---

[16]Tax burden is defined as the percentage of gross wage earnings of the average production worker that is spent on income tax plus employee social security contributions less cash benefits.

taxes) on social security programs such as old-age, disability, and survivors insurance; public health or sickness insurance; workers' compensation; unemployment insurance; and family allowance programs (U.S. Census Bureau, 1995).

CONCLUSIONS

Nine areas of health disadvantage are documented in Part I of this report:

- adverse birth outcomes;
- injuries, accidents, and homicides;
- adolescent pregnancy and sexually transmitted infections;
- HIV and AIDS;
- drug-related mortality;
- obesity and diabetes;
- heart disease;
- chronic lung disease; and
- disability.

There are policy implications for each of these. Although much is still to be learned, for many of these public health issues there are evidence-based policies that could address them at the national, state, and local levels.

Policy is also relevant to the unfavorable social, economic, and environmental conditions identified in this report as potential contributors to the U.S. health disadvantage. A variety of policies can contribute to high poverty rates, unemployment, inadequate educational achievement, low social mobility, and the absence of safety net programs to protect children and families from the consequences of these problems. However, identifying and implementing policy solutions is a formidable challenge. For example, national health objectives to address many of the conditions listed above were adopted decades ago by the federal government but only some have been achieved, a problem that global initiatives to improve public health have also encountered. Although there have been important public health successes in the United States and elsewhere, such as the remarkable progress in reducing the rate of tobacco use (Brownson et al., 2006), a variety of barriers have impeded progress on other fronts, such as stemming the obesity epidemic or reducing smoking among adolescents.

Other high-income countries with better health status, lower rates of poverty, and more impressive advances in education may owe their success to creative policies or strategies that could find application in the United States. These suppositions, however, amount only to informed speculation and are without empirical evidence. This panel did not undertake a systematic review of the policies and outcomes in other countries, but we believe that such an exercise would be worthwhile to identify useful lessons (see Chapter 9). Reports like the Transportation Research Board study (see Box 8-1) would be valuable for each of the leading causes of the U.S. health disadvantage. However, there are valid questions about the generalizability of "imported" models from overseas, and comparisons with other countries—even other high-income countries—may be seen as less applicable if the comparison countries are much smaller, have a more homogenous population, or have very different social or political systems.

The Measurement and Evidence Knowledge Network (Kelly et al., 2007, pp. 31-32) examined these issues in its final report to the World Health Organization Commission on the Social Determinants of Health. Its conclusions included the following challenges to implementation of such policies:

- [Social factors and other nonmedical determinants of health (SDH)] are multifaceted phenomena with multiple causes. [Although] conceptual models of SDH are useful, they do not necessarily provide policy makers with a clear pathway towards policy development and implementation. As specific policy initiatives tend to be targeted to a specific (population) group in certain circumstances and for prescribed time-periods, they can neglect the wider context within which the social and other determinants are generated and re-generated.
- . . . [R]ecent studies of SDH have emphasized the significance of the life-course perspective (Blane, 1999). Such a perspective poses serious challenges to policy-making processes whose time-scales are rarely measured over such long periods. The tenure of elected or appointed officials is measured in months and years rather than decades. Moreover, coalitions of interests in support of [these policies] may be unsustainable over the time periods necessary to [achieve] significant change. There have been some exceptions to this [general finding], especially in the field of public pension policies, but the general problem of time-scales remains important.
- . . . SDH necessarily imply policy action across a range of different sectors. It is increasingly recognized that action beyond health-care is essential and, as such, intersectoral partnerships are critical to formulating and implementing effective . . . [policies]. However, there is a significant body of evidence which shows that partnerships are hampered by cultural, organizational, and financial issues (Sullivan and Skelcher, 2002).
- Traditionally, government agencies have been organized vertically according to service delivery (Bogdanor, 2005; Ling, 2002) and

such "silo" or "chimney" approaches are not well equipped to tackle issues that cut across traditional structures and processes.

The report notes that silos within and across agencies make it difficult for leaders who address one social factor (e.g., education) to interact with health agencies. With the exception of some success stories (e.g., school health), meetings across agencies occur only occasionally except the Cabinet level. Looking at policies on social factors and other nonmedical determinants of health, the report notes:

- [They] must be viewed as only one of several competing priorities for policy makers' attention and resources. Economic policy or foreign affairs [often] take precedence over health concerns. More specifically, SDH may be over-shadowed . . . by [concerns over] health-care itself. However, this *health care* focus is often to the neglect of *health* and [its broader determinants].

The report further notes that a focus on health care also ignores the important connection between health and the economy: nonhealth policies that reduce disease burden and thus the costs of health care have enormous implications for medical spending and the economy itself (Milstein et al., 2011; Woolf, 2011). Unfortunately, the report notes, political realities often limit attention to "short-term [returns] rather than the long-term [ramifications] and on discrete interventions rather than coordinated, collaborative initiatives. . . ." Lastly, the report notes that globalization has been changing the role of national governments in shaping policy making:

- Governments' ability to shape and mould SDH with the goal of improving their population's health is becoming limited as many of the [upstream causes] no longer fall within their responsibility. There is a parallel argument that decentralization [of authority] to regions and cities has had a similar effect on the policy-making capacity of national governments.

Ultimately, meaningful initiatives to address the underlying causes of the U.S. health disadvantage may have to address the distribution of resources that are now directed to other categorical priorities—a change that is likely to engender political resistance. Is a shift in priorities warranted? This report documents that the United States is not keeping pace with other high-income countries in many areas of health and socioeconomic well-being, and the consequences to the nation can be measured not only in lives, but also in dollars. Understanding why this is occurring and identifying policies that could reverse these unfavorable trends are clearly important for the nation's future.

Part III:
Future Directions for Understanding the U.S. Health Disadvantage

The previous sections of this report have documented the scope of the U.S. health disadvantage (Part I) and have explored a range of potential factors that might explain why the United States is losing ground (Part II). We now turn to the question of what the nation should do about this situation.

As scientists, the panel is reluctant to answer the question without better evidence. Data are simply lacking to fully understand the causal factors responsible for each of the diseases and injuries that disproportionately affect the U.S. population. Even the scope of the U.S. health disadvantage is not fully understood because comparative data are lacking to fully compare population health across high-income countries. The statement of task from the study sponsor ended with this charge: "If insufficient evidence (data) exists currently to test new hypotheses, indicate the nature and extent of data that would be required." That research agenda is presented in Chapter 9.

The statement of task also instructed the panel to set out "more effective public health strategies" for the future. The research agenda outlined in Chapter 9 may take years to complete, and the panel is convinced by the scope and size of the U.S. health disadvantage that more immediate steps can, and should, be taken now. Chapter 10 provides specific recommendations about intensifying efforts to address national health objectives that target the areas in which the United States is experiencing a disadvantage and alerting the American public to the problem. Chapter 10 speaks to the need to stimulate a national discussion about the implications of the U.S. health disadvantage and proposes a comprehensive study to learn from policies that have been used by other nations that have better health outcomes.

Many proposals for rectifying the U.S. health disadvantage will remain "informed speculation" until sophisticated research and hypothesis testing convincingly establishes their effectiveness in the United States. The scientific community faces challenges in conducting such research because of limitations in access to comparable cross-national data and methodological challenges in conducting studies to identify the causes and solutions of the U.S. health disadvantage. To address those challenges, Chapter 9 proposes several specific and important actions that could be taken now by the scientific community, research and health statistics agencies, and funders to establish an infrastructure for vibrant research and scholarship on the causes of the U.S. health disadvantage.

9

Research Agenda

Although much is known about the U.S. health disadvantage, every chapter of this report has identified gaps in the evidence for documenting its scope, understanding its causes, and identifying effective solutions. It is clear that further research in epidemiology, demography, health services, economics, and political science is necessary to fully understand the root causes of the U.S. health disadvantage. The specific deficiencies in each topic area are too extensive to summarize here, and the literature in each field often includes articles that document specific research challenges. However, this report has identified several recurring themes that point to important priorities for research on cross-national differences in health.

This chapter focuses on advancing the science for understanding the U.S. health disadvantage, but we emphasize that such efforts should not come at the expense of vitally important fields of research that focus on how to reduce morbidity and mortality from today's leading health threats, including the nine conditions responsible for the U.S. health disadvantage documented in Part I. In general, cross-national comparisons provide only clues as to *why* one population is healthier than another. Investigating these important hypotheses should not detract from research on specific diseases and injuries and on effective interventions and policies to improve health outcomes in the United States. Research in the areas of biomedicine, health services, public health, social epidemiology, and the social, behavioral, and environmental sciences are all vital. Diverting support or funding for these important research endeavors to study the U.S. health disadvantage would be a fundamental mistake.

We also emphasize that our call for more research should not be used as an excuse for inaction on the part of the nation. As detailed in Chapter 10, the causes of the U.S. health disadvantage are clear enough, and there is sufficient evidence to justify an immediate response on the part of the nation, states, and local communities. The public health and social policy priorities are evident, and interventions of proven effectiveness are known. Delaying action to wait for more data will only allow the U.S. health disadvantage to grow (see Chapter 10).

Research can point to priority areas for action, but its value will depend on the quality of available cross-national data. This chapter focuses on how to build capacity for productive scholarship on international health differences and the need for an ongoing and coordinated commitment by research agencies, funding bodies, statistical agencies, and investigators. We focus on four areas: (1) data needs, (2) analytic methods development, (3) new lines of inquiry, and (4) stable research funding.

BACKGROUND

High-income countries currently collect extensive data on health indicators and a variety of factors that contribute to health and illness. The United States is among the countries with the most extensive efforts to collect health-related data using large population-based surveys. Large nationally representative population health surveys conducted by the U.S. Department of Health and Human Services include the National Health Information Survey (NHIS), the Behavioral Risk Factor and Surveillance System (BRFSS) survey, the National Health and Nutrition Examination Survey (NHANES), and the National Ambulatory Medical Care Survey (NAMCS). The BRFSS program has been systematically collecting risk factor data in a state-based system for nearly three decades. It is one of the few worldwide examples of a sustained systematic collection of data that tracks risk factors over time at the population level. Population surveys conducted by other agencies are also relevant to this topic, including the decennial U.S. census, the American Community Survey, the Current Population Survey, and surveys conducted by the U.S. Department of Education.

Data collection efforts in other countries range from local or provincial surveys to nationally representative survey methodologies and some cooperative multinational efforts to administer similar survey instruments to comparable populations in each country. An example of the latter is the Survey of Health, Ageing and Retirement in Europe (SHARE), the English Longitudinal Study on Ageing (ELSA), and the Health and Retirement Study (HRS). For a detailed list of data sets that are available for research (see Table 9-1).

In the United States, the National Institutes of Health (NIH)—and the National Institute on Aging in particular—has played a leadership role in

TABLE 9-1 Publicly Available Databases for Aging-Related Secondary Analyses in the Behavioral and Social Sciences

DATABASE	Baseline Sample Size	STUDY FEATURES						
		Start Year	End Year	Longitudinal	International	Minority Oversample(s)	Anthropometric Measures	Biospecimens
AVAILABLE DATA SETS								
Advanced Cognitive Training for Independent and Vital Elderly (ACTIVE)	2,802	1999	2011	X		X	X	X
Aging, Status, and the Sense of Control (ASOC)	2,593	1995	2001	X				
Alameda County Health and Ways of Living Study	6,928	1965	1999	X			X	
Americans Changing Lives Study (ACL)	3,617	1986	Pres	X				
Assessment of Doctor-Elderly Patient Encounters (ADEPT)	46	1998	2001	X		X		
The Australian Longitudinal Study of Ageing (ALSA)	2,087	1992	2010	X	X		X	X
Census Microdata Samples Project (also known as The Status of Older Persons in UNECE Countries)	~1 million	1989	1992		X	X		X
The Charleston Heart Study (CHS)	2,283	1960	Pres	X		X	X	
China Health and Retirement Longitudinal Study (CHARLS)	17,000	2008	Pres	X	X		X	X
Chinese Longitudinal Healthy Longevity Survey (CLHLS)	8,993	1998	Pres	X	X			X
Costa Rican Longevity and Healthy Aging Study (CRELES)	2,827	2004	Pres	X	X	X	X	X
The Cross National Equivalent Files (CNEF)	12,900–45,000+	1970	2010	X	X	X	X	
Cross-Sectional and Longitudinal Aging Study	2,891	1989	1992	X				
Early Indicators of Later Work Levels, Disease, and Death (EI)	39,338	1850	1930	X		X	X	

continued

TABLE 9-1 Continued

DATABASE	Baseline Sample Size	Start Year	End Year	Longitudinal	International	Minority Oversample(s)	Anthropometric Measures	Biospecimens
				STUDY FEATURES				
English Longitudinal Study of Ageing (ELSA)	12,100	2000	Pres	X	X		X	X
Epidemiology of Chronic Disease in the Oldest Old	1970-1979: 2,877 1980-1988: 3,113	1970	1988	X		X	X	
Established Populations for Epidemiologic Studies of the Elderly (EPESE)	14,458	1981	1993	X		X	X	
German Socio-Economic Panel (GSOEP)—see also Cross National Equivalent Files (CNEF) above	20,000+	1984	2011	X	X			
Health and Retirement Study (HRS)	22,000+	1992	Pres	X		X	X	X
Health Conditions of Elderly Puerto Ricans (PREHCO)	5,336	2000	2006	X	X	X	X	X
Human Mortality Database	37 countries	1751	Pres	X	X			
Indonesian Family Life Survey (IFLS)	22,000-44,000	1993	2008	X	X		X	X
International Database (IDB)	226 countries	1950	Pres		X			
Iowa 65+ Rural Health Study	3,673	1991	2001	X			X	X
Japanese Study on Aging and Retirement (JSTAR)	4,200	2007			X			
Irish Longitudinal Study on Ageing (TILDA)	8,000+	2009	Pres	X	X	X	X	X
Korean Longitudinal Study of Aging (KLoSA)	10,000+	2006	Pres	X	X			
Longitudinal Aging Study in India (LASI)	1,500+	2010	Pres	X	X		X	X
Longitudinal Employer-Household Dynamics (LEHD)	48 states or U.S. territories	1991	Pres	X				

Study	Sample	Start	End						
Longitudinal Studies of Aging (LSOAs)	1984-1990: 7,541 1994-2000: 9,447	1984	2000	X				X	X
Longitudinal Study of Generations	300+ three-generation families, 2,000+ individuals	1971	2001	X			X	X	X
Longitudinal Study of Mexican-American Elderly Health (The Hispanic EPESE)	3,050	1993	2006	X		X	X	X	X
Los Angeles Family and Neighborhood Survey (L.A.FANS)	2,548	2000	2008	X		X	X	X	
The Luxembourg Income Study (LIS)	30 countries	1968	Pres		X				
Matlab Health and Socioeconomic Survey (MHSS)	Main survey: 4,364 households	1996	Pres		X				
Mexican Family Life Survey (MxFLS)	35,000+	2002	2013	X		X	X	X	X
Mexican Health and Aging Study	15,186	2001	Pres	X		X	X	X	X
Midlife Development in Japan (MIDJA)	1,027	2008	Pres	X		X	X	X	X
Midlife Development in the United States (MIDUS)	1995-1996: 7,108 2004-2006: 5,555	1995	Pres	X		X	X	X	X
National Archive of Computerized Data on Aging (NACDA)									
National Health and Aging Trends Study (NHATS)	8,000+	2011	Pres	X		X	X		
National Long Term Care Survey (NLTCS)	20,485	1982	2004	X		X	X	X	
National Longitudinal Mortality Study (NLMS)	~3.3M	1973	2010	X					
National Longitudinal Survey: 1990 Resurvey of Older Males (NLS-Older Males)	5,020	1966	1990	X		X			
National Nursing Home Survey Follow-up (NNHSF)	6,001	1984	1990	X					
National Social Life, Health, and Aging Project (NSHAP)	3,004	2005	Pres	X		X	X	X	X
The National Survey of Families and Households (NSFH) Reinterview	13,017	2001	2003	X		X	X		
National Survey of Self-Care and Aging: Baseline and Follow-up	3,485	1990	1994	X					
The National Survey of the Japanese Elderly (NSJE)	3,990	1987	1993	X	X				

continued

246

TABLE 9-1 Continued

	STUDY FEATURES							
DATABASE	Baseline Sample Size	Start Year	End Year	Longitudinal	International	Minority Oversample(s)	Anthropometric Measures	Biospecimens
New Beneficiary Survey (NBS) and New Beneficiary Follow-up (NBF)	16,692	1982	1991	X				
New Immigrant Survey (NIS)	13,981	2003	2004	X		X		
Nihon University Japanese Longitudinal Study of Aging (NUJLSOA)	4,997	1999	2003	X	X			
Panel Study of Income Dynamics (PSID)	65,000+	1968	Pres	X		X		
PHSE Ten-Year Follow-up of the North Carolina EPESE	4,162	1996	1997	X		X		
Precursors of Premature Disease and Death	1,337	1946	2003	X				
Project TALENT	440,000	1960				X	X	
1990 Public Use Microdata Sample for the Older Population (PUMS-O)	Unknown	1990	2000					
Puerto Rican Elderly Health Conditions	4,291	2002	2007	X	X	X	X	
RAND Metadata Repository								
Religion, Aging, and Health Survey	1,500	2001	2004	X		X		
Resources for Enhancing Alzheimer's Caregiver Health (REACH) and REACH II	REACH: 600 REACH II: 642	1996	2004	X		X		
Sacramento Area Latino Study on Aging (SALSA)	1,789	1996	Pres	X		X	X	X
Seattle Longitudinal Study (SLS) of Adult Cognitive Development	6,000+	1956	Pres	X		X		X
The Second Malaysian Family Life Survey (MFLS-2)	Senior sample: 1,357	1988	1989		X			
Social Environment and Biomarkers of Aging Study (SEBAS) in Taiwan	1,023	1999	2006	X	X		X	X
Study of Dementia in Swedish Twins	3,838	1990	1999	X	X		X	X
Study on Global Ageing and Adult Health (SAGE)	90,000+	2002	Pres	X	X		X	X

Study	N	Start	End					
Survey of Health, Ageing and Retirement in Europe (SHARE)	31,115	2004	Pres	X	X		X	X
Survey of Health, Ageing and Retirement in Israel (SHARE-Israel)	2,598	2005	Pres	X	X		X	X
Swedish Adoption/Twin Study of Aging (SATSA)	2,020	1984	2005	X	X	X	X	X
Terman Life-Cycle Study, as supplemented	1,528	1922	2000	X	X			
The Wechsler Adult Intelligence Scale Archives (NIA-WAIS) on Aging and Multiple Cognitive Abilities	50,000+	1980	1998	X	X			
Whitehall II Study (also known as Stress and Health Study)	10,308	1985	Pres	X	X		X	X
Wisconsin Longitudinal Study (WLS) (graduates)	10,317	1957	Pres	X			X	X
Wisconsin Longitudinal Study (WLS) (siblings)	4,778	1975	Pres	X			X	X
The Women's Health and Aging Study (WHAS I, WHAS II, WHAS III)	1,002; 436; 1,438	1992	Pres	X			X	X
DATA SETS TO BE ARCHIVED IN THE FUTURE								
NIA Collaborative Studies on Dementia Special Care Units (SCUs)	3,000 Nursing Homes 1,500 SCUs	1991	1996	X				
Origins of Variance in the Old-Old: Octogenarian Twins (The OCTO Twin Study)	351 Same-Sex Twin Pairs	1997	2005	X	X		X	
Victoria Longitudinal Study	1,594	1987	Pres	X	X		X	X
DATA SETS AVAILABLE THROUGH PRINCIPAL INVESTIGATOR								
Epidemiology of Aging and Physical Functioning	2,092	1993	2006	X	X		X	X
The Longitudinal Study of Aging Danish Twins	4,371	1995	2005	X	X	X	X	X
Maine-Syracuse Longitudinal Study	~2,400	1975	Pres	X			X	X
The UAB Study of Aging: Mobility Among Older African-Americans and Whites	1,000	1999	2008			X		
The UNC Alumni Heart Study	6,340	1986	Pres	X				
Vietnam Era Twin Study of Aging (VETSA)	1,006	2002	Pres	X			X	X

SOURCE: National Institute on Aging (2012).

working with partners in other countries to coordinate the collection of comparable cross-national data to understand the epidemiology of health and aging (see Box 9-1). The National Institute of Child Health and Human Development (also part of NIH) and the National Science Foundation fund the University of Minnesota to maintain the Integrated Public Use Microdata Series-International (IPUMS-International), an effort to inventory, preserve, harmonize, and disseminate census microdata from around the world

---

BOX 9-1
International Health Studies
of the Population Age 50 and Older

International surveys of the population age 50 and older serve as good models of harmonization and collaboration in the production of comparable health information across a number of countries. The United States began this effort with the Health and Retirement Study (HRS) in 1992. HRS, which is administered by the University of Michigan, surveys a representative sample of more than 26,000 Americans over the age of 50 every 2 years. It now has 11 waves of data. It is supported by the National Institute on Aging and the Social Security Administration. The English Longitudinal Study on Ageing (ELSA) began in 2002 with a sample of about 12,000 people age 50 and older and now has had five waves of data collection. The Survey of Health, Ageing and Retirement in Europe (SHARE) is a cross-national panel database with four waves of data for more than 55,000 individuals age 50 or older from 20 European countries. A number of additional countries throughout the world are now undertaking studies that will be part of this family of surveys.

These surveys have been developed collaboratively across countries and across disciplines: there is significant overlap in the memberships of their monitoring committees and advisory groups and active investigators. Producing comparable international data has been an aim since their beginning. The multidisciplinary approach allows comparable cross-national examination across a wide variety of health outcomes, as well as comparison of the strength of associations with causal or related variables. All of these studies have collected information on health conditions, disability, physical functioning, cognitive functioning, risk factors, and health care use and expenditures. They also collect extensive socioeconomic, demographic, and life-style factors, as well as information on earlier life events. All three surveys measure functional status, and HRS and ELSA collect biomarkers. SHARE is now piloting biomarker collection.

(University of Minnesota, 2012b).[1] The project includes the Integrated Health Interview Series, which compiles a harmonized set of microdata and documentation based on material originally included in the public-use files of the National Health Interview Survey (University of Minnesota, 2012a).

Despite these impressive efforts, many of the data we sought for this report were scattered across disparate and sometimes obscure sources, and the data often did not exist, were inconsistently defined, or were contradictory. These limitations were often too great to allow us to reach definitive conclusions. We considered what steps the scientific community should take so that future analysts might benefit from a more robust body of evidence. Much of what can be done is beyond the scope of a single chapter to catalogue, but we offer three recommendations, beginning with one on the need for better data.

DATA NEEDS

RECOMMENDATION 1: **Acting on behalf of all relevant data-gathering agencies in the U.S. Department of Health and Human Services, the National Institutes of Health and the National Center for Health Statistics should join with an international partner (such as the OECD or the World Health Organization) to improve the quality and consistency of data sources available for cross-national comparisons. The partners should establish a data harmonization working group to standardize indicators and data collection methodologies. This harmonization work should explore opportunities for relevant U.S. federal agencies to add questions to ongoing longitudinal studies and population surveys that include various age groups—especially children and adolescents—and to replicate validated questionnaire items already in use by other high-income countries.**

A fundamental challenge to understanding the U.S. health disadvantage is a lack of data to identify, monitor, and analyze epidemiological changes over time. Ambiguities in how best to define and measure health outcomes and determinants of health and inconsistent measurements across countries plague any effort to compare countries on meaningful terms. In some cases, the epidemiological literature has not even confirmed a causal link between

---

[1]The database currently includes approximately 397 million people, from 185 censuses taken in 62 countries from 1960 to the present.

some putative causes and associated health outcomes.[2] Questions that have been validated in one country may perform differently in other countries, where life-styles, culture, or different meanings in translation can affect the results. Across topic areas, some countries participate in population surveys while others do not. These heterogeneities are an impediment to many efforts to make valid comparisons of health outcomes, determinants of health, and relevant contextual factors between one country and another. Thus, our first recommendation focuses on the need for better data, with an emphasis on the health outcomes discussed in this report, and on the determinants of health we reviewed, including health systems, personal behaviors, social and demographic factors, and physical and social environments.

Health Outcomes

Vital Statistics

Although vital statistics are available for nearly all of the populations of high-income countries and arguably measure the most precise endpoints imaginable—birth and death—attributions of cause of death still have inherent imprecision. For example, infant mortality comparisons are affected by differences in how countries register premature births and whether they are reported as live births. Cause-of-death attributions also may vary by country (e.g., "drug-related" deaths and suicides).

Physical Illnesses

Existing indicators do not go far enough to make meaningful cross-national comparisons of disease profiles of countries. For example, 30-day case-fatality rates assembled by the OECD focus on cross-national comparisons for only two conditions, acute myocardial infarction and stroke (OECD, 2011b). Similar data are needed for other major causes of death, and follow-up beyond 30 days is important to evaluate the quality of ambulatory and chronic illness care and readmission rates. Internationally comparable data are needed on the prevalence of ambulatory-sensitive conditions beyond the current focus on asthma and diabetes.

---

[2]Examples include precise quantification of the role of diet in causing cardiovascular disease, cancer, and other conditions; disentangling the effects of physical activity independent of obesity and diet; and which forms of "problem drinking" are predictive of disease or injury. Questions also surround the relative contribution of medical care to health outcomes and the causal role of stress, population-based services, and levels of public spending.

Mental Illnesses

As discussed in Chapter 2, deficiencies in available data and inconsistencies in diagnostic classifications prevented the panel from reaching conclusions about whether mental illness is more common in the United States than in other peer countries. Continued efforts are needed to standardize the collection of epidemiological data on mental illness based on established diagnostic instruments (see Chapter 2). An accepted international standard is lacking not only for established mental illnesses such as clinical depression, but also for relevant psychological factors, such as stress.

Determinants of Health

Health Systems

Cross-national comparisons of health system performance have been widely attempted (Davis et al., 2010), but they remain rudimentary. Validated indicators exist for delivery of specific services, such as those used for performance measures, but not for other dimensions of care important to outcomes, including measurements of the quality of care coordination for chronic illnesses or the quality of communication between providers and with patients. As discussed in Chapter 4, the only currently available systematic data to compare the quality of health care in countries come from surveys administered by the Commonwealth Fund, which are fielded in only 7-11 countries and rely on the perceptions of patients and primary care providers. Differences in cultural norms and expectations may skew patients' responses to questions about whether doctors "spend enough time," make mistakes, or communicate well. Developing more objective measures of quality, coordination, and communication that can be administered consistently across countries and can also account for contextual factors (e.g., differences in health systems and social policy) will require a collaborative effort among health services researchers from high-income countries.

For medical and public health systems among countries, no established (let alone validated) measures for access or quality are in use, apart from efforts to define the core content of public health, such as the core functions or the accreditation criteria recently developed in the United States by the Public Health Accreditation Board (Institute of Medicine, 2012) (see Chapter 4). Even national spending in these areas is difficult to compare: countries differ in both how they track spending on public health or social programs and how they classify these spending categories.[3] Community-

---

[3]For example, countries differ in what is classified as public health, prevention, health promotion, primary care, and social transfers.

level financial data to assess population health investments are generally lacking. The Institute of Medicine (IOM) has made recommendations about the need to develop validated measures of public health services, including systems to track quality and accountability in public health departments (Institute of Medicine, 2011e, 2012). This work could inform the development of metrics to compare the core competencies of countries in providing public health services to their populations.

Personal Behaviors

Inconsistencies and ambiguous indicators also plague comparisons of how people behave in different countries. As noted in Chapter 5, a variety of questions have been asked about physical activity, diet, sexual practices, drinking, driving practices, and violence. Some questions have not undergone adequate scientific validation, rely on different sampling or survey administration protocols, or are interpreted differently across populations. For example, "leisure-time activity" and "total physical activity" are not constants across cultures.[4] In addition, the degree to which such behavior predicts health outcomes among countries is unclear.

Similarly, international differences in dietary habits, menus, and even norms for portion sizes make it difficult to compare food consumption patterns. To our knowledge, comparable international data are lacking to compare injurious behaviors, such as driving while intoxicated; the failure to use seatbelts or child safety seats; or reckless acts that lead to poisoning, falls, drowning, fires, or burns. Few indicators exist for measuring the prevalence of injurious behaviors (e.g., fights), making it difficult to disentangle the role of weapons (e.g., firearms) in understanding high homicide rates in the United States. The leading drugs of abuse may vary across countries, and the focus of substance abuse surveys may therefore differ.

Social and Demographic Factors

Demographic and socioeconomic data (e.g., race and ethnicity, income) would have to be collected more systematically to make valid cross-national comparisons, such as comparing people of a given income level in one country with people of a similar income level in another country. Many U.S. surveys do not even ask about income or education, lack consistency in response options, or list broad income ranges that fail to meaningfully differentiate the purchasing power of different populations.

---

[4]Survey instruments like the International Physical Activity Questionnaire (Bauman et al., 2009) hold promise as a standardized tool for assessing physical activity.

Physical Environment

Chapter 7 discusses the limited availability of environmental data to compare physical and chemical exposures across countries or to document differences in land use and the built environment. Accepted metrics of the built environment that are applicable across different countries will be challenging to devise.

Social Environment

Even greater challenges to generalizability affect research efforts to compare countries in terms of social capital, social cohesiveness, and social participation. To begin such research will require agreement on accepted metrics, as well as the capacity to pose such questions in population studies.

Age Groups

Many of the data sources for the seminal studies on the U.S. health disadvantage—SHARE, ELSA, and HRS—were studies of aging that focused on adults age 50 and older. A pressing research priority is to define indicators and capture data for younger people and for each stage of the life cycle. Particular priority should be given to data on the health and behavior of children and adolescents (e.g., diet, exercise, alcohol and other drug use, driving, sexual activity) and related contextual variables. Such data are often lacking for children and adolescents (Institute of Medicine and National Research Council, 2011). Just as surveys such as SHARE provide morbidity data to measure the prevalence of diseases in adults over age 50, surveys are needed to make cross-national comparisons of morbidity, risk factors, and biomarkers in children, adolescents, and young adults.

The Maternal and Child Health Life Course Research Network (2012) is a recently established network of investigators who are committed to such an effort. Such development work is important to identify appropriate indicators to measure at each phase of the life cycle and to begin expanding understanding of the important influences that foster (or prevent) the development of risk behaviors, pathophysiological disease processes, and emotional turmoil. Children may be affected by a host of different environmental factors than are adults, and those effects may have different levels of significance in different life stages and settings.

Some of the important determinants of health are not conventional public health measures. For example, the Early Development Inventory and community "dashboards" created by Halfon and colleagues (e.g., Halfon et al., 2010) are data tools that help evaluate children in terms of social and emotional development, approaches to learning, language skills, and cognition. Such indicators will need to be validated and used consistently across countries so that surveys can capture data on how differences in exposures may explain cross-national differences. Surveys will need to ask age-appropriate questions to create a temporal understanding of the major health determinants for each stage of the life course. As shown in Figure 9-1, a social-ecological interaction is occurring at age each stage of life, and the challenge of sophisticated longitudinal study of this process is to evaluate the effects of an evolving cast of potential influences at each stage.

Measures of adolescent health deserve particular attention because this is the life-cycle stage when so many nonmedical determinants of health come into play and when many life-long health behaviors are established. The infant mortality rate has long been regarded as an indicator of human development, as well as of the state of a nation's public health (Lee et al.,

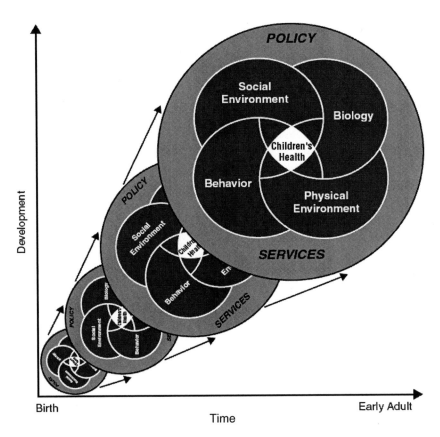

FIGURE 9-1 Social-ecologic influences on children's health over time
SOURCE: National Research Council and Institute of Medicine (2004, Figure 2-2).

1997). Given that infant and early childhood mortality have fallen in all countries and that adolescent mortality exceeds child mortality in all high-income countries, adolescent mortality may be emerging as a stronger indicator of healthy human development for these countries (Viner, 2012).

Challenges and Opportunities

International Collaboration

Although harmonized data collection across multiple countries faces a variety of challenges, some inconsistencies are surmountable simply by agreeing to coordinate methods and to agree on harmonization proactively ("input harmonization") rather than after the fact (Hoffmeyer-Zlotnik and Harkness, 2005). Relatively immediate steps (pursuing the "low-hanging fruit") and long-term planning would be possible through a consortium of statistical officials from major countries who meet regularly and are committed to data harmonization. Interest in such efforts is increasing, and a number of harmonization efforts are already under way, including some with a particular focus on social determinants of health. A partial list of such global efforts is provided in Box 9-2. These efforts have received enthusiastic endorsement from world leaders, including a landmark September 2011 decision by the United Nations (2012a) General Assembly.[5]

A logical U.S. partner for international collaboration is the NIH, which has a deep interest in understanding the U.S. health disadvantage and in establishing a common data set that all countries could use for investigating and monitoring cross-national health differences. Such data are important to NIH not only to explain the U.S. health disadvantage, but also to expand knowledge of the biological and social determinants of disease over the life course. Other agencies in the U.S. Department of Health and Human Services share this interest—such as the Centers for Disease Control and Prevention (CDC)—and bring unique resources to the task of collecting and analyzing health data in the United States and overseas. The panel therefore

---

[5]The United Nations (2012a) General Assembly voted for the following action item in its resolution on noncommunicable diseases (e.g., heart disease, cancer): "Call upon WHO, with the full participation of Member States, informed by their national situations, through its existing structures, and in collaboration with United Nations agencies, funds and programmes, and other relevant regional and international organizations, as appropriate, building on continuing efforts to develop before the end of 2012, a comprehensive global monitoring framework, including a set of indicators, capable of application across regional and country settings, including through multisectoral approaches, to monitor trends and to assess progress made in the implementation of national strategies and plans on non-communicable diseases . . .," see Rio Declaration in Box 9-2.

BOX 9-2
International Efforts to Harmonize Data

The European Commission (EC) has a commitment to "produce comparable data on health and health-related behaviour, diseases and health systems . . . to be based on common EU [European Union] health indicators, for which there is Europe-wide agreement regarding definitions, collection and use" (European Commission, 2012). Of 88 health indicators, the EC has identified more than 40 core indicators on demographic and socioeconomic conditions, health status, health determinants, and interventions (including health services and health promotion) for which data are "readily available and reasonably comparable" (European Commission, 2012). As part of the Public Health Program of its Directorate for Health and Consumers, the EC has made a long-standing commitment to data harmonization in Europe. The Public Health Program has a separate strand, with a separate budget, for improving health monitoring systems, from which many initiatives have received funding (European Commission, 2012). Examples of such activities include the European Health Survey System (which aims to harmonize health interview surveys) and the European Health Examination Survey (which prepares a harmonized health examination survey in 14 European countries).

The OECD is also active in promoting data harmonization, with a focus on the creation of comparable health care accounts and other indicators. This report, which cites OECD data extensively on a broad variety of issues, has been a beneficiary of the organization's efforts to compile extensive cross-national data sets.

The European Observatory on Health Systems and Policies promotes evidence-based health policy making by studying health systems in more than 50 countries in Europe and elsewhere. This project represents a collaboration between the governments of nine European countries, the EC, the European Investment Bank, the World Bank, UNCAM (French National Union of Health Insurance Funds), the Regional Office for Europe of the World Health Organization (WHO), the London School of Economics and Political Science, and the London School of Hygiene and Tropical Medicine (European Observatory on Health Systems and Policies, 2012).

The Eurothine project, based at Erasmus University in Rotterdam (Erasmus, 2012), is "collecting and analyzing information from different European countries that will help policy-makers at the European and national level to develop rational strategies for tackling socioeconomic inequalities in health." It is developing and collecting indicators of health inequalities to provide benchmarking data on inequalities to participating countries, along with assessing evidence and making recommendations on policy interventions. It intends to disseminate the results and to "develop a proposal for a permanent European clearinghouse on tackling health inequalities."

The World Alliance for Risk Factor Surveillance is working to finalize a definition and conceptual framework for behavioral risk factor surveillance that "can be shared and discussed globally" and "serve as a reference for researchers, practitioners, and countries that are developing behavioral risk factor surveillance" (International Union for Health Promotion and Education, 2009).

The Washington Group on Disability Statistics was formed as a result of the UN International Seminar on Measurement of Disability, which occurred in New York in 2001. An outcome of that meeting was the recognition that statistical and methodological work was needed at an international level to facilitate the comparison of disability data cross-nationally. Consequently, the UN Statistical Commission authorized the formation of a "city group" to address some of these issues and invited the National Center for Health Statistics in the United States to host the group's first meeting. The city group format typically involves three or four working meetings at which representatives from national statistical agencies address selected problems in statistical methods. The Washington Group on Disability Statistics takes its name from the location of the group's first meeting (Centers for Disease Control and Prevention, 2009e).

In 2011, WHO issued the Rio Declaration on Social Determinants of Health, which included a commitment to 12 strategies of data collection, listed below. The declaration (World Health Organization (2011c, p. 6) noted:

> that monitoring of trends in health inequities and of impacts of actions to tackle them is critical to achieving meaningful progress, that information systems should facilitate the establishment of relationships between health outcomes and social stratification variables and that accountability mechanisms to guide policy-making in all sectors are essential, taking into account different national contexts.

The Rio Declaration was adopted by WHO Member States in 2012 at the Sixty-fifth World Health Assembly in Geneva, Switzerland. It includes 12 pledges to monitor progress and increase accountability (World Health Organization (2011c, pp. 6-7):

continued

1. Establish, strengthen and maintain monitoring systems that provide disaggregated data to assess inequities in health outcomes as well as in allocations and use of resources;
2. Develop and implement robust, evidence-based, reliable measures of societal well-being, building where possible on existing indicators, standards and programmes and across the social gradient, that go beyond economic growth;
3. To promote research on the relationships between social determinants and health equity outcomes with a particular focus on evaluation of effectiveness of interventions;
4. Systematically share relevant evidence and trends among different sectors to inform policy and action;
5. Improve access to the results of monitoring and research for all sectors in society;
6. Assess the impacts of policies on health and other societal goals, and take these into account in policy making;
7. Use intersectoral mechanisms such as a Health in All Policies approach for addressing inequities and social determinants of health; enhance access to justice and ensure accountability, which can be followed up;

believes that the role of coordinating these efforts should logically occur at the departmental level and not any one agency.

Data Access

Another opportunity for overcoming the limitations of data is to broaden access to existing data sets. The open government movement in the United States shares this aim. A secure publicly accessible data warehouse platform to enable investigators to access data for cross-national comparisons is a research priority well within the capabilities of modern information systems architecture. The widely used Luxembourg Income Study (LIS) has been identified as a particularly strong model worthy of replication in the health field (Bambra and Beckfield, 2012, pp. 31-32):

Several key characteristics of the LIS make it a perfect model for a comparative database on population health: researchers can access individual-level

8. Support the leading role of the World Health Organization in its collaboration with other United Nations agencies in strengthening the monitoring of progress in the field of social determinants of health and in providing guidance and support to Member States in implementing a Health in All Policies approach to tackling inequities in health;
9. Support the World Health Organization on the follow-up to the recommendations of the Commission on Information and Accountability for Women's and Children's Health;
10. Promote appropriate monitoring systems that take into consideration the role of all relevant stakeholders including civil society, nongovernmental organizations as well as the private sector, with appropriate safeguard against conflict of interests, in the monitoring and evaluation process;
11. Promote health equity in and among countries, monitoring progress at the international level and increasing collective accountability in the field of social determinants of health, particularly through the exchange of good practices in this field; and
12. Improve universal access to and use of inclusive information technologies and innovation in key social determinants of health.

data (critical for examining social inequality), access is via remote server (LIS requires application for permission to access the data, but a researcher never "owns" the data, which allows for the free dissemination of sensitive information), and the LIS team harmonizes the data to aid in international comparison (just as the LIS developed an "income concept" to facilitate comparison, likewise a "health concept" could be developed for comparative analysis). . . . [W]e think the development of such a rich resource for the scientific community is the top priority for new science on these critical questions. The sort of detailed comparative research on health that Elo (2009) and others have called for simply cannot be conducted in a way that allows knowledge to cumulate without such data infrastructure.

Expanding Current Surveys

Another important opportunity to enhance data is to add new questions (or replace outdated questions) on existing longitudinal or cohort studies and on national population surveys, such as the National Health and Nutrition Examination Survey and BRFSS,[6] with a focus on variables relevant to health determinants and life-cycle influences at all ages that may help to explain the U.S. health disadvantage. Such efforts would serve two important purposes. First, expanding the data collected would provide an opportunity for U.S. surveys to include questions already in use in Europe and elsewhere, thereby enabling prevalence rates in the United States to be accurately compared with those of other countries. For example, only minor modifications are needed to align questions about sexual behavior on the Youth Risk Behavioral Survey (YRBS) and the Health Behaviour in School-Aged Children (HBSC) survey. Similar modifications could occur with other surveys, such as harmonizing the National Longitudinal Survey of Youth (NLSY) with questions asked in other countries. Second, an expansion would provide data that investigators could use to study the causal pathways responsible for health in general and the U.S. health disadvantage in particular.

Cross-National Surveillance

In addition to expanding the collection of survey data in the United States, the National Center for Health Statistics and the CDC could maintain an ongoing effort to look abroad to track how efforts to improve public health in the United States compare with those in other countries. Regular reports—such as those that appear in the *Morbidity and Mortality Weekly Reports* (*MMWR*) and are reprinted in the *Journal of the American Medical Association* and by many news outlets—would be an appropriate vehicle for publishing updates on U.S. health rankings and health statistics relative to other high-income countries. This ongoing surveillance would help to monitor progress and identify opportunities to learn from successes in other high-income countries. For example, the Transportation Research Board (TRB) gleaned important insights from other countries on strategies that could be used to reduce traffic fatalities in the United States (see Chapter 8). This is not to say that other countries have all the answers; but their experiences suggest strategies and approaches that might also work in (parts of) the United States or that can be adapted and piloted in the United States.

It will be important to continue to compare health rankings for population subgroups (e.g., by socioeconomic status, race and ethnicity, access to health care) to help differentiate the contribution (or lack thereof) of

---

[6]Supplementary modules administered by BRFSS provide a system that can adapt to needs and could be used to explore causal hypotheses about cross-national health differences. For example, questions added to these modules could collect data on such factors as injurious behaviors and the social and physical environment.

different factors in accounting for cross-national differences. Comparisons with England have already demonstrated that the U.S. health disadvantage appears to persist across racial and ethnic groups in the United States and among college-educated and upper income populations (Banks et al., 2006; Martinson et al., 2011a, 2011b), but the studies to date have their limitations. For example, the study by Banks et al. (2006) was restricted to adults aged 55-64. The stratified data described by Martinson and colleagues, which did not appear in their original article (2011a), could be analyzed more carefully in future studies that apply multifactorial regression analyses. Such analyses could be informed by new studies and better data, as they become available, to assign coefficients that more precisely quantify the relative contribution of different variables to the U.S. health disadvantage.

Logistical Challenges

Logistical and resource constraints may make it difficult to take these steps or to sustain data collection activities across countries or even within countries. Most data agencies are plagued by tenuous funding streams, and they often face other organizational and technological barriers to data collection. For example, the United States lacks the capacity to evaluate the public health services it provides to populations because those services are fragmented across federal health agencies, state health departments, and more than 3,000 local health departments. Because BRFSS is administered by states, which can elect which modules to administer, it is also an example of how decentralization (see Chapter 8) can limit the ability of the U.S. government to gather consistent national health statistics. A previous Institute of Medicine (2011e) study made extensive recommendations, directed to the National Center for Health Statistics and CDC, to undertake bold reforms in the scope and quality of public health data collected by the agencies. The recommendations included the need for a greater focus on social determinants of health, on community conditions that affect health, and on linking data systems in the public and private sector to create more informative "dashboard" data on health conditions at the national, state, and local level.

Methodological Challenges

Research efforts that rely on surveys must contend with different methods for sampling and administration and low, or at least inconsistent, response rates. For example, ELSA had a lower response rate than HRS, which can introduce a response bias. Some countries participating in SHARE had response rates below 50 percent at baseline. Nonrespondents may be more likely to be disadvantaged (and less healthy).

Much of the data compiled in this report relies on self-reported information in population surveys (e.g., of diet, drinking habits, physical activity, medical errors). Methodological research should explore the use of biomarkers and objective reference standards to establish the validity of self-reported observations, such as the use of pedometers and global positioning system (GPS) data to validate self-reports of physical activity and the use of administrative and medical record data to validate reports of medical errors.

Cross-national comparisons treat countries as units of analysis, but from a policy and statistical perspective they are not truly independent because of cross-national ties, such as drug policies enforced by the European Union. The important distinctions may be less about geographic national borders than the relevant regions or populations within countries (Hans, 2009). For example, some argue that U.S. states are more appropriate units of comparison for cross-national studies because of the large size of the United States and significant state-level disparities in health status and other health-related variables.[7]

A further nuance for researchers to tease apart in cross-national comparisons is whether a risk factor, such as poverty or inadequate education, may have different "toxicity" (health implications) depending on the contextual circumstances in each country. For example, as discussed in Chapter 6, there is some evidence that the absence of a college education may have greater effects on employment and health in the United States than in other countries.

ANALYTIC METHODS DEVELOPMENT

RECOMMENDATION 2: The National Institutes of Health and other research funding agencies should support the development of more refined analytic methods and study designs for cross-national health research. These methods should include innovative study designs, creative uses of existing data, and novel analytical approaches to better elucidate the complex causal pathways that might explain cross-national differences in health.

A daunting methodological challenge is how to design studies to understand the causes of the U.S. health disadvantage. Randomized controlled trials, which are considered the strongest evidence of effectiveness

---

[7]Similarly, like the United States, comparison countries also experience important health disparities by province or canton. It could be argued that an "apples to apples" comparison with the United States would contrast similar regions or, perhaps more meaningfully, would compare similar populations in each country, along with appropriate adjustments for relevant covariates affecting health.

in much medical research, are hardly the answer for this field (Anderson and McQueen, 2009; Black, 1996; Braveman et al., 2011c; Glasgow et al., 2006; McQueen, 2009; Petticrew and Roberts, 2003; Victora et al., 2004). Innovations in study designs require thoughtful methodological research and support from funding agencies to sponsor such efforts. The National Institutes of Health, other grant-making institutions, and public-private partnerships could play important roles in funding pilot studies and innovative methodological research to develop such designs in collaboration with colleagues in other countries.

Recent advances in analytic techniques enable researchers to answer more complex questions about the explanation of between-country differences in levels and trends of health indicators. These advances include techniques that allow improved causal inferences regarding population differences, such as multilevel analysis methods that account for factors at different levels of organization (e.g., countries, regions, and individuals) (see Diez Roux, 2011); instrumental variable and other approaches that improve causal inferences regarding population-level differences (see Ahern et al., 2009; Hernán and Robins, 2006); and systems modeling tools that allow one to integrate information from different sources and consider dynamic relations. Counterfactual analysis techniques (such as techniques based on population attributable fractions, which allow an assessment of the contribution of specific risk factors to variations in health outcomes between populations [Northridge, 1995]), fixed effects models, complex systems theory (Diez Roux, 2011), and econometric techniques are among the important tools available. In lieu of longitudinal studies, data fusion through statistical matching has been explored as a way of combining data for life-course research (D'Orazio et al., 2006; Rässler, 2002).

Longitudinal Research

Much of the evidence presented in this report consists of cross-sectional comparisons of mortality and prevalence rates in recent years, along with some historical trend data that span several decades for some indicators. The more important question in understanding the U.S. health disadvantage is to explore the relationship between antecedent factors and health outcomes, some of which occur relatively soon after a risk exposure (e.g., unintended pregnancies) and some of which transpire over years or decades. The life-course perspective embraced by this panel places a premium on understanding how the health problems experienced by Americans are shaped by their early life circumstances—and perhaps even the circumstances of their parents before they were born—and this approach obviously necessitates either prospective longitudinal studies or creative uses of historical data to retrospectively examine causal factors. Clues to the health disadvantages experienced by today's senior citizens may lie in (archived) post–World War II data on living conditions generations ago. Data to track all age groups are important, but longitudinal data that includes middle-aged adults appear to be especially lacking.

The United States is responsible for important longitudinal research, both historically (e.g., the Framingham Heart Study) and in ongoing studies. One example is the National Children's Study (Hirschfeld et al., 2010), an effort to study the long-term experience of 100,000 children and their families that exemplifies such a commitment but has faced its own challenges. Other examples include the Panel Study of Income Dynamics, the longest running longitudinal household survey in the world (Institute for Social Research, 2011), and the National Longitudinal Surveys (U.S. Bureau of Labor Statistics, 2012).

However, the United States generally lags behind the investments other countries have made in conducting longitudinal research. Sweden has maintained an impressive longitudinal data set for more than a century. The Whitehall Studies in the United Kingdom, which began in the late 1960s, produced seminal data on the role of social determinants of health (Marmot et al., 1991). In New Zealand, the Dunedin Multidisciplinary Health and Development Study has been following 1,037 individuals for four decades, since their birth in 1972-1973. Other classic British cohort studies include the National Study of Health and Development (launched in 1946), the National Child Development Study (launched in 1958), and the British Cohort Study (launched in 1970).

Worldwide interest in identifying and measuring early life precursors of both health and socioeconomic outcomes has spawned the launch of "Millennium Cohort Studies" in many countries. These studies include the Danish National British Cohort (100,418 pregnant women, launched in 1996), the Norwegian Mother and Child Cohort (108,000 pregnant women, launched in 1999), the Dutch Generation R Study (9,748 pregnant women, launched in 2001), Born in Bradford, UK (14,000 pregnant women, launched in 2007), Growing Up in Australia (10,000 children, launched in 2004), Millennium Cohort Study U.K. (20,000 children, launched in 2001), the Étude Longitudinale Français Depuis L'Enfance (20,000 children across France, launched in 2011), and Growing Up in Scotland (8,000 children, launched in 2003, with a second wave of 8,000 children enrolled in 2008).[8]

The United States has opportunities to expand this genre of research by collecting prospective data on the influence of social factors on health

---

[8]The U.S. Department of Defense also launched a Millennium Cohort Study in 2001, which now involves approximately 150,000 participants, but its focus is on the long-term effects of military deployment.

outcomes. For example, for many years the research community has advocated the collection of longitudinal data by NHANES. Another option is to expand the Early Childhood Longitudinal Study (ECLS) by collecting data on health outcomes. The ECLS-K of 1998-1999 followed a nationally representative sample of more than 21,000 children who were kindergartners in 1998-1999 through the eighth grade. The ECLS-K program provides national data on children's transition to school and their experiences and growth through the eighth grade. The ECLS program also provides data to analyze the relationships among a wide range of family, school, community, and child characteristics with children's development, learning, and performance in school (National Center for Education Statistics, 2012a).

Mining Currently Available Data

Although a new wave of long-term longitudinal research is certainly needed, more can be done with the available data already in existence. At a time of limited resources for new research, funding agencies can achieve important economies by funding investigator-initiated studies that propose secondary analysis of existing data sets. For example, BRFSS data could be more effectively mined to uncover more valuable information than currently occurs. Important insights could be obtained by examining historical data because the U.S. health disadvantage has been growing over time and the important antecedents occurred decades ago. An important avenue the panel identified, which it did not have time to explore, would be to map historical trends in social factors, such as poverty and social capital, from post–World War II and to track their evolution over time as the U.S. health disadvantage has grown more pronounced.

In some cases, useful data for international comparisons may exist in disparate databases that are familiar only to specialists in certain disciplines. This report demonstrates the broad spectrum of disciplines relevant to understanding the U.S. health disadvantage, spanning health, social science, economics, and the environment. Even in the digital information era, an exhaustive search of all databases that provide a basis for making international comparisons is challenging. For example, this panel was not confident that it had adequately identified all possible data sources for comparing the physical and social environments in high-income countries. It is possible, for example, that specialists in air quality, climate, social networks, or other disciplines are aware of differences between the United States and other countries that we did not discover in our review.

Other opportunities exist in exploiting linked data sets, such as the National Death Index, which combines death records and survey data (Centers for Disease Control and Prevention, 2009c), and in conducting secondary data analysis across multiple studies. For example, the Emerging Risk Factors Collaboration has established a central database on more than 2 million people from more than 125 prospective population-based studies (Emerging Risk Factors Collaboration, 2012). That collaboration's focus is on risk factors for cardiovascular disease, but a similar model could be pursued to stratify the data by country in order to study determinants of a broader array of cross-national health differences.

Another opportunity exists in reinvigorating successful surveys that have not been fielded for many years and could be updated. For example, the most recent European data from HBSC survey are from 2005-2006 (Roberts et al., 2009). Data from that survey on children's activity levels may not reflect current behaviors in an era in which use of electronic devices, video games, and smart phones has probably supplanted many outdoor play activities. In the United States, the National Maternal and Infant Health Survey (NMIHS) has not been conducted since 1988. Other similarly dormant surveys could be revived in a new effort to understand cross-national differences.

Discordant family designs are another intriguing model, because they enable statistical comparisons to be made within pairs of twins or clusters of siblings and can help quantify the role of environmental factors in health disparities. This design implicitly controls for confounding from factors that are constant within families and helps to disentangle the different contributions from adult socioeconomic position, childhood socioeconomic position, and genetic factors. Such studies have their limitations, including requirements for large sample sizes, the persistent inability to prove causation or eliminate confounding (e.g., from "ability bias"), random measurement error, and international generalizability (Madsen et al., 2010; McGue et al., 2010). However, these study designs remain useful to consider for appropriate research questions.

Funding agencies can support efforts by researchers to devise creative strategies to design and conduct such studies and to explore alternative study design concepts that can yield more immediate results. Such efforts have led to important findings. For example, important information about the role of smoking in explaining the U.S. health disadvantage among adults aged 50 and older (see Chapter 6) was derived from sophisticated modeling studies that relied on evidence based on smoking-attributable fractions (Preston et al., 2010a, 2010b). Further research could explore the validity of this method of estimating contribution to mortality for a variety of conditions and age groups.

NEW LINES OF INQUIRY

RECOMMENDATION 3: The National Institutes of Health and other research funding agencies should commit to a coordinated portfolio of investigator-initiated and invited research devoted to understanding the factors responsible for the U.S. health disadvantage and potential solutions, including lessons that can be learned from other countries.

For reasons noted throughout this report, a meaningful effort to understand the U.S. health disadvantage requires multiple lines of inquiry and a long time horizon, which in turn will require ongoing financial and political support. Long-term cohort studies, and even elaborate retrospective studies, require an investment of time and committed funding to sustain data collection over many years. Although financial constraints make long-term commitments to research very difficult, the persistence (and worsening) of the U.S. health disadvantage over decades, with its profound human and economic implications, may justify such an investment. The knowledge gleaned from such research has the potential not only to help the United States regain its footing as a leader in health and improve its long-term economic outlook but also to broaden universal understanding of the factors responsible for cross-national health differences.

The cause-effect relationships for some aspects of the social and other nonmedical determinants of health are not yet well established. Knowing and understanding causal pathways is a first step in devising appropriate policies, but the question of attribution remains. As Deaton (2002, p. 15) argues: "policy cannot be intelligently conducted without an understanding of mechanisms; correlations are not enough."[9] A particular suite of questions is especially important in understanding why the United States is experiencing a health disadvantage relative to other countries. We resist the temptation to prioritize the research questions—such as arguing that the study of social determinants is more important than comparing the role of health care systems—because we recognize the fallacy of predicting which lines of scientific inquiry will yield the greatest insights and propel the most important improvements in public health. To fill the gaps in scholarship on the subject, the panel envisions a portfolio of research supported by the National Institutes of Health and other funding entities, including the following:

- international tracking studies that maintain a current epidemiologic dashboard on cross-national patterns in the prevalence of diseases,

---

[9]In circumstances in which a clear cause-effect relationship cannot be linked with a discrete policy intervention, there may still be a valid policy argument for pursuing interventions based on other values, with causal mechanisms only partially understood.

biomarkers, and risk factors; all-cause and cause-specific mortality rates; and the incidence of injuries for key age groups (especially for people under age 50), by administering the same instrument in a standard group of high-income countries;

- further research on how the U.S. health disadvantage is distributed by income and education and what factors may be responsible for the differential influence of income on health;

- long-term prospective cohort studies and other innovative designs (e.g., twin and family studies) that could document the role of antecedent factors (policy, the environment, social factors, behaviors, and health systems) on the U.S. health disadvantage;

- questions about past experiences and exposures (retrospective questions)[10] on population surveys, which can facilitate research on life-course influences (although validation of such questions may require longitudinal studies);

- retrospective studies of historical data and time-series analyses to better elucidate how past conditions in the United States might help explain current health patterns;

- environmental measurement to understand place-based influences on cross-national health disparities, including the effect of land use and urban planning decisions in cities and contextual factors in the large rural areas of the United States;[11] and

- area-based research using geocodable data, geographic information systems (GIS) technology, and a variety of newer approaches based on global positioning.

This report has identified a series of important research questions that deserve exploration, too many to list here. However, it is important to note some of the crucial unanswered research questions about the U.S. health disadvantage:

- What specific factors explain the unfavorable birth outcomes (e.g., high infant mortality rates) experienced in the United States, which exist even after adjusting for race, ethnicity, and maternal education?

---

[10]Examples include the National Survey of Family Growth and National Longitudinal Study of Youth.

[11]Important areas for future research include characterizing levels and distributions of environmental risk factors using comparable measures across countries, documenting inequalities in the distribution of these environmental factors, and identifying the extent to which these environmental factors affect health and the extent to which their effects are modulated by individual-, community-, or country-level factors.

- To what extent does inadequate health care explain why Americans are more likely than their counterparts in peer countries to die from:
  o transportation-related injuries
  o violence
  o noncommunicable diseases
  o communicable diseases
- What specific factors explain why the United States has a higher obesity rate than any other high-income country?
- Do firearms alone explain the excessively high homicide rate in the United States, including murders of children?
- What accounts for the "drug-related" deaths that claim a large proportion of excess years of life lost in the United States?
- What specific factors explain why U.S. adolescents have a higher rate of pregnancies and abortions than their counterparts in other high-income countries?
- Why does the United States have the highest rate of AIDS among OECD countries?
- Is mental illness generally, and are specific mental illnesses, more common in the United States than in other peer countries?
- Why are Americans more likely than people elsewhere to describe their health as good or excellent?
- Why do Americans have a health advantage for certain conditions (e.g., stroke), and can the answers to that phenomenon help explain the causes of health disadvantages for other conditions?
- To what extent do social and economic inequality and low social mobility, independent of absolute poverty, contribute to the aggregate disadvantage in U.S. health?
- To what extent do working conditions in the United States differ from those in peer countries, and how might these contribute to the U.S. health disadvantage?
- Are advantaged Americans—by race (e.g., non-Hispanic whites), education, income, insurance status, and risk factors (e.g., non-smokers, nonobese)—in worse health than their counterparts in other countries as some research suggests?[12] If true, does the find-

---

[12]Four studies have now reported this pattern (Avendano et al., 2009, 2010; Banks et al., 2006; Martinson et al., 2011a), but some of them looked only at education and not other variables (Avendano et al., 2010), or are restricted to comparisons of a narrow age group in only two countries (Banks et al., 2006). Replication with more focused criteria would help confirm the finding.

ing apply to only some of the health conditions for which the United States is experiencing a disadvantage?[13]

- To what extent do epigenetic processes help explain the links between environmental factors and the biological outcomes observed in the U.S. health disadvantage?

- To what extent does the United States lack "protective factors" that buffer the effects of adverse factors, from poverty to adolescent pregnancy?

Stable Funding

Conditions that encourage investigators to commit themselves to long-term lines of inquiry require confidence in a stable source of funding for competitive applications. Solicitations for proposals that are expected to achieve specific aims within 3-5 years discourage ambitious enterprises, like the National Children's Study, which follow individuals over their life courses. Another limitation is lack of "ownership" of the problem: no entity in government is responsible for studying the U.S. health disadvantage.

Creating an institutional home for such research in the Office of the Director of NIH would give the effort a strategic position to coordinate the research portfolios across the agency's 27 institutes and centers, to support inclusion of measures from OECD and European studies in relevant NIH-funded studies, and to solicit proposals for developing measures, data collection instruments, and sampling and administration methodologies. Other potential institutional homes include the Office of Global Health at CDC and the U.S. Department of State. Academic centers of excellence, such as the CDC prevention centers, or academic institutions with expertise on social determinants of health, could be further expanded with appropriate funding to study the underlying causes of the U.S. health disadvantage.

CONCLUSIONS

Perhaps the largest impediments to furthering research on comparisons of cross-national health are the data collection capacities of countries. Not all countries are positioned, or can afford, to administer large-scale population surveys on an annual basis and to maintain publicly available data repositories for research use by the scientific community. Impediments include limited budgets, pragmatic and bureaucratic constraints, and the absence of international collaborative arrangements. In the United

---

[13]Further research is needed to know whether the higher rates of disease for certain conditions in the United States, such as HIV infection, are differentially distributed across social classes.

States, for example, funding for the National Children's Study has met with numerous challenges (Wadman, 2012). Countries can sometimes field large surveys, but not on a regular basis. Political considerations can also influence research by dictating what gets measured and who decides.

Another challenge is the shortage of qualified investigators to conduct such research. Tenuous funding opportunities affect the training pipeline and discourage young investigators from pursuing research careers directed at the issues covered in this report. Dedicated funding for a research portfolio that includes career development awards and support for investigator-initiated studies would help to change this climate, not to mention the motivation to undertake protracted longitudinal studies.

Calls for research in this field by NIH, in particular, would be very persuasive in shifting attitudes in academia, and universities' promotion and tenure committees might adopt more enlightened policies on scholarship that draw more young faculty into careers focused on these issues. Establishing special emphasis panels and study sections composed of reviewers with appropriate expertise and knowledge of the special methodological challenges in conducting cross-national comparative research (and who consider a wider range of analytic approaches, appropriate to the subject, than randomized controlled trials) would also cultivate career growth in this field. These changes in research culture might then find their way to professional journals and persuade editors to involve reviewers with appropriate expertise to make judgments about the worthiness of manuscripts for publication.

For their part, scientists and researchers who conduct studies of the U.S. health disadvantage—a topic with profound public policy implications—need to enhance their skills in communicating scientific findings to general audiences, to policy makers, and to stakeholders. These audiences do not generally read peer-reviewed journals or attend scientific meetings, and findings need to be presented in understandable formats and venues that are relevant to their decisions. Work by Marmot (e.g., 2010), Cutler (e.g., Meara et al., 2008), McGinnis (e.g., McGinnis et al., 2002), Woolf (e.g., Woolf, 2011), Lantz (e.g., Lantz et al., 2010), Schoeni (e.g., Schoeni et al., 2011), and Kindig (e.g., Kindig et al., 2010) among others, illustrates how to present scientific evidence about population health interventions—including returns on investment—to nonscientific policy audiences and decision makers who shape public policies.

The diversity of disciplines needed for studying and understanding the U.S. health disadvantage, ranging from biomedicine to political science, should compel scientists to enthusiastically embrace interdisciplinary collaborations. Medicine, public health, epidemiology, sociology, demography, behavioral science, economics, marketing, and other diverse disciplines and fields can all contribute to the study of the U.S. health disadvantage, but they will need to work through differences in nomenclature, research traditions, data repositories, top journals, and key scientific meetings. Yet the collaboration of experts with different perspectives and skills produces insights that are greater than the sum of the parts and holds the greatest promise in solving the mysteries of the U.S. health disadvantage.

10

Next Steps

The United States and many other nations should take pride in the dramatic gains in life expectancy and disease survival rates that they have achieved in the past century, a credit to major advances in medicine and public health. However, as documented throughout this report, advances in the United States have generally not kept pace with those of many other high-income countries. Using data from a wide range of sources, Part I details these elements of the U.S. health disadvantage:

- Americans have shorter life expectancy than people in almost all other high-income countries.
- This disadvantage has been growing for the past three decades, especially among women.
- This disadvantage is pervasive—it affects all age groups up to the oldest ages and is observed for multiple diseases, biological and behavioral risk factors, and injuries.
- More specifically, when compared with the average of other high-income countries, the United States fares worse in nine health domains:
  o adverse birth outcomes (e.g., low birth weight and infant mortality);
  o injuries, accidents, and homicides;
  o adolescent pregnancy and sexually transmitted infections;
  o HIV and AIDS;
  o drug-related mortality;

o   obesity and diabetes;
o   heart disease;
o   chronic lung disease; and
o   disability.

Part II considers potential explanations for this disadvantage and documents that important *antecedents* of good health are also frequently problematic in the United States:

- The U.S. health system is highly fragmented, with weak public health and primary care components and a large uninsured population. Compared with people in other high-income countries, Americans are more likely to find care inaccessible or unaffordable and to report lapses in the quality and safety of ambulatory care.
- Americans are less likely to smoke and may drink less heavily than their counterparts in other countries; however, they consume the most calories per capita, abuse more prescription and illicit drugs, are less likely to use seatbelts, have more traffic accidents involving alcohol, and own more firearms. U.S. adolescents seem to become sexually active at an earlier age, have more sexual partners, and are more likely to engage in riskier sexual practices than adolescents in other high-income countries.
- The United States has higher rates of poverty and income inequality than do most rich democracies. U.S. children, especially, are more likely than children in many other affluent countries to grow up in poverty, and they are less likely to surpass their parents socio-economically. In addition, although the United States was once the world leader in education, it has not kept pace with many other countries for several decades.
- There are stark differences in land use patterns and transportation systems between the United States and other high-income countries. Americans are less likely than people in other high-income countries to live close to sources of healthy foods. There is also some evidence that residential segregation by socioeconomic position is greater in the United States than in some European countries.

In this chapter we turn to the question of what else the nation should do about the U.S. health disadvantage. We believe that there is sufficient evidence for the country to act now, without waiting for additional research.

The pervasiveness of the U.S. health disadvantage and the fact that it has been worsening for decades leads us to recommend that the nation and its leaders act now in three areas: (1) intensify efforts to pursue existing national health objectives that already target the specific areas in which the United States is lagging behind other high-income countries, (2) alert the public about the problem and stimulate a national discussion about inherent tradeoffs in a range of actions to begin to match the achievements of other high-income nations, and (3) undertake analyses of policy options by studying the policies used by other high-income countries with better health outcomes and their adaptability to the United States.

PURSUE NATIONAL HEALTH OBJECTIVES

RECOMMENDATION 4: The nation should intensify efforts to achieve established national health objectives that are directed at the specific disadvantages documented in this report and that use strategies and approaches that reputable review bodies have identified as effective.

Although the panel was not tasked with evaluating *specific* policies or programs that could address the U.S. health disadvantage we document in this report, the broad outlines are clear enough. The list of factors that may be responsible for the U.S. health disadvantage is daunting, but it is also very familiar to experts in public health and social policy. The list of specific health problems have been long-standing concerns: infant mortality, injuries, violence, adolescent pregnancy, sexually transmitted infections and HIV, drug abuse, obesity, diabetes, heart and lung disease, and disability. Similarly, the underlying contributors are familiar explanations: smoking and other unhealthy behaviors, education, poverty, and the physical and social environment. Many evidence-based strategies to address these specific public health challenges have been identified, and the United States has set national objectives to address them.

Indeed, the very areas in which the United States is deficient relative to other high-income countries are outlined in *Healthy People 2020* (U.S. Department of Health and Human Services, 2012a) (see Table 10-1). The problem areas identified in this report align fully with the 12 priority areas in that report that were subsequently singled out as "critical to the nation's health needs" (Institute of Medicine, 2011g, p. 2). For example, high U.S. transportation-related injury or violent deaths could be ameliorated by efforts that reduce traffic fatalities or homicides. The U.S. ranking as world leader in obesity and the high prevalence of diseases related to obesity (e.g., diabetes) could be helped by initiatives that succeed in lowering the average body mass index of the population.

Similarly, the national prevention strategy of the Surgeon General's

TABLE 10-1 National Health Objectives That Address Specific U.S. Health Disadvantages

Disadvantages Relative to Other High-Income Countries	Examples of Relevant *Healthy People 2020* Objectives
Chapters 1-2: **Shorter Lives, Poorer Health**	
Higher prevalence and death rates from cardiovascular disease	*HDS-2: Reduce coronary heart disease deaths.* *HDS-16: Increase the proportion of adults age 20 and older who are aware of, and respond to, early warning symptoms and signs of a heart attack.*
Higher prevalence and death rates from diabetes	*D-1: Reduce the annual number of new cases of diagnosed diabetes in the population.* *D-3: Reduce the diabetes death rate.*
Higher prevalence and death rates from chronic lung diseases	*RD-10: Reduce deaths from chronic obstructive pulmonary disease (COPD) among adults.*
Higher homicide rates	*IVP-29: Reduce homicides.*
Higher transportation injury fatality rates	*SA-17: Decrease the rate of alcohol-impaired driving (.08 + blood alcohol content [BAC]) fatalities.*
Higher transportation and non-transportation injury fatality rates	*IVP-1: Reduce fatal and nonfatal injuries.*
Higher rate of drug-related deaths	*SA-12: Reduce drug-induced deaths.*
Higher death rates from communicable diseases	
Higher death rates from AIDS	*HIV-3: Reduce the rate of HIV transmission among adolescents and adults.* *HIV-4: Reduce the number of new AIDS cases among adolescents and adults.* *HIV-12: Reduce deaths from HIV infection.*
Higher prevalence of obesity	*NWS-9: Reduce the proportion of adults who are obese.* *NWS-10: Reduce the proportion of children and adolescents who are considered obese.*
Higher prevalence of hypertension	*HDS-5: Reduce the proportion of persons in the population with hypertension.*
Higher prevalence of asthma	*RD-1: Reduce asthma deaths.* *RD-2: Reduce hospitalizations for asthma.*
Higher infant mortality rate	*MICH-1: Reduce the rate of fetal and infant deaths.*
Higher prevalence of low birth weight and prematurity	*MICH-8: Reduce low birth weight (LBW) and very low birth weight (VLBW).* *MICH-9: Reduce preterm births.*
Higher maternal mortality ratio	*MICH-5: Reduce the rate of maternal mortality.*

TABLE 10-1 Continued

Disadvantages Relative to Other High-Income Countries	Examples of Relevant *Healthy People 2020* Objectives
Higher adolescent pregnancy rates	*FP-1: Increase the proportion of pregnancies that are intended.* *FP-8: Reduce pregnancy rates among adolescent females.*
Higher prevalence of sexually transmitted diseases	*STD-1: Reduce the proportion of adolescents and young adults with chlamydia trachomatis infections.* *STD-6: Reduce gonorrhea rates.*
Higher prevalence of mental illness	*MHMD-4: Reduce the proportion of persons who experience major depressive episodes (MDE).*
Chapter 4: **Public Health and Medical Care Systems**	
Low childhood immunization rates	*IID-7: Achieve and maintain effective vaccination coverage levels for universally recommended vaccines among young children.*
Lower health insurance coverage	*AHS-1: Increase the proportion of persons with health insurance.*
Greater difficulties with affordability	*AHS-6: Reduce the proportion of individuals who are unable to obtain or delay in obtaining necessary medical care, dental care, or prescription medicines.*
Less access to primary care/ regular physician	*AHS-3: Increase the proportion of persons with a usual primary care provider.* *AHS-5: Increase the proportion of persons who have a specific source of ongoing care.*
Greater deficiencies in ambulatory care, such as care of diabetes	*HDS-24: Reduce hospitalizations of older adults with heart failure as the principal diagnosis.* *D-5: Improve glycemic control among the population with diagnosed diabetes.* *D-9: Increase the proportion of adults with diabetes who have at least an annual foot examination.* *D-10: Increase the proportion of adults with diabetes who have an annual dilated eye examination.* *D-11: Increase the proportion of adults with diabetes who have a glycosylated hemoglobin measurement at least twice a year.* *D-12: Increase the proportion of persons with diagnosed diabetes who obtain an annual urinary microalbumin measurement.*
Fewer electronic medical records	*HC/HIT-10: Increase the proportion of medical practices that use electronic health records.*

continued

Disadvantages Relative to Other High-Income Countries	Examples of Relevant *Healthy People 2020* Objectives
Fewer registry capacities	C-12: *Increase the number of central, population-based registries from the 50 states and the District of Columbia that capture case information on at least 95 percent of the expected number of reportable cancers.*
Chapter 5: Individual Behaviors	
Higher consumption of calories and dietary fat	NWS-17: *Reduce consumption of calories from solid fats and added sugars in the population age 2 and older.*
Higher prevalence of sedentary activity	PA-1: *Reduce the proportion of adults who engage in no leisure-time physical activity.*
Higher rates of screen time	PA-8: *Increase the proportion of children and adolescents who do not exceed recommended limits for screen time.*
Higher use of drugs	SA-2: *Increase the proportion of adolescents never using substances.* SA-19: *Reduce the past-year nonmedical use of prescription drugs.*
Earlier initiation of adolescent sexual activity and more sexual partners	FP-9: *Increase the proportion of adolescents age 17 and under who have never had sexual intercourse.*
Less use of oral contraceptives and condoms, especially among adolescents	FP-6: *Increase the proportion of females or their partners at risk of unintended pregnancy who used contraception at most recent sexual intercourse.* FP-10: *Increase the proportion of sexually active persons aged 15-19 who use condoms to both effectively prevent pregnancy and provide barrier protection against disease.* FP-11: *Increase the proportion of sexually active persons aged 15 to 19 years who use condoms and hormonal or intrauterine contraception to both effectively prevent pregnancy and provide barrier protection against disease.*
Less use of front seatbelts	IVP-15: *Increase use of safety belts.*
Less use of motorcycle helmets	
More traffic deaths attributable to alcohol	SA-1: *Reduce the proportion of adolescents who report that they rode, during the past 30 days, with a driver who had been drinking alcohol.*

continued

TABLE 10-1 Continued

Disadvantages Relative to Other High-Income Countries	Examples of Relevant *Healthy People 2020* Objectives
Greater access to firearms	*IVP-34: Reduce physical fighting among adolescents.* *IVP-36: Reduce weapon carrying by adolescents on school property.* *AH-11: Reduce adolescent and young adult perpetration of, as well as victimization by, crimes.*
Chapter 6: **Social Factors**	
Higher poverty	
Higher social inequality	
Lower educational performance	*AH-5: Increase educational achievement of adolescents and young adults.* *ECBP-6: Increase the proportion of the population that completes high school education.*
Lower social mobility	
Chapter 7: **Physical and Social** **Environmental Factors**	
Heavier reliance on automobiles	*EH-2: Increase use of alternative modes of transportation for work.*
Lower public transit and non-motorized travel mode shares	
Longer work hours and less employment protection	
Greater residential segregation	
Higher prevalence of food deserts	*NWS-3: Increase the number of states that have state-level policies that incentivize food retail outlets to provide foods that are encouraged by the dietary guidelines.*

NOTES: Examples of the objectives are from the U.S. Department of Health and Human Services (2012). The codes in the table refer to theme areas identified by *Healthy People 2020*.

National Prevention Council targets the same issues responsible for the U.S. health disadvantage (see Box 10-1). Appendix A catalogues the specific policy solutions to address these problems and the supporting evidence and citations provided by the National Prevention Council. Although further research (as outlined in Chapter 9) can help prioritize this list, the largest obstacle to addressing the U.S. health disadvantage is not a lack of evidence

BOX 10-1
Recommendations of U.S. Surgeon General's
National Prevention Council

HEALTHY AND SAFE COMMUNITY ENVIRONMENTS

- Improve quality of air, land, and water.
- Design and promote affordable, accessible, safe, and healthy housing.
- Strengthen state, tribal, local, and territorial public health departments to provide essential services.
- Integrate health criteria into decision making, where appropriate, across multiple sectors.
- Enhance cross-sector collaboration in community planning and design to promote health and safety.
- Expand and increase access to information technology and integrated data systems to promote cross-sector information exchange.
- Identify and implement strategies that are proven to work and conduct research where evidence is lacking.
- Maintain a skilled, cross-trained, and diverse prevention workforce.

CLINICAL AND COMMUNITY PREVENTIVE SERVICES

- Support the National Quality Strategy's focus on improving cardiovascular health.
- Use payment and reimbursement mechanisms to encourage delivery of clinical preventive services.
- Expand use of interoperable health information technology.
- Support implementation of community-based preventive services and enhance linkages with clinical care.
- Reduce barriers to accessing clinical and community preventive services, especially among populations at greatest risk.
- Enhance coordination and integration of clinical, behavioral, and complementary health strategies.

EMPOWERED PEOPLE

- Provide people with tools and information to make healthy choices.
- Promote positive social interactions and support healthy decision making.
- Engage and empower people and communities to plan and implement prevention policies and programs.
- Improve education and employment opportunities.

ELIMINATION OF HEALTH DISPARITIES

- Ensure a strategic focus on communities at greatest risk.
- Reduce disparities in access to quality health care.

- Increase the capacity of the prevention workforce to identify and address disparities.
- Support research to identify effective strategies to eliminate health disparities.
- Standardize and collect data to better identify and address disparities.

TOBACCO-FREE LIVING
- Support comprehensive tobacco-free policies and other evidence-based tobacco control policies.
- Support full implementation of the 2009 Family Smoking Prevention and Tobacco Control Act (Tobacco Control Act).
- Expand use of tobacco cessation services.
- Use media to educate and encourage people to live tobacco free.

PREVENTING DRUG ABUSE AND EXCESSIVE ALCOHOL USE
- Support state, tribal local, and territorial implementation and enforcement of alcohol control policies.
- Create environments that empower young people not to drink or use other drugs.
- Identify alcohol and other drug abuse disorders early and provide brief intervention, referral, and treatment.
- Reduce inappropriate access to, and use of, prescription drugs.

HEALTHY EATING
- Increase access to healthy and affordable foods in communities.
- Implement organizational and programmatic nutrition standards and policies.
- Improve nutritional quality of the food supply.
- Help people recognize and make healthy food and beverage choices.
- Support policies and programs that promote breastfeeding.
- Enhance food safety.

ACTIVE LIVING
- Encourage community design and development that supports physical activity.
- Promote and strengthen school and early learning policies and programs that increase physical activity.
- Facilitate access to safe, accessible, and affordable places for physical activity.
- Support workplace policies and programs that increase physical activity.
- Assess physical activity levels and provide education, counseling, and referrals.

continued

BOX 10-1 Continued

INJURY AND VIOLENCE FREE LIVING

- Implement and strengthen policies and programs to enhance transportation safety.
- Support community and streetscape design that promotes safety and prevents injuries.
- Promote and strengthen policies and programs to prevent falls, especially among older adults.
- Promote and enhance policies and programs to increase safety and prevent injury in the workplace.
- Strengthen policies and programs to prevent violence.
- Provide individuals and families with the knowledge, skills, and tools to make safe choices that prevent violence and injuries.

REPRODUCTIVE AND SEXUAL HEALTH

- Increase utilization of preconception and prenatal care.
- Support reproductive and sexual health services and support services for pregnant and parenting women.
- Provide effective sexual health education, especially for adolescents.
- Enhance early detection of HIV, viral hepatitis, and other sexually transmitted infections and improve linkage to care.

MENTAL AND EMOTIONAL WELL-BEING

- Promote positive early childhood development, including positive parenting and violence-free homes.
- Facilitate social connectedness and community engagement across the lifespan.
- Provide individuals and families with the support necessary to maintain positive mental well-being.
- Promote early identification of mental health needs and access to quality services.

NOTE: See Appendix A for specific policy recommendations and supporting evidence cited by the National Prevention Council.
SOURCE: Adapted from Appendix 5, National Prevention Council (2011).

or uncertainty about effective interventions[1] but limited political support among both the public and policy makers to enact the policies and commit the necessary resources to implement them. As this report is being written, the major debate relevant to this issue is whether to reduce or eliminate discretionary spending on public health and social policy initiatives in an effort to balance budgets and limit the size of government.

Setting aside ideological arguments about whether such curtailments are right or wrong, the evidence reviewed in this report suggests that reduced attention to public health priorities will exacerbate the U.S. health disadvantage, resulting in both the human and economic consequences of excess loss of life. The disturbing findings in this report about the relative disadvantages affecting American youth suggest that inattention to these problems will claim the lives of infants, children, and adolescents and shape the health trajectories of those who survive to adulthood. Evidence from tobacco control efforts and other examples in this report (e.g., German unification; see Chapter 8) underscore that interventions with middle-aged and older adults can also be very instrumental in improving the health of a nation. Thus, all age groups—young and old—are important in reversing the U.S. health disadvantage. It is important to add that the solutions are not to be found solely at the national level. As the discussion in Box 10-2 emphasizes, meaningful solutions to the nation's health disadvantage requires the involvement of states and local communities.

ALERT THE PUBLIC

RECOMMENDATION 5: The philanthropy and advocacy communities should organize a comprehensive media and outreach campaign to inform the general public about the U.S. health disadvantage and to stimulate a national discussion about its implications for the nation.

---

[1]The panel acknowledges that the quality of supporting evidence for the listed interventions varies. Some of the policy solutions have been the subject of randomized trials and other useful scientific study designs that document their effectiveness in improving outcomes. Both U.S. and international review groups have conducted numerous systematic reviews and rated the strength of evidence for these strategies: see, especially, Campbell Collaboration (2012), Cochrane Library (2012), and Community Preventive Services Task Force (2012). However, the evidence that other policy solutions are effective is less developed. Some evidence is circumstantial or ecological: health outcomes may have improved in a country after the introduction of a policy, but evidence of a causal relationship may be lacking. And debates continue about proper outcomes for measuring health: for example, some critics argue that mortality rates or life expectancy are less meaningful than measures of health-related quality of life, such as quality-adjusted life years (Institute of Medicine, 2011e), and they fault national health objectives that lack such metrics and do not set specific goals for reducing disparities.

BOX 10-2
Roles for Governments and
Nongovernment Actors at All Levels

The steps advocated by the panel to meet the health objectives that address areas of the U.S. health disadvantage and to stimulate a national discussion on these issues are not activities for the federal government alone. Quite to the contrary, productive discussion, design, and implementation of on-the-ground strategies to address the U.S. health disadvantage often require action at the regional, state, and local levels and involvement of local employers, health care institutions, public health officials, school boards, park authorities, civic groups, retailers, restaurants, developers, media, and other such stakeholders (see Institute of Medicine, 2009b).

In the United States, the statutory authority for government to address a variety of contributing factors, from motor vehicle safety to education policy, rests with state and local governments. For some years, in fact, states and localities throughout North America have emerged as laboratories for devising and testing solutions within a "health in all policies" framework. For example, important efforts are under way in the Bay Area of California, Denver, Seattle, Vancouver, New York City, Somerville (MA), and Atlanta, where health officials are collaborating with community partners to address a range of social and economic factors that affect health. The federal government is recognizing this work with Community Transformation Grants and Communities Putting Prevention to Work grants, funded by the Centers for Disease Control and Prevention (CDC) to encourage pursuit and testing of creative solutions to health problems. At the same time, the federal government is making its own inroads by forging cross-Cabinet collaborations aimed at achieving these vital goals, such as healthy housing and combating childhood obesity.

Of particular concern to the panel is whether the public is fully aware of the U.S. health disadvantage. The depth and breadth of the problem, as documented in this report, came as a surprise to many of us. Although we do not know of survey or poll data that gauge Americans' awareness of their poor health rankings relative to other high-income countries, we suspect that the information detailed in this report is not widely known.

Although people are increasingly aware that the U.S. health care system is costly, inefficient, and out of reach for many Americans (Pew Research Center, 2009), many people may still believe that their own health—if not their health care—is the best in the world. The public likely has little awareness that the United States ranks unfavorably on so many antecedents of disease. For example, the average American may not realize that the country has one of the highest child poverty rates of developed countries and has less success in promoting social mobility (see Chapter 6). Many people may also mistakenly attribute unfavorable health statistics to the conditions of poor, unemployed, or uninsured Americans, when several studies now suggest that even advantaged Americans are in poorer health than their counterparts in other countries. In short, we believe that most Americans do not realize that their expensive, world-class health care system—and the very large economy that supports it—has not enabled them to keep pace with the health gains achieved by people in other high-income countries.

With this in mind, the panel believes it is critically important to share our findings not only with relevant professional audiences, but also with the public at large. We believe that doing so will serve to build knowledge of the facts, correct misperceptions, and raise awareness of the health and economic consequences of the nation's current course.

To that end, although this publication will be widely distributed and made available online, a broader, concerted effort will also be needed to reach the general public and policy makers. Such an effort could include a comprehensive communications strategy[2] that identifies a broad range of target audiences and packages the report's key messages in formats that are appropriate and accessible. To broadly spread the word, it could focus on traditional media (e.g., newspaper articles and television and radio coverage), as well as new media (e.g., social networking sites, community listservs, and information-sharing vehicles, such as blogs, Facebook, and Twitter).

The panel believes that a national discussion on the implications of the U.S. health disadvantage is an important step, and one that is long overdue. U.S. rankings on many health indicators have been deteriorating

---

[2]Although the government has considerable resources that could be devoted to a communication effort on this scale, the panel believes that it may be more appropriate and effective for independent, objective, nonpartisan organizations to organize a communications effort on this topic. For example, this topic speaks to deficiencies the United States faces relative to other countries, a message that may be politically awkward for an administration to disseminate to a domestic or international audience. Yet the public deserves the facts. Thus, the panel believes an independent scientific body, with support from one or more foundations or advocacy organizations concerned with public health (perhaps collaborating as a consortium to share resources), should spearhead a communications campaign. We also think the National Institutes of Health (NIH) would be an ideal entity to take responsibility for disseminating the findings of this report to colleagues and leaders on the NIH campus, to other agencies in the U.S. Department of Health and Human Services, and to the scientific community more broadly. We hope this effort would spur discussion of how to revise solicitations for future research and the composition of study sections to advance scholarship in this field. The National Institute on Aging has been a leader at NIH in studying cross-national health differences.

for decades. As shown by the morbidity and mortality data in this report, this information has not yet been sufficient to arrest or reverse the decline. The panel believes that a national discussion aimed at building consensus is a critical step. Because the factors and determinants underlying the U.S. health disadvantage are far reaching and complex, they raise important questions about national strategies, governance, and policies. A concerted effort is needed to present the evidence to the public and policy makers in a way that is accurate, engaging, and convincing and that stimulates thoughtful discussion of the implications.

The goal of a national discussion would be to publicly consider a wide range of tradeoffs. For example, making meaningful progress on our health rankings might require the adoption of policies and practices that give greater priority to public health but impose restrictions on individuals or businesses. As described in Chapter 8, such steps—which some other countries have used successfully—may be at odds with traditional American beliefs (e.g., limited government, free enterprise, individual rights and freedoms); they might be seen as undermining constitutional protections (e.g., the right to bear arms), or as contravening religious and moral beliefs (e.g., the use of birth control).

A national discussion could help determine whether the American people deem such tradeoffs acceptable. It could explore whether this poses a false choice, whether models and practices used overseas could be adapted ("Americanized") or, better yet, whether new solutions could be devised that better conform to American sensibilities. In situations where individual liberties or societal values are in conflict with policies that can produce better health outcomes, a thoughtful national discussion could help Americans consider what investments and compromises they are willing to make to begin to overcome the U.S. health disadvantage.

EXPLORE INNOVATIVE POLICY OPTIONS

RECOMMENDATION 6: The National Institutes of Health or another appropriate entity should commission an analytic review of the available evidence on (1) the effects of policies (including social, economic, educational, urban and rural development and transportation, health care financing and delivery) on the areas in which the United States has an established health disadvantage, (2) how these policies have varied over time across high-income countries, and (3) the extent to which those policy differences may explain cross-national health differences in one or more health domains. This report should be followed by a series of issue-focused investigative studies to explore why the United States experiences poorer outcomes than other countries in the specific areas documented in this report.

As noted throughout this report, the areas in which the United States has a health disadvantage are familiar challenges that the nation has been trying to address for decades. There is no shortage of good ideas on how to address the obesity epidemic and control diabetes, to control violent crime and homicides, to create jobs and enhance the economic stability of American families, and to improve the quality of education in the United States. There have been many blue-ribbon reports, strategic plans, and even international charters that list best practices and policy recommendations—too many to cite here.

Yet the panel believes that the United States can learn more by studying the policies that have been used by those countries that have been outpacing the United States on both health outcomes and social factors related to health. Chapter 8 engaged in "informed speculation" about whether the health advantages enjoyed by these countries can be traced to styles of governance or policies adopted in those countries and offered suppositions about dominant values in those societies and their potential links to observed outcomes. However, the panel lacked the time and was not charged to undertake a systematic examination of the nature and history of the policies that exist in the 16 peer countries with which the United States was compared.

Nor did this panel have the appropriate qualifications for such a study. This panel was composed primarily of demographers, epidemiologists, physicians, and social scientists. Although it did include several European and foreign-born experts, its members did not include authorities from outside the United States with extensive knowledge of the policy landscape in comparable countries.[3]

The panel therefore recommends that an appropriate organization or federal or international agency undertake a follow-up effort that involves appropriate experts from many of the high-income countries considered in this report.[4] In some ways, what we envision would amount to the third report in a trilogy. The first report by the National Research Council (2011) drew attention to the growing U.S. mortality disadvantage among adults age 50 and older. This second report documents the significant health disadvantage for Americans under age 50 and offers a systematic examination of some of the potential causes. It moves "upstream" and highlights the potential importance of policy influences on health, but the panel was unable to examine in any detail whether specific cross-national

---

[3]For example, the panel did not include officials from health ministries or political scientists from Japan or Europe.

[4]The effort should also include U.S. experts who understand the opportunities and challenges that come with translating policies from one place to another—whether cross-nationally, across states, or from one local area to another.

policy differences might help explain the cross-national health differences documented in Part I of this report.

A third report could complete this analysis by evaluating the evidence for health-promoting policies that other top performing countries have adopted and identifying strategies that offer promise in the United States. These strategies could then be assessed for their feasibility or adaptability to the U.S. context. The research community can also adapt and test the effectiveness of these strategies in U.S. settings, through demonstration projects, policy research, and intervention studies. The panel notes that the scope of the proposed exercise would not be trivial if it is to cover policies from a life-course perspective. Besides health, the report would need to examine specific policies related to education, family support, workplace benefits, and other social factors that affect health outcomes, as well as contextual factors and other secular trends that bear on all countries' health patterns (e.g., globalization, population aging).

Whereas the proposed report would focus on cross-cutting policies that appear to improve a country's health outcomes (across multiple diseases and conditions), we believe it would also be of great value to launch a series of issue-focused reports on the specific conditions (diseases and injuries) for which the United States has a health disadvantage to identify useful policies to address those health conditions. The panel was impressed with the value of the 2011 report issued by the Transportation Research Board (TRB), *Achieving Traffic Safety Goals in the United States: Lessons from Other Nations* (discussed in Chapter 8). For one of the prime areas of U.S. health disadvantage—traffic fatalities—the TRB study considered how other countries have achieved lower death rates. The TRB authoring committee included experts in safety research, public policy, evaluation, and public administration, as well as members of state legislatures. That committee included a transportation specialist from the World Bank, current and former officials of federal and state transportation agencies in the United States, a state police commissioner, economists, and others with special knowledge of how other countries achieve lower traffic fatalities. As noted in Chapter 8, the TRB report's analytic approach and findings mirror those of this panel, but the report also provides specific guidance that the U.S. transportation community, policy makers, and traffic safety advocates can use to improve conditions in the United States.

Thus, the panel recommends a series of similar issue-focused investigative studies to seek explanations for the nine specific health disadvantages identified in this report: (1) adverse birth outcomes; (2) injuries, accidents, and homicides; (3) adolescent pregnancy and sexually transmitted infections; (4) HIV and AIDS; (5) drug-related mortality; (6) obesity and diabetes; (7) cardiovascular disease; (8) chronic lung disease; and (9) disability. The panels commissioned for each report would be composed of experts on the topic, with knowledge of relevant data sources, clinical practices, and policy strategies for addressing the conditions in other rich nations (or knowledgeable contacts in each country for obtaining this information). Like the TRB report, such studies would seek to find discrete explanations for how and why other high-income countries are achieving lower morbidity and mortality rates for the specific conditions under study and perhaps model or estimate the predicted health and economic effects of alternative policy strategies that target different components of the causal chain.

These issue-focused inquiries are likely to uncover many of the same general themes raised in this report. For example, it is likely that social factors or the lack of universal health insurance in the United States will be found to interfere with access to health care for many of the above conditions. But these focused inquiries will also be able to "unpack" the specifics. They can examine, for example, whether strategies for treating drug abuse or controlling access to prescription opioids account for lower drug-related deaths in other countries. The inquiry into adverse birth outcomes can attempt to tease out the specific reasons that U.S. infant mortality rates have not kept pace with other countries for decades by examining differences in not only prenatal or newborn care, but also preconception and prenatal efforts in public health or social policy to lessen maternal risks for adverse birth outcomes.

Our vision is a published series of issue-specific reports that would be released over several years, with each study building on the findings and insights of those coming before it. The first report could be commissioned immediately. The series would support a critical ongoing cycle of evidence production, guidance regarding effective policies and practices, implementation and evaluation, and learning from practice. The rollout of these reports over time will not only deliver practical solutions to enable the United States to begin to turn the tide in specific domains in which there is a disadvantage, but it will also provide a basis for steady and continued public attention on this issue. It is important to this panel that the public and the nation's leaders maintain awareness of the U.S. health disadvantage and not lose momentum in efforts to find solutions.

LOOKING AHEAD

Although the evidence reviewed in this report documents a U.S. health disadvantage that spans decades and continues to trend downward, no one knows for certain what will come next. The health trajectory of the United States and many other countries will be affected by known global trends—such as climate change, dwindling sources of energy, military conflicts, and overcrowding—but also by unforeseen influences yet to emerge. However, almost all trend lines indicate that, in the absence of corrective action, the

U.S. health disadvantage relative to other high-income countries will continue to worsen, as it has for years.

A number of factors support the prediction that the health of Americans will continue to slip behind that of people in other countries. For example, to the extent that education of today's youth predicts the health of tomorrow's adults, the failure of the United States to keep pace with the educational advances occurring in other countries is a discouraging sign. So is the continuing rise of income inequality in the United States, the persistence of poverty (especially child poverty) at rates that exceed those of most other rich nations, and the relative lack of social mobility. The increasing prevalence of obesity and diabetes among U.S. children at rates that exceed those of other countries is certainly an ominous trend in a country whose adults already suffer from high rates of cardiovascular disease.

Other factors, however, could mitigate these trends and perhaps improve the rankings of the United States relative to other countries. For example, there is some evidence that the obesity epidemic is beginning to stabilize in the United States (Ogden et al., 2012a) while it is continuing to spread globally (Finucane et al., 2011). The prevalence of smoking in the United States has fallen considerably while rates in other countries continue to increase (OECD, 2011b).[5] These trends might temper the excessive burden of chronic disease in the United States relative to other countries, especially as today's middle-aged adults (the beneficiaries of lower smoking rates) become older adults.[6] And as these behaviors begin to affect morbidity and mortality in other countries, it is possible that they may "catch up" with the United States, in a negative sense, and so improve the country's relative ranking. However, such an "improvement" would mean only that progress in safeguarding public health is faltering globally, and that would hardly be good news for the United States.

Indeed, the important point about the U.S. health disadvantage is not that the United States is losing a competition with other countries, but that Americans are dying and suffering at rates that are demonstrably unnecessary. The fact that other high-income countries have better health outcomes

---

[5]The rate of decrease in tobacco use among young adults has decreased in recent years in the United States, and smokeless tobacco use has increased (U.S. Department of Health and Human Services, 2012b). These trends could diminish the salutary effects of tobacco control on the U.S. health disadvantage of the next generation. Nonetheless, it bears noting that smoking rates among U.S. youth are generally lower than rates among their peers in other high-income countries.

[6]As noted in Chapter 5, Wang and Preston (2009) predicted that deaths attributable to smoking among men would decline relatively soon but that improvements for women would come later. Other authors, however, have questioned whether the obesity epidemic will outweigh any gains in life expectancy achieved by lower smoking rates (Stewart et al., 2009). Furthermore, specific aspects of the U.S. health disadvantage, such as the high prevalence of low-birth-weight babies, may persist if smoking rates remain high for women of childbearing age.

is evidence that better health is achievable for Americans. The same lesson will apply to other countries if epidemiologic trends cause health improvements in their societies to falter, because they too will know that they are capable of achieving better health outcomes for their populations.

That the health of Americans does not meet the standard that now exists in other rich nations is a tragedy for all age groups, but especially for children. Behind the statistics detailed in this report are the faces of young people—infants, children, and adolescents—who are unwell and dying early because conditions in this country are not as favorable as those in other countries. Overall, young Americans are entering adulthood in poorer health than their counterparts in other countries and therefore face a future with greater risks of disease and the other life challenges they bring than did their parents.

This alone is reason enough for concern, but the nation's leaders—in government and business—also understand what the nation can expect from a future generation of workers, executives, and military recruits whose illnesses and socioeconomic disadvantages compromise their productivity and require more intensive health care. This forecast has obvious implications for national security and for the economy—the price tag of the U.S. health disadvantage is unlikely to be small.

With this many lives and dollars at stake, we believe the U.S. health disadvantage is a problem the country can no longer afford to ignore.

References and Bibliography

Aarts, L., and van Schagen, I. (2006). Driving speed and the risk of road crashes: A review. *Accident, Analysis and Prevention, 38*(2), 215-224.

Acemoglu, D. and Johnson, S. (2007). Disease and development: The effect of life expectancy on economic growth. *Journal of Political Economy, 115*(6), 925-985.

Acemoglu, D., and Robinson, J.A. (2012). *Why nations fail: The origins of power, prosperity and poverty.* New York: Crown.

Acevedo-Garcia, D., Osypuk, T.L., McArdle, N., and Williams, D.R. (2008). Toward a policy-relevant analysis of geographic and racial/ethnic disparities in child health. *Health Affairs, 27*(2), 321-333.

Acs, G. (2011). *Downward mobility from the middle class: Waking up from the American dream.* Washington, DC: PEW Charitable Trusts, Economic Mobility Project.

Adler, N.E., and Rehkopf, D.H. (2008). U.S. disparities in health: Descriptions, causes, and mechanisms. *Annual Review of Public Health, 29*, 235-252.

Adler, N.E., and Stewart, J. (2010). Health disparities across the lifespan: Meaning, methods, and mechanisms. *Annals of the New York Academy of Sciences, 1,186*(1), 5-23.

Agency for Healthcare Research and Quality. (2011). *National healthcare disparities report: 2010.* Rockville, MD: Agency for Healthcare Research and Quality.

Agency for Healthcare Research and Quality. (2012). *National healthcare quality report: 2011.* Rockville, MD: Agency for Healthcare Research and Quality.

Ahern, J., and Galea, S. (2011). Collective efficacy and major depression in urban neighborhoods. *American Journal of Epidemiology, 173*(12), 1,453-1,462.

Ahern, J., Hubbard, A., and Galea, S. (2009). Estimating the effects of potential public health interventions on population disease burden: A step-by-step illustration of causal inference methods. *American Journal of Epidemiology, 169*(9), 1,140-1,147.

Alderson, A.S., and Nielsen F. (2002). Globalization and the great U-turn: Income inequality trends in 16 OECD countries. *American Journal of Sociology, 107*(5), 1,244-1,299.

Aliyu, M.H., Luke, S., Kristensen, S., Alio, A.P., and Salihu, H.M. (2010). Joint effect of obesity and teenage pregnancy on the risk of preeclampsia: A population-based study. *Journal of Adolescent Health, 46*(1), 77-82.

292

Alley, D.E., Lloyd, J., and Shardell, M. (2010). Can obesity account for cross-national differences in life expectancy trends? In National Research Council, *International differences in mortality at older ages: Dimensions and sources* (pp. 164-192). E.M. Crimmins, S.H. Preston, and B. Cohen (Eds.), Panel on Understanding Divergent Trends in Longevity in High-Income Countries. Committee on Population. Division of Behavioral and Social Sciences and Education. Washington, DC: The National Academies Press.

Allison, D.B., Fontaine, K.R., Manson, J.E., Stevens, J., and VanItallie, T.B. (1999). Annual deaths attributable to obesity in the United States. *Journal of the American Medical Association, 282*, 1,530-1,538.

Altman, S.H., and Shactman, D. (2011). *Power, politics, and universal health care: The inside story of a century-long battle.* Amherst, NY: Prometheus Books.

American Academy of Family Physicians. (2009). *Family physician workforce reform: Recommendations of the American Academy of Family Physicians.* Leawood, KS: American Academy of Family Physicians.

American Anthropological Association. (1998). *Statement on "race."* Available: http://www.aaanet.org/stmts/racepp.htm [April 2012].

An, J., Braveman, P., Dekker, M., Egerter, S., and Grossman-Kahn, R. (2011). *Work matters for health.* Princeton, NJ: Robert Wood Johnson Foundation.

Anand, S., and Ravallion, M. (1993). Human development in poor countries: On the role of private incomes and public services. *Journal of Economic Perspectives, 7*(1), 133-150.

Anda, R., Felitti, V., Bremner, J., Walker, J., Whitfield, C., Perry, B., Dube, S., and Giles, W. (2006). The enduring effects of abuse and related adverse experiences in childhood. *European Archives of Psychiatry and Clinical Neuroscience, 256*(3), 174-186.

Anderson, G., and Squires, D. (2010). *Measuring the U.S. health care system: A cross-national comparison.* New York: Commonwealth Fund.

Anderson, L., and McQueen, D. (2009). Informing public health policy with the best available evidence. In A. Killoran and M. Kelly (Eds.), *Evidence-based public health: Effectiveness and efficiency* (pp. 436-447). Oxford, UK: Oxford University Press.

Anderson, L.M., Quinn, T.A., Glanz, K., Ramirez, G., Kahwati, L.C., Johnson, D.B., Buchanan, L.R., Archer, W.R., Chattopadhyay, S., Kalra, G.P., Katz, D.L., and Task Force on Community Preventive Services. (2009). The effectiveness of worksite nutrition and physical activity interventions for controlling employee overweight and obesity: A systematic review. *American Journal of Preventive Medicine, 37*(4), 340-357.

Andrade, L., Caraveo-Anduaga, J.J., Berglund, P., Bijl, R.V., De Graaf, R., Vollebergh, W., Dragomirecka, E., Kohn, R., Keller, M., Kessler, R.C., Kawakami, N., Kilic, C., Offord, D., Ustun, T.B., and Wittchen, H.U. (2003). The epidemiology of major depressive episodes: Results from the International Consortium of Psychiatric Epidemiology (ICPE) surveys. *International Journal of Methods in Psychiatric Research, 12*(1), 3-21.

Aneshensel, C.S., and Sucoff, C.A. (1996). The neighborhood context of adolescent mental health. *Journal of Health and Social Behavior, 37*(4), 293-310.

Armingeon, K., Careja, A., Weisstanner, D., Engler, S., Potolidis, P., and M. Gerber. (2012). *Comparative political data sets III: 1990-2010.* Institute for Political Science, University of Berne. Available: http://www.ipw.unibe.ch/content/team/klaus_armingeon/comparative _political_data_sets/index_ger.html [June 2012].

Arnesen, E., and Forsdahl, A. (1985). The Tromsø heart study: Coronary risk factors and their association with living conditions during childhood. *Journal of Epidemiology and Community Health, 39*(3), 210-214.

Aron, L.Y. (2006). Rural homelessness in the United States. In P. Milborne and P.J. Cloke (Eds.). *International perspectives on rural homelessness* (pp. 9-24). New York: Routledge.

Aron, L.Y., Honberg, R., Duckworth, K., et al. (2009). *Grading the states 2009: A report on America's health care system for adults with serious mental illness,* Arlington, VA: National Alliance on Mental Illness.

Ashcraft, A., and Lang, K. (2006). *The consequences of teenage childbearing.* Cambridge, MA: National Bureau of Economic Research.

Aspalter, C. (2006). The East Asian welfare model. *International Journal of Social Welfare, 15*(4), 290-301.

Auchincloss, A.H., and Diez Roux, A.V. (2008). A new tool for epidemiology: The usefulness of dynamic-agent models in understanding place effects on health. *American Journal of Epidemiology, 168*(1), 1-8.

Avendano, M., and Kawachi, I. (2011). Invited commentary: The search for explanations of the American health disadvantage relative to the English. *American Journal of Epidemiology, 173*(8), 866-869.

Avendano, M., Glymour, M.M., Banks, J., and Mackenbach, J.P. (2009). Health disadvantage in U.S. adults aged 50 to 74 years: A comparison of the health of rich and poor Americans with that of Europeans. *American Journal of Public Health, 99*(3), 540-548.

Avendano, M., Kok, R., Glymour, M., Berkman, L., Kawachi, I., Kunst, A., Mackenbach, J., and Eurothine Consortium. (2010). In National Research Council, *International differences in mortality at older ages: Dimensions and sources* (pp. 313-332). E.M. Crimmins, S.H. Preston, and B. Cohen (Eds.), Panel on Understanding Divergent Trends in Longevity in High-Income Countries. Committee on Population. Division of Behavioral and Social Sciences and Education. Washington, DC: The National Academies Press.

Babey, S.H., Diamant, A.L., Hastert, T.A., Harvey, S., Goldstein, H., Flournoy, R., Banthia, R., Rubin, V., and Treuhaft, S. (2008). *Designed for disease: The link between local food environments and obesity and diabetes.* Los Angeles: University of California, Center for Health Policy Research.

Baer, J.D., and McGrath, D. (2007). *The reading literacy of U.S. fourth-grade students in an international context: Results from the 2001 and 2006 Progress in International Reading Literacy Study (PIRLS).* Washington, DC: National Center for Education Statistics.

Baker, D.W., Einstadter, D., Husak, S.S., and Cebul, R.D. (2004). Trends in postdischarge mortality and readmissions: Has length of stay declined too far? *Archives of Internal Medicine, 164*(5), 538-544.

Baldasano, J.M., Valera, E., and Jimenez, P. (2003). Air quality data from large cities. *Science of the Total Environment, 307*(1-3), 141-165.

Baldwin, W.H., and Cain, V.S. (1980). The children of teenage parents. *Family Planning Perspectives, 12*(1), 34-43.

Bambra, C. (2005). Cash versus services: "Worlds of welfare" and the decommodification of cash benefits and health care services. *Journal of Social Policy, 34*(2), 195-213.

Bambra, C. (2006). Decommodification and the worlds of welfare revisited. *Journal of European Social Policy, 16*(1), 73-80.

Bambra, C. (2007). Going beyond the three worlds of welfare capitalism: Regime theory and public health. *Journal of Epidemiology and Community Health, 61,* 1,098-1,102.

Brambra, C. (2011). *Work, worklessness, and the political economy of health.* New York: Oxford University Press.

Bambra, C., and Beckfield, J. (2012). *Institutional arrangements as candidate explanations for the U.S. mortality disadvantage.* Unpublished paper prepared for the NAS/IOM Panel on Understanding Cross-National Health Differences Among High-Income Countries. Department of Geography, Durham University and Department of Sociology, Harvard University.

Bambra, C., and Eikemo, T.A. (2009). Welfare state regimes, unemployment and health: A comparative study of the relationship between unemployment and self-reported health in 23 European countries. *Journal of Epidemiology and Community Health, 63*(2), 92-98.

Bambra, C., Gibson, M., Sowden, A., Wright, K., Whitehead, M., and Petticrew, M. (2010). Tackling the wider social determinants of health and health inequalities: Evidence from systematic reviews. *Journal of Epidemiology and Community Health, 64*(4), 284-291.

Banks, J., and Smith, J.P. (2011). *International comparisons in health economics: Evidence from aging studies*. Santa Monica, CA: RAND Corporation.

Banks, J., Marmot, M., Oldfield, Z., and Smith, J.P. (2006). Disease and disadvantage in the United States and in England. *Journal of the American Medical Association, 295*(17), 2,037-2,045.

Banks, J., Berkman, L., Smith, J.P., Avendano, M., and Glymour, M. (2010). Do cross-country variations in social integration and social interactions explain differences in life expectancy in industrialized countries? In National Research Council, *International differences in mortality at older ages: Dimensions and sources* (pp. 217-256). E.M. Crimmins, S.H. Preston, and B. Cohen (Eds.), Panel on Understanding Divergent Trends in Longevity in High-Income Countries. Committee on Population. Division of Behavioral and Social Sciences and Education. Washington, DC: The National Academies Press.

Barefoot, J.C., Maynard, K.E., Beckham, J.C., Brummett, B.H., Hooker, K., and Siegler, I.C. (1998). Trust, health, and longevity. *Journal of Behavioral Medicine, 21*(6), 517-526.

Barker, D.J.P. (1998). *Mothers, babies and health in later life* (2nd ed.). Edinburgh, UK: Churchill Livingstone.

Barker, D.J.P. (1999). Early growth and cardiovascular disease. *Archives of Disease in Childhood, 80*(4), 305-307.

Barker, D.J.P., Godfrey, K.M., Gluckman, P.D., Harding, J.E., Owens, J.A., and Robinson, J.S. (1993). Fetal nutrition and cardiovascular disease in adult life. *Lancet, 341*(8,850), 938-941.

Bartley, M., and Owen, C. (1996). Relation between socioeconomic status, employment, and health during economic change, 1973-1993. *British Medical Journal, 313*, 445-449.

Bartley, M., Blane, D., and Montgomery, S. (1997). Health and the life course: Why safety nets matter. *British Medical Journal, 314*, 1,194-1,196.

Bassett, D.R., Jr., Pucher, J., Buehler, R., Thompson, D.L., and Crouter, S.E. (2008). Walking, cycling, and obesity rates in Europe, North America, and Australia. *Journal of Physical Activity and Health, 5*(6), 795-814.

Bauman, A., Bull, F., Chey, T., Craig, C.L., Ainsworth, B.E., Sallis, J.F., Bowles, H.R., Hagstromer, M., Sjostrom, M., Pratt, M., and IPS Group. (2009). The international prevalence study on physical activity: Results from 20 countries. *International Journal of Behavioral Nutrition and Physical Activity, 6*(21).

Bay Area Regional Health Inequities Initiative. (2008). *Health inequalities in the Bay Area*. Oakland, CA: Bay Area Regional Health Inequities Initiative. Available: http://www.barhii.org/press/download/barhii_report08.pdf [June 2012].

Beaulac, J., Kristjansson, E., and Cummins, S. (2009). A systematic review of food deserts, 1966-2007. *Preventing Chronic Disease, 6*(3), A105.

Beckett, M. (2000). Converging health inequalities in later life—An artifact of mortality selection. *Journal of Health and Social Behavior, 41*, 106-119.

Beckfield, J. (2004). Does income inequality harm health? New cross-national evidence. *Journal of Health and Social Behavior, 45*(3), 231-248.

Beckfield, J., and Krieger, N. (2009). Epi + demos + cracy: Linking political systems and priorities to the magnitude of health inequities—Evidence, gaps, and a research agenda. *Epidemiologic Reviews, 31*(1), 152-177.

Bellinger, D.C. (2008). Very low lead exposures and children's neurodevelopment. *Current Opinion in Pediatrics, 20*(2), 172-177

Bergen, G., Chen, L.H., Warner, M., and Fingerhut, L.A. (2008). *Injury in the United States: 2007 chartbook*. National Center for Health Statistics. Washington, DC: U.S. Department of Health and Human Services.

Bernard, S.J., Paulozzi, L.J., and Wallace, D.L. (2007). Fatal injuries among children by race and ethnicity—United States, 1999-2002. *Morbidity and Mortality Weekly Report Surveillance Summaries, 56*, 1-16.

Berwick, D.M., and Hackbarth, A.D. (2012). Eliminating waste in U.S. health care. *Journal of the American Medical Association, 307*(14), 1,513-1,516.

Berwick, D.M., Nolan, T.W., and Whittington, J. (2008). The triple aim: Care, health, and cost. *Health Affairs, 27*(3), 759-769.

Billari, F.C. (2009). The life course is coming of age. *Advances in Life Course Research, 14*(3), 83-86.

Black, N. (1996). Why we need observational studies to evaluate the effectiveness of health care. *British Medical Journal, 312*(7,040), 1,215-1,218.

Blakely, T.A., Kennedy, B.P., and Kawachi, I. (2001). Socioeconomic inequality in voting participation and self-rated health. *American Journal of Public Health, 91*(1), 99-104.

Blanden, J., Gregg, P., and Machin, S. (2005). *Intergenerational mobility in Europe and North America: A report supported by the Sutton Trust.* London, UK: Centre for Economic Performance, London School of Economics and Political Science.

Bleich, S.N., Jarlenski, M.P., Bell, C.N., and LaVeist, T.A. (2012). Health inequalities: Trends, progress, and policy. *Annual Review of Public Health, 33*, 7-40.

Blencowe, H., Cousens, S., Oestergaard, M.Z., Chou, D., Moller, A.B., Narwal, R., Adler, A., Vera Garcia, C., Rohde, S., Say, L., Lawn, J.E. (2012). National, regional, and worldwide estimates of preterm birth rates in the year 2010 with time trends since 1990 for selected countries: a systematic analysis and implications. *Lancet, 379*(9,832), 2,162-2,172.

Bloom, D.E., Canning, D., and Sevilla, J. (2004). The effect of health on economic growth: A production approach. *World Development, 32*(1), 1-13.

Boehm, J., and Kubzansky, L. (2012). The heart's content: The association between positive psychological well-being and cardiovascular health. *Psychological Bulletin, 138*(4), 655-691.

Bogdanor V. (Ed.). (2005). *Joined-up government.* Oxford, UK: Oxford University Press.

Bonczar, T.P. (2003). *Prevalence of imprisonment in the U.S. Population, 1974-2001.* Office of Justice Programs. Washington, DC: U.S. Department of Justice.

Bongaarts, J. (2006). How long will we live? *Population and Development Review, 32*(4), 605-628.

Bonnefoy, J., Morgan, A., Kelly, M.P., Butt, J., and Bergman, V. (2007). *Constructing the evidence base on the social determinants of health: A guide.* Geneva, Switzerland: WHO Commission on Social Determinants of Health, Measurement and Evidence Knowledge Network.

Bonoli, G. (1997). Classifying welfare states: A two-dimension approach. *Journal of Social Policy, 26*, 351-372.

Bots, M.L., and Grobbee, D.E. (1996). Decline of coronary heart disease mortality in the Netherlands from 1978 to 1985: Contribution of medical care and changes over time in presence of major cardiovascular risk factors. *Journal of Cardiovascular Risk, 3*(3), 271-276.

Bowles, S., Gintis, H., and Osborne Groves, M. (Eds.). (2005). *Unequal chances: Family background and economic success.* New York: Russell Sage Foundation.

Bradley, E.H., Elkins, B.R., Herrin, J., and Elbel, B. (2011). Health and social services expenditures: Associations with health outcomes. *British Medical Journal Quality and Safety, 20*(10), 826-831.

Brady, D. (2005). The welfare state and relative poverty in rich western democracies, 1967-1997. *Social Forces, 83*(4), 1,329-1,364.

Braveman, P.A., and Barclay, C. (2009). Health disparities beginning in childhood: A life-course perspective. *Pediatrics, 124*(Suppl. 3), S163-S175.

Braveman, P.A., and Egerter, S. (2008). *Overcoming obstacles to health.* Report from the Robert Wood Johnson Foundation to the Commission to Build a Healthier America. Princeton, NJ: Robert Wood Johonson Foundation. Available: http://www.rwjf.org/files/research/obstaclestohealth.pdf [June 2012].

Braveman, P.A., Cubbin, C., Egerter, S., Chideya, S., Marchi, K.S., Metzler, M., and Posner, S. (2005). Socioeconomic status in health research: One size does not fit all. *Journal of the American Medical Association, 294*(22), 2,879-2,888.

Braveman, P.A., Cubbin, C., Egerter, S., Williams, D.R., and Pamuk, E. (2010a). Socioeconomic disparities in health in the United States: What the patterns tell us. *American Journal of Public Health, 100*(Suppl. 1), S186-S196.

Braveman, P.A., Marchi, K., Egerter, S., Kim, S., Metzler, M., Stancil, T., and Libet, M. (2010b). Poverty, near-poverty, and hardship around the time of pregnancy. *Maternal and Child Health Journal, 14*(1), 20-35.

Braveman, P.A., Egerter, S.A., and Mockenhaupt, R.E. (2011a). Broadening the focus: The need to address the social determinants of health. *American Journal of Preventive Medicine, 40*(Suppl. 1), S4-S18.

Braveman, P.A., Egerter, S.A., and Williams, D.R. (2011b). The social determinants of health: Coming of age. *Annual Review of Public Health, 32*(1), 381-398.

Braveman, P.A., Egerter, S.A., Woolf, S.H., and Marks, J.S. (2011c). When do we know enough to recommend action on the social determinants of health? *American Journal of Preventive Medicine, 40*(1), S58-S66.

Bringolf-Isler, B., Grize, L., Mäder, U., Ruch, N., Sennhauser, F.H., and Braun-Fahrländer, C. (2010). Built environment, parents' perception, and children's vigorous outdoor play. *Preventive Medicine, 50*(5-6), 251-256.

Bromet, E., Andrade, L., Hwang, I., Sampson, N., Alonso, J., de Girolamo, G., de Graaf, R., Demyttenaere, K., Hu, C., Iwata, N., Karam, A., Kaur, J., Kostyuchenko, S., Lepine, J.-P., Levinson, D., Matschinger, H., Mora, M., Browne, M., Posada-Villa, J., Viana, M., Williams, D., and Kessler, R. (2011). Cross-national epidemiology of DSM-IV major depressive episode. *BMC Medicine, 9*(1), 90.

Brook, R.D., Rajagopalan, S., Pope, C.A., Brook, J.R., Bhatnagar, A., Diez Roux, A.V., Holguin, F., Hong, Y., Luepker, R.V., Mittleman, M.A., Peters, A., Siscovick, D., Smith, S.C., Whitsel, L., Kaufman, J.D., on behalf of the American Heart Association Council on Epidemiology Prevention, Council on the Kidney in Cardiovascular Disease, and Council on Nutrition, Physical Activity and Metabolism. (2010). Particulate matter air pollution and cardiovascular disease. *Circulation, 121*(21), 2,331-2,378.

Brook, R.H. (2011a). Health services research and clinical practice. *Journal of the American Medical Association, 305*(15), 1,589-1,590.

Brook, R.H. (2011b). The role of physicians in controlling medical care costs and reducing waste. *Journal of the American Medical Association, 306*(6), 650-651.

Brownell, K.D., and Warner, K.E. (2009). The perils of ignoring history: Big tobacco played dirty and millions died. How similar is big food? *Milbank Quarterly, 87*(1), 259-294.

Brownell, K.D., Kersh, R., Ludwig, D.S., Post, R.C., Puhl, R.M., Schwartz, M.B., and Willett, W.C. (2010). Personal responsibility and obesity: A constructive approach to a controversial issue. *Health Affairs, 29*(3), 379-387.

Brownson, R.C., Haire-Joshu, D., and Luke, D.A. (2006). Shaping the context of health: A review of environmental and policy approaches in the prevention of chronic diseases. *Annual Review of Public Health, 27,* 341-370.

Brugger, H., Elsensohn, F., Syme, D., Sumann, G., and Falk, M. (2005). A survey of emergency medical services in mountain areas of Europe and North America: Official recommendations of the International Commission for Mountain Emergency Medicine (ICAR Medcom). *High Altitude Medicine & Biology, 6*(3), 226-237.

Brulle, R.J., and Pellow, D.N. (2006). Environmental justice: Human health and environmental inequalities. *Annual Reviews of Public Health, 27,* 103-124.

Brunello, G., Michaud, P., and Sanz-de-Galdeano, A. (2008). *The rise in obesity across the Atlantic: An economic perspective.* Bonn, Germany: Institute for the Study of Labor.

Buehler, R., Pucher, J., Merom, D., and Bauman, A. (2011). Active travel in Germany and the
    U.S.: Contributions of daily walking and cycling to physical activity. *American Journal
    of Preventive Medicine, 41*(3), 241-250.

Bunker, J.P. (2001). The role of medical care in contributing to health improvements within
    societies. *International Journal of Epidemiology, 30*(6), 1,260-1,263.

Bunker, J.P., Frazier, H.S., and Mosteller, F. (1994). Improving health: Measuring effects of
    medical care. *Milbank Quarterly, 72*(2), 225-258.

Burström, B., Macassa, G., Oberg, L., Bernhardt, E., and Smedman, L. (2005). Equitable child
    health interventions: The impact of improved water and sanitation on inequalities in child
    mortality in Stockholm, 1878 to 1925. *American Journal of Public Health, 95*(2), 208-216.

Byrne, D. (2004). *Enabling good health for all: A reflection process for a new EU health
    strategy.* European Commissioner for Health and Consumer Protection. Available: http://
    www.anme.info/ttcms/uploads/en/ENCommissioner.pdf [August 2012].

California Newsreel. (2008). *Unnatural causes: Is inequality making us sick?* San Francisco,
    CA. Available: http://www.unnaturalcauses.org [August 2012].

Campbell, C.A., Hahn, R.A., Elder, R., Brewer, R., Chattopadhyay, S., Fielding, J., Naimi,
    T.S., Toomey, T., Lawrence, B., and Middleton, J.C. (2009). The effectiveness of limiting
    alcohol outlet density as a means of reducing excessive alcohol consumption and alcohol-
    related harms. *American Journal of Preventive Medicine, 37*(6), 556-569.

Campbell Collaboration. (2012). *The Campbell Collaboration.* Available: http://www.camp
    bellcollaboration.org/ [August 2012].

Capewell, S., Morrison, C.E., and McMurray, J.J. (1999). Contribution of modern cardiovas-
    cular treatment and risk factor changes to the decline in coronary heart disease mortality
    in Scotland between 1975 and 1994. *Heart, 81*(4), 380-386.

Capewell, S., Beaglehole, R., Seddon, M., and McMurray, J. (2000). Explanation for the decline
    in coronary heart disease mortality rates in Auckland, New Zealand, between 1982 and
    1993. *Circulation, 102*(13), 1,511-1,516.

Card, D.E., and Freeman, R.B. (1993). Introduction. In D.E. Card and R.B. Freeman (Eds.), *Small
    differences that matter: Labor markets and income maintenance in Canada and the United
    States.* Chicago: University of Chicago Press.

Card, J.J. (1981). Long-term consequences for children of teenage parents. *Demography,
    18*(2), 137-156.

Case, A., and Paxson, C. (2010). Causes and consequences of early-life health. *Demography,
    47*(Suppl.), S65-S85.

Case, A., and Paxson, C. (2011). The long reach of childhood health and circumstance: Evi-
    dence from the Whitehall II study. *Economic Journal, 121*(554), F183-F204.

Case, A., Fertig, A., and Paxson, C.H. (2005). The lasting impact of childhood health and
    circumstances. *Journal of Health Economics 24*(2), 365-389.

Caspi, A., (2000). The child is father of the man: Personality continuities from childhood to
    adulthood. *Journal of Personality and Social Psychology, 78*(1), 158-172.

Caspi, A., and Moffitt, T.E. (1995). The continuity of maladaptive behavior: From descrip-
    tion to understanding in the study of antisocial behavior. In D. Cicchetti and D.J. Cohen
    (Eds.), *Developmental psychopathology* (vol. 2, pp. 472-511). New York: Wiley.

Caspi, A., McClay, J., Moffitt, T.E., Mill, J., Martin, J., Craig, I.W., Taylor, A., and Poulton,
    R. (2002). Role of genotype in the cycle of violence in maltreated children. *Science,
    297*(5,582), 851-854.

Caspi, A., Sugden, K., Moffitt, T.E., Taylor, A., Craig, I.W., Harrington, H., McClay, J., Mill,
    J., Martin, J., Braithwaite, A., and Poulton, R. (2003). Influence of life stress on depres-
    sion: Moderation by a polymorphism in the 5-htt gene. *Science, 301*(5,631), 386-389.

Cassel, C.K., and Guest, J.A. (2012). Choosing wisely: Helping physicians and patients make
    smart decisions about their care. *Journal of the American Medical Association, 307*(17),
    1,801-1,802.

Castles, F.G., and Mitchell D. (1993). Worlds of welfare and families of nations. In F.G. Casltes (Ed.), *Families of nations: Patterns of public policy in Western democracies* (pp. 93-128). Aldershot: Dartmouth.

Catalano, R., Goldman-Mellor, S., Saxton, K., Margerison-Zilko, C., Subbaraman, M., LeWinn, K., and Anderson, E. (2011). The health effects of economic decline. *Annual Review of Public Health, 32*, 431-450.

Cavelaars, A.E.J.M., Kunst, A.E., Geurts, J.J.M., Crialesi, R., Grötvedt, L., Helmert, U., Lahelma, E., Lundberg, O., Matheson, J., Mielck, A., Rasmussen, N.K., Regidor, E., Rosário-Giraldes, M.d., Spuhler, T., and Mackenbach, J. (2000). Educational differences in smoking: International comparison. *British Medical Journal, 320*(7,242), 1,102-1,107.

Cawley, J., and Maclean, J. (2010). *Unfit for service: The implications of rising obesity for U.S. military recruitment.* (NBER Working Paper No. 16408). Cambridge, MA: National Bureau of Economic Research.

Center on Human Needs. (2012a). *Project on societal distress.* Available: http://www.human needs.vcu.edu/PoSD.aspx [August 2012].

Center on Human Needs. (2012b). *Social capital and health outcomes in Boston.* E. Zimmerman, B.F. Evans, S.H. Woolf and A.D. Haley. Richmond: Virginia Commonwealth University.

Center on the Developing Child at Harvard University. (2010). *The foundations of lifelong health are built in early childhood.* Available: http://www.developingchild.harvard.edu [June 2012].

Centers for Disease Control and Prevention. (1999). Ten great public health achievements—United States, 1900-1999. *Morbidity and Mortality Weekly Report, 48*(12), 241-243.

Centers for Disease Control and Prevention. (2004). Trends in intake of energy and macronutrients—United States, 1971-2000. *Morbidity and Mortality Weekly Report, 53*(4), 80-82.

Centers for Disease Control and Prevention. (2006). *Preventing injuries in America: Public health in action.* In National Center for Injury Prevention and Control. *CDC Injury Fact Book* (pp. 35-101). Atlanta, GA: Centers for Disease Control and Prevention.

Centers for Disease Control and Prevention. (2008). Annual smoking-attributable mortality, years of potential life lost, and productivity losses—United States, 2000-2004. *Morbidity and Mortality Weekly Report, 57*(45), 1,226-1,228.

Centers for Disease Control and Prevention. (2009a). *Behavioral risk factor surveillance system, 2009.* Available: http://apps.nccd.cdc.gov/brfss/list.asp?cat=FV&yr=2009&qkey=4415&state=All [June 2012].

Centers for Disease Control and Prevention. (2009b). *National maternal and infant health survey.* Available: http://www.cdc.gov/nchs/nvss/nmihs.htm [April 2012].

Centers for Disease Control and Prevention. (2009c). *NCHS data linked to mortality files.* Available: http://www.cdc.gov/nchs/data_access/data_linkage/mortality.htm [April 2012].

Centers for Disease Control and Prevention. (2009d). *State indicator report on fruits and vegetables, 2009.* Available: http://www.cdc.gov/nutrition/downloads/StateIndicatorReport 2009.pdf [August 2012].

Centers for Disease Control and Prevention. (2009e). *Washington group on disability statistics.* Available: http://www.cdc.gov/nchs/washington_group.htm [August 2012].

Centers for Disease Control and Prevention. (2011a). *Community Preventive Services Task Force: First annual report to Congress.* Atlanta, GA: U.S. Department of Health and Human Services. Available: http://www.thecommunityguide.org/library/ARC2011/congress-report-exec.pdf [May 2011].

Centers for Disease Control and Prevention. (2011b). *The guide to community preventive services.* Available: http://www.thecommunityguide.org/index.html [June 2012].

Centers for Disease Control and Prevention. (2011c). *Health effects of cigarette smoking.* Atlanta, GA: U.S. Department of Health and Human Services.

Centers for Disease Control and Prevention. (2011d). *National hospital ambulatory medical care survey: 2008 emergency department summary tables.* Available: http://www.cdc.gov/nchs/data/ahcd/nhamcs_emergency/nhamcsed2008.pdf [November 2011].

Centers for Disease Control and Prevention. (2011e). *Suicide and self-inflicted injury.* Available: http://www.cdc.gov/nchs/fastats/suicide.htm [November 2011].

Centers for Disease Control and Prevention. (2011f). *Vital Signs: Alcohol-impaired driving among adults—United States, 2010.* Available: http://www.cdc.gov/mmwr/preview/mmwrhtml/mm6039a4.htm?s_cid=mm6039a4_w [August 2012].

Centers for Disease Control and Prevention. (2011g). *Vital Signs: Prescription painkiller overdoses in the U.S.* Available: http://www.cdc.gov/features/vitalsigns/painkilleroverdoses [August 2012].

Centers for Disease Control and Prevention. (2011h). *Web-based injury statistics query and reporting system (WISQARS).* Available: http://www.cdc.gov/injury/wisqars/index.html [June 2012].

Centers for Disease Control and Prevention. (2012a). *Behavioral risk factor surveillance system, prevalence and trends data, fruits and vegetables-2009. Adults who have consumed fruits and vegetables five or more times per day.* Available: http://apps.nccd.cdc.gov/brfss/list.asp?cat=FV&yr=2009&qkey=4415&state=All [June 2012]

Centers for Disease Control and Prevention. (2012b). *Deaths and mortality statistics.* Available: http://www.cdc.gov/nchs/fastats/deaths.htm [June 2012].

Centers for Disease Control and Prevention. (2012c). *National Vital Statistics System (NVSS).* Available: http://www.cdc.gov/nchs/deaths.htm [June 2012].

Centers for Disease Control and Prevention. (2012d). Vital Signs: Binge drinking prevalence, frequency, and intensity among adults—United States, 2010. *Morbidity and Mortality Weekly Report, 61*(1), 14-19.

Chang, C.H., Stukel, T.A., Flood, A.B., and Goodman, D.C. (2011). Primary care physician workforce and Medicare beneficiaries' health outcomes. *Journal of the American Medical Association, 305*(20), 2,096-2,104.

Charbit, C. (2011). *Governance of public policies in decentralised contexts: The multi-level approach.* Paris: OECD.

Chen, C.Y., and Lin, K.M. (2009). Health consequences of illegal drug use. *Current Opinion in Psychiatry, 22*(3), 287-292.

Chisholm, D., Rehm, J., Van Ommeren, M., and Monteiro, M. (2004). Reducing the global burden of hazardous alcohol use: A comparative cost-effectiveness analysis. *Journal of Studies on Alcohol, 65*(6), 782-793.

Christakis, N.A., and Fowler, J.H. (2007). The spread of obesity in a large social network over 32 years. *New England Journal of Medicine, 357*(4), 370-379.

Christensen, K., Davidsen, M., Juel, K., Mortensen, L., Rau, R., and Vaupel, J.W. (2010). The divergent life-expectancy trends in Denmark and Sweden—and some potential explanations. In National Research Council, *International differences in mortality at older ages: Dimensions and sources* (pp. 385-408). E.M. Crimmins, S.H. Preston, and B. Cohen (Eds.), Panel on Understanding Divergent Trends in Longevity in High-Income Countries. Committee on Population. Division of Behavioral and Social Sciences and Education. Washington, DC: The National Academies Press.

Chuang, Y., Cubbin, C., Ahn, D., and Winkleby, M.A. (2005). Effects of neighbourhood socioeconomic status and convenience store concentration on individual-level smoking. *Tobacco Control, 14*(5), 568-573.

Chung, H., and Muntaner, C. (2007). Welfare state matters: A typological multilevel analysis of wealthy countries. *Health Policy, 80*(2), 328-339.

Ciccolallo, L., Capocaccia, R., Coleman, M.P., Berrino, F., Coebergh, J.W.W., Damhuis, R.A.M., Faivre, J., Martinez-Garcia, C., Møller, H., Ponz de Leon, M., Launoy, G., Raverdy, N., Williams, E.M.I., and Gatta, G. (2005). Survival differences between European and U.S. patients with colorectal cancer: Role of stage at diagnosis and surgery. *Gut, 54*(2), 268-273.

Clougherty, J.E., Souza, K., and Cullen, M.R. (2010). Work and its role in shaping the social gradient in health. *Annals of the New York Academy of Sciences, 1,186*(1), 102-124.

Cnattingius, S., Bergström, R., Lipworth, L., and Kramer, M.S. (1998). Prepregnancy weight and the risk of adverse pregnancy outcomes. *New England Journal of Medicine, 338*, 147-152.

Coburn, D. (2004). Beyond the income inequality hypothesis: Class, neo-liberalism, and health inequalities. *Social Science and Medicine, 58*(1), 41-56.

Cochrane Library. (2012). *Home page.* Available: http://www.thecochranelibrary.com/view/0/index.html [June 2012].

Cohen, R.A., Gindi, R.M., and Kirzinger, W.K. (2012). *Burden of medical care cost: Early release of estimates from the National Health Interview Survey, January-June 2011.* National Center for Health Statistics. Available: http://www.cdc.gov/nchs/nhis/releases.htm [June 2012].

Cohen, S., and Wills, T.A. (1985). Stress, social support, and the buffering hypothesis. *Psychological Bulletin, 98*(2), 310-357.

Cohen, S., Janicki-Deverts, D., Chen, E., and Matthews, K.A. (2010). Childhood socioeconomic status and adult health. *Annals of the New York Academy of Sciences, 1,186*(1), 37-55.

Cole, B.L., and Fielding, J.E. (2007). Health impact assessment: A tool to help policy makers understand health beyond health care. *Annual Review of Public Health, 28*, 393-412.

Collins, J.W., Wu, S.Y., and David, R.J. (2002). Differing intergenerational birth weights among the descendants of U.S.-born and foreign-born whites and African Americans in Illinois. *American Journal of Epidemiology, 155*(3), 210-216.

Commission on Social Determinants of Health. (2008). *Closing the gap in a generation: Health equity through action on the social determinants of health.* Geneva, Switzerland: World Health Organization.

Commonwealth Fund Commission on a High Performance Health System. (2008). *Why not the best? Results from a national scorecard on U.S. health system performance, 2008.* New York: Commonwealth Fund.

Commonwealth Fund Commission on a High Performance System. (2011). *Why not the best? Results from the national scorecard on U.S. health system performance, 2011.* New York: Commonwealth Fund.

Community Preventive Services Task Force. (2012). *The guide to community preventive services.* Available: http://www.thecommunityguide.org/index.html [October 2011].

Congressional Budget Office. (1992). *CBO staff memorandum: Factors contributing to the infant mortality ranking of the United States.* Available: http://www.cbo.gov/sites/default/files/cbofiles/ftpdocs/62xx/doc6219/doc05b.pdf [June 2012].

Conley, D., and Springer, K.W. (2001). Welfare state and infant mortality. *American Journal of Sociology, 107*(3), 768-807.

Cooper, H.L.F., Friedman, S.R., Tempalski, B., and Friedman, R. (2007). Residential segregation and injection drug use prevalence among black adults in U.S. metropolitan areas. *American Journal of Public Health, 97*(2), 344-352.

Corak, M. (2004). *Generational income mobility in North America and Europe.* New York: Cambridge University Press.

Courtemanche, C.J., and Zapata, D. (2012). *Does universal coverage improve health? The Massachusetts experience.* Cambridge, MA: National Bureau of Economic Research.

Crimmins, E., and Solé-Auró, A. (2011). *The health of those near age 50 in the U.S. relative to Europeans.* Unpublished paper prepared for the NAS/IOM Panel on Understanding Cross-National Health Differences Among High-Income Countries. Davis School of Gerontology, University of Southern California and USC/UCLA Center on Biodemography and Population Health.

Crimmins, E.M., Vasunilashorn, S., Kim, J.K., Hagedorn, A., and Saito, Y. (2008). A comparison of biological risk factors in two populations: The United States and Japan. *Population and Development Review, 34*(3), 457-482.

Crimmins, E.M., Garcia, K., and Kim, J.K. (2010). Are international differences in health similar to international differences in life expectancy? In National Research Council, *International differences in mortality at older ages: Dimensions and sources* (pp. 68-101). E.M. Crimmins, S.H. Preston, and B. Cohen (Eds.), Panel on Understanding Divergent Trends in Longevity in High-Income Countries. Committee on Population. Division of Behavioral and Social Sciences and Education. Washington, DC: The National Academies Press.

Crocombe, L.A., Mejia, G.C., Koster, C.R., and Slade, G.D. (2009). Comparison of adult oral health in Australia, the USA, Germany and the UK. *Australian Dental Journal, 54*(2), 147-153.

Croissant, A. (2004). Changing welfare regimes in East and Southeast Asia: Crisis, change and challenge. *Social Policy & Administration, 38*(5), 504-524.

Crosnoe, R. (2007). Gender, obesity, and education. *Sociology of Education, 80*(3), 241-260.

Cubbin, C., LeClere, F.B., and Smith, G.S. (2000). Socioeconomic status and injury mortality: Individual and neighbourhood determinants. *Journal of Epidemiology and Community Health, 54*(7), 517-524.

Cudnik, M.T., Newgard, C.D., Sayre, M.R., and Steinberg, S.M. (2009). Level I versus level II trauma centers: An outcomes-based assessment. *Journal of Trauma and Acute Care Surgery, 66*(5), 1,321-1,326.

Cullen, M.R., Cummins, C., and Fuchs, V.R. (2012). Geographic and racial variation in premature mortality in the U.S.: Analyzing the disparities. *PLoS ONE, 7*(4), e32930.

Culyer, A.J., and Lomas, J. (2006). Deliberative processes and evidence-informed decision making in healthcare: Do they work and how might we know? *Evidence and Policy: A Journal of Research, Debate and Practice, 2*(3), 357-371.

Cummins, S., Petticrew, M., Higgins, C., Findlay, A., and Sparks, L. (2005). Large-scale food retailing as an intervention for diet and health: Quasi-experimental evaluation of a natural experiment. *Journal of Epidemiology and Community Health, 59*(12), 1,035-1,040.

Cunradi, C.B., Mair, C., Ponicki, W., and Remer, L. (2012). Alcohol outlet density and intimate partner violence-related emergency department visits. *Alcoholism: Clinical and Experimental Research, 36*(5), 847-853.

Currie, C., Gabhainn, S.N., Godeau, E., Roberts, C., Smith, R., Currie, D., Picket, W., Richter, M., Morgan, A., and Barnekow, V. (2008). *Inequalities in young people's health: HBSC international report from the 2005/2006 survey.* Geneva, Switzerland: World Health Organization.

Currie, J., and Widom, C.S. (2010). Long-term consequences of child abuse and neglect on adult economic well-being. *Child Maltreatment, 15*(2), 111-120.

Cutler, D.M., and Glaeser, E.L. (2006). *Why do Europeans smoke more than Americans?* Cambridge, MA: National Bureau of Economic Research.

Cutler, D.M., and Miller, G. (2005). The role of public health improvements in health advances: The twentieth-century United States. *Demography, 42*(1), 1-22.

Cutler, D.M., Richardson, E., Keeler, T.E., and Staiger, D. (1997) *Measuring the health of the U.S. population.* Brookings Papers on Economic Activity. Available: http://www.brookings.edu/~/media/Projects/BPEA/1997%20micro/1997_bpeamicro_cutler.PDF [October 2012].

Cutler, D.M., Deaton, A., and Lleras-Muney, A. (2006a). The determinants of mortality. *Journal of Economic Perspectives, 20*(3), 97-120.

Cutler, D.M., Rosen, A.B., and Vijan, S. (2006b). The value of medical spending in the United States, 1960-2000. *New England Journal of Medicine, 355*(9), 920-927.

Cutler, D.M., Lange, F., Meara, E., Richards-Shubik, S., and Ruhm, C.J. (2011). Rising educational gradients in mortality: The role of behavioral risk factors. *Journal of Health Economics, 30*(6), 1,174-1,187.

Cutrona, C.E., Russell, D.W., Hessling, R.M., Brown, P.A., and Murry, V. (2000). Direct and moderating effects of community context on the psychological well-being of African American women. *Journal of Personal and Social Psychology, 79*(6), 1,088-1,101.

Cutrona, C.E., Wallace, G., and Wesner, K.A. (2006). Neighborhood characteristics and depression: An examination of stress processes. *Current Directions in Psychological Science, 15*(4), 188-192.

Dahl, E., Fritzell, J., Lahelma, E., Martikainen, P., Kunst, A.E., and Mackenbach, J.P. (2006). Welfare state regimes and health inequalities. In J. Siegrist and M. Marmot (Eds.), *Social inequalities in health: New evidence and policy implications.* New York: Oxford University Press.

Dahlgren, G., and Whitehead, M. (1991). *Policies and strategies to promote social equity in health. Background document to WHO—strategy paper for Europe.* Arbetsrapport: Institute for Futures Studies.

Danaei, G., Finucane, M.M., Lin, J.K., Singh, G.M., Paciorek, C.J., Cowan, M.J., Farzadfar, F., Stevens, G.A., Lim, S.S., Riley, L.M., and Ezzati, M. (2011a). National, regional, and global trends in systolic blood pressure since 1980: Systematic analysis of health examination surveys and epidemiological studies with 786 country-years and 5.4 million participants. *Lancet, 377*(9,765), 568-577.

Danaei, G., Finucane, M.M., Lu, Y., Singh, G.M., Cowan, M.J., Paciorek, C.J., Lin, J.K., Farzadfar, F., Khang, Y.-H., Stevens, G.A., Rao, M., Ali, M.K., Riley, L.M., Robinson, C.A., and Ezzati, M. (2011b). National, regional, and global trends in fasting plasma glucose and diabetes prevalence since 1980: Systematic analysis of health examination surveys and epidemiological studies with 370 country-years and 2.7 million participants. *Lancet, 378*(9,785), 31-40.

Danese, A., Pariante, C.M., Caspi, A., Taylor, A., and Poulton, R. (2007). Childhood maltreatment predicts adult inflammation in a life-course study. *Proceedings of the National Academy of Sciences, 104*(4), 1,319-1,324.

Daniels, N., Kennedy, B., and Kawachi, I. (2000). *Is inequality bad for our health?* Boston, MA: Beacon Press.

Danziger, S., and Gottschalk, P. (1986). *Do rising tides lift all boats? The impact of secular and cyclical changes on poverty.* AEA Papers and Proceedings. Pittsburgh, PA: American Economic Association.

Danziger, S., and Gottschalk, P. (1995). *America unequal.* London, UK: Cambridge University Press.

Darroch, J.E., Singh, S., and Frost, J.J. (2001). Differences in teenage pregnancy rates among five developed countries: The roles of sexual activity and contraceptive use. *Family Planning Perspectives, 33*(6), 244-250, 281.

David, R.J., and Collins, J.W. (1997). Differing birth weight among infants of U.S.-born blacks, African-born blacks, and U.S.-born whites. *New England Journal of Medicine, 337*(17), 1,209-1,214.

Davis, K., Schoen, C., and Stremikis, K. (2010). *Mirror, mirror on the wall: How the performance of the U.S. health care system compares internationally, 2010 update.* New York: Commonwealth Fund.

Davison, K., and Lawson, C. (2006). Do attributes in the physical environment influence children's physical activity? A review of the literature. *International Journal of Behavioral Nutrition and Physical Activity, 3*(1), 19.

Dawwas, M.F., Gimson, A.E., Lewsey, J.D., Copley, L.P., and van der Meulen, J.H.P. (2007). Survival after liver transplantation in the United Kingdom and Ireland compared with the United States. *Gut, 56*(11), 1,606-1,613.

De Silva, M.J., McKenzie, K., Harpham, T., and Huttly, S.R.A. (2005). Social capital and mental illness: A systematic review. *Journal of Epidemiology and Community Health, 59*(8), 619-627.

Deaton, A. (2002). Policy implications of the gradient of health and wealth. *Health Affairs, 21*(2), 13-30.

Deaton, A., and Lubotsky, D. (2009). Income inequality and mortality in U.S. cities: Weighing the evidence. A response to Ash. *Social Science and Medicine, 68*(11), 1,914-1,917.

Dehlendorf, C., Marchi, K., Vittinghoff, E., and Braveman, P. (2010). Sociocultural determinants of teenage childbearing among Latinas in California. *Maternal and Child Health Journal, 14*(2), 194-201.

Delaney, L., and Smith, J.P. (2012). Childhood health: Trends and consequences over the life course. *Future Child, 22*(1), 43-63.

Demyttenaere, K., Bruffaerts, R., Posada-Villa, J., Gasquet, I., et al., and the WHO World Mental Health Survey Consortium. (2004). Prevalence, severity, and unmet need for treatment of mental disorders in the World Health Organization World Mental Health Surveys. *Journal of the American Medical Association, 291*(21), 2,581-2,590.

DeNavas-Walt, C., Proctor, B.D., and Smith, J.C. (2011). *Income, poverty, and health insurance coverage in the United States: 2010.* Washington, DC: U.S. Census Bureau.

Diez Roux, A.V. (2000). Multilevel analysis in public health research. *Annual Review of Public Health, 21*, 171-192.

Diez Roux, A.V. (2011). Complex systems thinking and current impasses in health disparities research. *American Journal of Public Health, 101*(9), 1,627-1,634.

Diez Roux, A.V., and Mair, C. (2010). Neighborhoods and health. *Annals of the New York Academy of Sciences, 1,186*(1), 125-145.

Diez Roux, A.V., Merkin, S.S., Arnett, D., Chambless, L., Massing, M., Nieto, F.J., Sorlie, P., Szklo, M., Tyroler, H.A., and Watson, R.L. (2001). Neighborhood of residence and incidence of coronary heart disease. *New England Journal of Medicine, 345*(2), 99-106.

Ding, D., Sallis, J.F., Kerr, J., Lee, S., and Rosenberg, D.E. (2011). Neighborhood environment and physical activity among youth: A review. *American Journal of Preventive Medicine, 41*(4), 442-455.

Do, D.P., Diez Roux, A.V., Hajat, A., Auchincloss, A.H., Merkin, S.S., Ranjit, N., Shea, S., and Seeman, T. (2011). Circadian rhythm of cortisol and neighborhood characteristics in a population-based sample: The multi-ethnic study of atherosclerosis. *Health & Place, 17*(2), 625-632.

Dong, M., Anda, R.F., Felitti, V.J., Dube, S.R., Williamson, D.F., Thompson, T.J., Loo, C.M., and Giles, W.H. (2004). The interrelatedness of multiple forms of childhood abuse, neglect, and household dysfunction. *Child Abuse & Neglect, 28*(7), 771-784.

D'Orazio, M., Di Zio, M., and Scanu, M. (2006). *Statistical matching: Theory and practice.* Hoboken, NJ: Wiley.

Douglas, M.J., Watkins, S.J., Gorman, D.R., and Higgins, M. (2011). Are cars the new tobacco? *Journal of Public Health, 33*(2), 160-169.

Dow, W.H., and Rehkopf, D.H. (2010). Socioeconomic gradients in health in international and historical context. *Annals of the New York Academy of Sciences, 1186*(1), 24-36.

Dow, W.H., Schoeni, R.F., Adler, N.E., and Stewart, J. (2010). Evaluating the evidence base: Policies and interventions to address socioeconomic status gradients in health. *Annals of the New York Academy of Sciences, 1,186*(1), 240-251.

Downey, L., and Van Willigen, M. (2005). Environmental stressors: The mental health impacts of living near industrial activity. *Journal of Health and Social Behavior, 46*(3), 289-305.

Downing, S.R., Oyetunji, T.A., Greene, W.R., Jenifer, J., Rogers, S.O., Jr., Haider, A.H., Cornwell, E.E., III, and Chang, D.C. (2011). The impact of insurance status on actuarial survival in hospitalized trauma patients: When do they die? *Journal of Trauma, 70*(1), 130-134; discussion 134-135.

Dozier, M., and Peloso, E. (2006). The role of early stressors in child health and mental health outcomes. *Archives of Pediatrics and Adolescent Medicine, 160*(12), 1,300-1,301.

Drösler, S.E., Romano, P.S., Tancredi, D.J., and Klazinga, N.S. (2012). International comparability of patient safety indicators in 15 OECD member countries: A methodological approach of adjustment by secondary diagnoses. *Health Services Research, 47*(1 Pt. 1), 275-292.

Dunn, J.R. (2010). Health behavior vs the stress of low socioeconomic status and health outcomes. *Journal of the American Medical Association, 303*(12), 1,199-1,200.

Durand, C.P., Andalib, M., Dunton, G.F., Wolch, J., and Pentz, M.A. (2011). A systematic review of built environment factors related to physical activity and obesity risk: Implications for smart growth urban planning. *Obesity Reviews, 12*(5), e173-e182.

Durkin, M.S., Davidson, L.L., Kuhn, L., O'Connor, P., and Barlow, B. (1994). Low-income neighborhoods and the risk of severe pediatric injury: A small-area analysis in northern Manhattan. *American Journal of Public Health, 84*(4), 587-592.

Durkin, M.S., McElroy, J., Guan, H., Bigelow, W., and Brazelton, T. (2005). Geographic analysis of traffic injury in Wisconsin: Impact on case fatality of distance to level I/II trauma care. *WMJ: Official Publication of the State Medical Society of Wisconsin, 104*(2), 26-31.

Eaker, E.D., Sullivan, L.M., Kelly-Hayes, M., D'Agostino, R.B., and Benjamin, E.J. (2004). Does job strain increase the risk for coronary heart disease or death in men and women? *American Journal of Epidemiology, 159*(10), 950-958.

Eaton, D.K., Kann, L., Kinchen, S., Ross, J., Hawkins, J., Harris, W.A., Lowry, R., McManus, T., Chyen, D., Shanklin, S., Lim, C., Grunbaum, J.A., and Wechsler, H. (2006). Youth risk behavior surveillance—United States, 2005. *Journal of School Health, 76*(7), 353-372.

Eaton, D.K., Kann, L., Kinchen, S., Shanklin, S., Ross, J., Hawkins, J., Harris, W.A., Lowry, R., McManus, T., Chyen, D., Lim, C., Brener, N.D., and Wechsler, H. (2008). Youth risk behavior surveillance—United States, 2007. *Morbidity and Mortality Weekly Report Surveillance Summaries, 57*(4), 1-131.

Edwards, V.J., Holden, G.W., Felitti, V.J., and Anda, R.F. (2003). Relationship between multiple forms of childhood maltreatment and adult mental health in community respondents: Results from the Adverse Childhood Experiences Study. *American Journal of Psychiatry, 160*(8), 1,453-1,460.

Eikemo, T.A., and Bambra, C. (2008). The welfare state: A glossary for public health. *Journal of Epidemiology and Community Health, 62*, 3-6.

Eikemo, T.A., Bambra, C., Joyce, K., and Dahl, E. (2008a). Welfare state regimes and income-related health inequalities: A comparison of 23 European countries. *European Journal of Public Health, 18*(6), 593-599.

Eikemo, T.A., Huisman, M., Bambra, C., and Kunst, A.E. (2008b). Health inequalities according to educational level in different welfare regimes: A comparison of 23 European countries. *Sociology of Health and Illness, 30*(4), 565-582.

Eisenberg, L. (2005). Violence and the mentally ill: Victims, not perpetrators. *Archives of General Psychiatry, 62*(8), 825-826.

Elder, G., and Shanahan, M. (2007). The life course and human development. In W. Damon and R. Lerner (Eds.), *Handbook of child psychology* (6th ed., vol. 1). New York: Wiley.

Elo, I.T. (2009). Social class differentials in health and mortality: Patterns and explanations in comparative perspective. *Annual Review of Sociology, 35*, 553-572.

Elo, I.T., and Preston, S.H. (1996). Educational differentials in mortality: United States, 1979-1985. *Social Science and Medicine, 42*(1), 47-57.

Emerging Risk Factors Collaboration, Wormser, D., Kaptoge, S., Di Angelantonio, E., Wood, A.M., Pennells, L., Thompson, A., Sarwar, N., Kizer, J.R., Lawlor, D.A., Nordestgaard, B.G., Ridker, P., Salomaa, V., Stevens, J., Woodward, M., Sattar, N., Collins, R., Thompson, S.G., Whitlock, G., and Danesh, J. (2012). Separate and combined associations of body-mass index and abdominal adiposity with cardiovascular disease: Collaborative analysis of 58 prospective studies. *Lancet, 377*(9,771), 1,085-1,095.

Epstein, A.M. (2007). Pay for performance at the tipping point. *New England Journal of Medicine, 356*(5), 515-517.

Erasmus. (2012). *Eurothine: Objectives.* Available: http://survey.erasmusmc.nl/eurothine/index.php?id=5,0,0,1,0,0 [April 2012].

Erinosho, T.O., Thompson, O.M., Moser, R.P., and Yaroch, A.L. (2011). Fruit and vegetable intake of U.S. adults: Comparing intake by mode of survey administration. *Journal of the American Dietetic Association, 111*(3), 408-413.

Ermisch, J., Jäntti, M., and Smeeding, T.M. (2012). *From parents to children: The intergenerational transmission of advantage.* New York: Russell Sage Foundation.

Esping-Andersen, G. (1987). Citizenship and socialism: Decommodification and solidarity in the welfare state. In G. Esping-Andersen and L. Rainwater (Eds.), *Stagnation and renewal in social policy: The rise and fall of policy regimes.* London, UK: Sharpe.

Esping-Andersen, G. (1990). *The three worlds of welfare capitalism.* London, UK: Polity.

Esping-Andersen, G., and Myles, J. (2009). The welfare state and redistribution. In *The Oxford handbook of economic inequality* (pp. 639-664). W. Salverda, B. Nolan, and T.M. Smeeding (Eds.). Oxford, UK: Oxford University Press.

Etzel, R.A. (2003). How environmental exposures influence the development and exacerbation of asthma. *Pediatrics, 112*(1 Pt. 2), 233-239.

European Commission. (2012). *Data collection: Policy.* Available: http://ec.europa.eu/health/data_collection/policy/index_en.htm [April 2012].

European Food Information Council. (2012). *Fruit and vegetable consumption in Europe—Do Europeans get enough?* Available: http://www.eufic.org/article/en/expid/Fruit-vegetable-consumption-Europe/ [June 2012].

European Observatory on Health Systems and Policies. (2012). *About us.* Available: http://www.euro.who.int/en/who-we-are/partners/observatory/about-us [August 2012].

Evans, G.W., and Kantrowitz, E. (2002). Socioeconomic status and health: The potential role of environmental risk exposure. *Annual Review of Public Health, 23*(May), 303-331.

Evans, G.W., and Kim, P. (2010). Multiple risk exposure as a potential explanatory mechanism for the socioeconomic status-health gradient. *Annals of the New York Academy of Sciences, 1,186*(1), 174-189.

Evans, G.W., Fuller-Rowell, T.E., and Doan, S.N. (2012). Childhood cumulative risk and obesity: The mediating role of self-regulatory ability. *Pediatrics, 129*(1), e68-e73.

Evans, R.G., and Stoddart, G.L. (1990). Producing health, consuming health care. *Social Science and Medicine, 31*(12), 1,347-1,363.

Ezzati, M., Friedman, A.B., Kulkarni, S.C., and Murray, C.J.L. (2008). The reversal of fortunes: Trends in county mortality and cross-county mortality disparities in the United States. *PLoS Medicine, 5*(4), e66. doi:10.1371/journal.pmed.0050066.

Fairchild, A.L., Rosner, D., Colgrove, J., Bayer, R., and Fried, L.P. (2010). The EXODUS of public health: What history can tell us about the future. *American Journal of Public Health, 100*(1), 54-63.

Farzadfar, F., Finucane, M.M., Danaei, G., Pelizzari, P.M., Cowan, M.J., Paciorek, C.J., Singh, G.M., Lin, J.K., Stevens, G.A., Riley, L.M., and Ezzati, M. (2011). National, regional, and global trends in serum total cholesterol since 1980: Systematic analysis of health examination surveys and epidemiological studies with 321 country-years and 3.0 million participants. *Lancet, 377*(9,765), 578-586.

Felitti, V.J., Anda, R.F., Nordenberg, D., Williamson, D.F., Spitz, A.M., Edwards, V., Koss, M.P., and Marks, J.S. (1998). Relationship of childhood abuse and household dysfunction to many of the leading causes of death in adults. The adverse childhood experiences (ACE) study. *American Journal of Preventive Medicine, 14*(4), 245-258.

Felker, B., Yazel, J.J., and Short, D. (1996). Mortality and medical comorbidity among psychiatric patients: A review. *Psychiatric Service, 47,* 1,356-1,363.

Fell, J. C., Lacey, J. H., and Voas, R. B. (2004). Sobriety checkpoints: Evidence of effectiveness is strong but use is limited. *Traffic Injury Prevention, 5*(3), 220-227.

Feng, J., Glass, T.A., Curriero, F.C., Stewart, W.F., and Schwartz, B.S. (2010). The built environment and obesity: A systematic review of the epidemiologic evidence. *Health & Place, 16*(2), 175-190.

Ferrera, M. (1996). The "southern model" of welfare in social Europe. *Journal of European Social Policy, 6*(1), 17-37.

Fielding, J.E., and Teutsch, S.M. (2009). Integrating clinical care and community health. *Journal of the American Medical Association, 302*(3), 317-319.

Fielding, J.E., and Teutsch, S.M. (2011). An opportunity map for societal investment in health. *Journal of the American Medical Association, 305*(20), 2,110-2,111.

Finch, C.E. (2010). Evolution of the human lifespan and diseases of aging: Roles of infection, inflammation, and nutrition. *Proceedings of the National Academy of Sciences, 107*(Suppl. 1), 1,718-1,724.

Finch, C.E., and Crimmins, E.M. (2004). Inflammatory exposure and historical changes in human life-spans. *Science, 305*(5,691), 1,736-1,739.

Fine, L.J., Philogene, G.S., Gramling, R., Coups, E.J., and Sinha, S. (2004). Prevalence of multiple chronic disease risk factors. 2001 National Health Interview Survey. *American Journal of Preventive Medicine, 27*(Suppl. 2), S18-S24.

Fineberg, H.V. (2012). A successful and sustainable health system—How to get there from here. *New England Journal of Medicine, 366*(11), 1,020-1,027.

Finkelstein, E.A., Brown, D.S., Wrage, L.A., Allaire, B.T., and Hoerger, T.J. (2010). Individual and aggregate years-of-life-lost associated with overweight and obesity. *Obesity, 18*(2), 333-339.

Finucane, M.M., Stevens, G.A., Cowan, M.J., Danaei, G., Lin, J.K., Paciorek, C.J., Singh, G.M., Gutierrez, H.R., Lu, Y., Bahalim, A.N., Farzadfar, F., Riley, L.M., and Ezzati, M. (2011). National, regional, and global trends in body-mass index since 1980: Systematic analysis of health examination surveys and epidemiological studies with 960 country-years and 9.1 million participants. *Lancet, 377*(9,765), 557-567.

Fisher, E.S., Wennberg, D.E., Stukel, T.A., Gottlieb, D.J., Lucas, F.L., and Pinder, E.L. (2003). The implications of regional variations in Medicare spending. Part 2: Health outcomes and satisfaction with care. *Annuals of Internal Medicine, 138,* 288-298.

Fisher, E.S., McClellan, M.B., and Safran, D.G. (2011). Building the path to accountable care. *New England Journal of Medicine, 365*(26), 2,445-2,447.

Fix, B.V., Hyland, A., Rivard, C., McNeill, A., Fong, G.T., Borland, R., Hammond, D., and Cummings, K.M. (2011). Usage patterns of stop smoking medications in Australia, Canada, the United Kingdom, and the United States: Findings from the 2006-2008 international tobacco control (ITC) four country survey. *International Journal of Environmental Research and Public Health, 8*(1), 222-233.

Fletcher, J.M., and Richards, M.R. (2012). Diabetes's "health shock" to schooling and earnings: Increased dropout rates and lower wages and employment in young adults. *Health Affairs, 31*(1), 27-34.

Foege, W.H. (2004). Redefining public health. *Journal of Law, Medicine & Ethics, 32*(Suppl. 4), S23-S26.

Fong, I.W. (2000). Emerging relations between infectious diseases and coronary artery disease and atherosclerosis. *Canadian Medical Association Journal, 163*(1), 49-56.

Fonseca, R., and Zheng, Y. (2011). *The effect of education on health: Cross-country evidence.* Santa Monica, CA: RAND Corporation.

Food and Agricultural Organization of the United Nations. (2010). *FAO statistical yearbook 2010.* Available: http://www.fao.org/economic/ess/ess-publications/ess-yearbook/ess-yearbook2010/en/ [December 2012].

Ford, E.S., and Capewell, S. (2011). Proportion of the decline in cardiovascular mortality disease due to prevention versus treatment: Public health versus clinical care. *Annual Review of Public Health, 32*, 5-22.

Ford, E.S., Ajani, U.A., Croft, J.B., Critchley, J.A., Labarthe, D.R., Kottke, T.E., Giles, W.H., and Capewell, S. (2007). Explaining the decrease in U.S. deaths from coronary disease, 1980-2000. *New England Journal of Medicine, 356*(23), 2,388-2,398.

Forey, B., Hamling, J., Lee, P., and Wald, N. (2002). *International smoking statistics: A collection of historical data from 30 economically developed countries* (2nd ed.). Oxford, UK: Oxford University Press.

Förster, M., and d'Ercole, M. (2005). *Income distribution and poverty in OECD countries in the second half of the 1990s.* Paris: OECD.

Foster, S., and Giles-Corti, B. (2008). The built environment, neighborhood crime, and constrained physical activity: An exploration of inconsistent findings. *Preventive Medicine, 47*(3), 241-251.

Franco, M., Diez Roux, A.V., Glass, T.A., Caballero, B., and Brancati, F.L. (2008). Neighborhood characteristics and availability of healthy foods in Baltimore. *American Journal of Preventive Medicine, 35*(6), 561-567.

Frankel, S., Davey Smith, G., and Gunnell, D. (1999). Childhood socioeconomic position and adult cardiovascular mortality: The Boyd Orr cohort. *American Journal of Epidemiology, 150*(10), 1,081-1,084.

Freedman, V.A., Martin, L.G., Schoeni, R.F., and Cornman, J.C. (2008). Declines in late-life disability: The role of early- and mid-life factors. *Social Science and Medicine, 66*(7), 1,588-1,602.

Freeman, J.D., Kadiyala, S., Bell, J.F., and Martin, D.P. (2008). The causal effect of health insurance on utilization and outcomes in adults: A systematic review of U.S. studies. *Medical Care, 46*(10), 1,023-1,032

Freeman, R., and Marmor, T. (2003). Making sense of health services politics through cross-national comparison. *Journal of Health Services Research & Policy, 8*(3), 180-182.

Friese, B., and Grube, J.W. (2001). *Youth drinking rates and problems: A comparison of European countries and the United States.* Berkeley, CA: Pacific Institute for Research and Evaluation.

Frumkin, H. (2005). Health, equity, and the built environment. *Environmental Health Perspectives, 113*(5), A290-A291.

Fujishiro, K., Diez Roux, A.V., Landsbergis, P., Baron, S., Barr, R.G., Kaufman, J.D., Polak, J.F., and Stukovsky, K.H. (2011). Associations of occupation, job control and job demands with intima-media thickness: The multi-ethnic study of atherosclerosis (MESA). *Occupational and Environmental Medicine, 68*(5), 319-326.

Galama, T., and Kapteyn, A. (2011). Grossman's missing health threshold. *Journal of Health Economics. 30*(5), 1,044-1,056.

Galea, S., Tracy, M., Hoggatt, K.J., DiMaggio, C., and Karpati, A. (2011). Estimated deaths attributable to social factors in the United States. *American Journal of Public Health, 101*(8), 1,456-1,465.

Galobardes, B., Lynch, J.W., and Davey Smith, G. (2008). Is the association between childhood socioeconomic circumstances and cause-specific mortality established? Update of a systematic review. *Journal of Epidemiology and Community Health, 62*(5), 387-390.

Galvez, M.P., Pearl, M., and Yen, I.H. (2010). Childhood obesity and the built environment. *Current Opinion in Pediatrics, 22*(2), 202-207.

Garber, A.M., and Skinner, J. (2008). *Is American health care uniquely inefficient?* (NBER Working Paper No. 14257). Cambridge, MA: National Bureau of Economic Research.

Garrett, B.E., Dube, S.R., Trosclair, A., Caraballo, R.S., Pechacek, T.F., and Centers for Disease Control and Prevention. (2011). Cigarette smoking—United States, 1965-2008. *Morbidity and Mortality Weekly Report Surveillance Summaries, 60*(Suppl.), 109-113.

Gatta, G., Capocaccia, R., Coleman, M.P., Gloeckler Ries, L.A., Hakulinen, T., Micheli, A., Sant, M., Verdecchia, A., and Berrino, F. (2000). Toward a comparison of survival in American and European cancer patients. *Cancer, 89*(4), 893-900.

Gavin, A.R., Hill, K.G., Hawkins, J.D., and Maas, C. (2011). The role of maternal early-life and later-life risk factors on offspring low birth weight: Findings from a three-generational study. *Journal of Adolescent Health, 49*(2), 166-171.

Gay, J., Devaux, M., de Looper, M., and Paris, V. (2011). *Mortality amenable to health care in 31 OECD countries: Estimates and methodological issues.* (No. 55, OECD Health Working Papers). Paris: OECD.

Gaziano, T.A., Young, C.R., Fitzmaurice, G., Atwood, S., and Gaziano, J.M. (2008). Laboratory-based versus non-laboratory-based method for assessment of cardiovascular disease risk: The NHANES I follow-up study cohort. *Lancet, 371*(9,616), 923-931.

Glantz, L.H., and Annas, G.J. (2009). Handguns, health, and the second amendment. *New England Journal of Medicine, 360*(22), 2,360-2,365.

Glanz, K., Rimer, B.K., and Viswanath, K. (Eds.). (2008). *Health behavior and health education: theory, research, and practice* (4th ed.). San Francisco, CA: Jossey-Bass.

Glasgow, R.E., Green, L.W., Klesges, L.M., Abrams, D.B., Fisher, E.B., Goldstein, M.G., Hayman, L.L., Ockene, J.K., and Orleans, C.T. (2006). External validity: We need to do more. *Annals of Behavioral Medicine, 31*(2), 105-108.

Glaze, L.E. (2011). *Correctional population in the United States, 2010.* Washington, DC: U.S. Department of Justice.

Global Burden of Metabolic Risk Factors of Chronic Diseases Collaborating Group. (2012). *Global burden of metabolic risk factors of chronic diseases.* Available: http://www1.imperial.ac.uk/publichealth/departments/ebs/projects/eresh/majidezzati/healthmetrics/metabolicriskfactors/ [June 2012].

Global Education Project. (2004). *Earth: A graphic look at the state of the world.* Available: http://www.theglobaleducationproject.org/earth/human-conditions.php#4 [April 2012].

Global Health Observatory Data Repository. (2011). *Table 3. Age-standardized death rates per 100,000 by cause, sex and member state, 2008.* Geneva, Switzerland: World Health Organization. Available: http://www.who.int/gho/mortality_burden_disease/global_burden_disease_death_estimates_sex_2008.xls [August 2012].

Gluckman, P.D., and Hanson, M.A. (2006). The developmental origins of health and disease: An overview. In P.D. Gluckman and M.A. Hanson (Eds.), *Developmental origins of health and disease* (pp. 6-32). New York: Cambridge University Press.

Gluckman, P.D., Hanson, M.A., Cooper, C., and Thornburg, K.L. (2008). Effect of in utero and early-life conditions on adult health and disease. *New England Journal of Medicine, 359*(1), 61-73.

Gluckman, P.D., Hanson, M., and Mitchell, M. (2010). Developmental origins of health and disease: Reducing the burden of chronic disease in the next generation. *Genome Medicine, 2*(2), 14.

Godeau, E., Nic Gabhainn, S., Vignes, C., Ross, J., Boyce, W., and Todd, J. (2008). Contraceptive use by 15-year-old students at their last sexual intercourse: Results from 24 countries. *Archives of Pediatrics and Adolescent Medicine, 162*(1), 66-73.

Godfrey, K. (2006). The "developmental origins" hypothesis: Epidemiology. In P.D. Gluckman and M.A. Hanson (Eds.), *Developmental origins of health and disease* (pp. 6-32). New York: Cambridge University Press.

Gohmann, S.F. (2010). A comparison of health care in Canada and the United States: The case of pap smears. *Medical Care, 48*(11), 1,036-1,040.

Gold, M.R., Siegel, J.E., Russell, L.B. and Weinstein, M.C. (Eds). (1996). *Cost-effectiveness in health and medicine.* New York: Oxford University Press.

Goldberg, D.S. (2011). Against the very idea of the politicization of public health policy. *American Journal of Public Health, 102*(1), 44-49.

Goldberg, R.J., Spencer, F.A., Fox, K.A.A., Brieger, D., Steg, P.G., Gurfinkel, E., Dedrick, R., and Gore, J.M. (2009). Prehospital delay in patients with acute coronary syndromes (from the Global Registry of Acute Coronary Events [GRACE]). *American Journal of Cardiology, 103*(5), 598-603.

Goldman, L., and Cook, E.F. (1984). The decline in ischemic heart disease mortality rates. An analysis of the comparative effects of medical interventions and changes in lifestyle. *Annals of Internal Medicine, 101*(6), 825-836.

Gonzales, P.A. (2008). *Highlights from TIMSS 2007: Mathematics and science achievement of U.S. fourth- and eighth-grade students in an international context.* Washington, DC: National Center for Education Statistics, Institute of Education Sciences, U.S. Department of Education.

Goodman, A., Joyce, R., and Smith, J.P. (2011). The long shadow cast by childhood physical and mental problems on adult life. *Proceedings of the National Academy of Sciences, 108*(15), 6,032-6,037.

Goodman, L.A., Smyth, K.F., Borges, A.M., and Singer, R. (2009). When crises collide: How intimate partner violence and poverty intersect to shape women's mental health and coping? *Trauma, Violence, and Abuse, 10*(4), 306-329.

Gordon, R., MacKintosh, A.M., and Moodie, C. (2010). The impact of alcohol marketing on youth drinking behaviour: A two-stage cohort study. *Alcohol and Alcoholism, 45*(5), 470-480.

Gordon-Larsen, P., Nelson, M.C., Page, P., and Popkin, B.M. (2006). Inequality in the built environment underlies key health disparities in physical activity and obesity. *Pediatrics, 117*(2), 417-424.

Gore, F.M., Bloem, P.J.N., Patton, G.C., Ferguson, J., Joseph, V., Coffey, C., Sawyer, S.M., and Mathers, C.D. (2011). Global burden of disease in young people aged 10-24 years: A systematic analysis. *Lancet, 377*(9,783), 2,093-2,102.

Gornick, J.C., and Jäntti, M. (2010). Child poverty in upper-income countries: Lessons from the Luxembourg Income Study. In S. Kamerman, S. Phipps, and A. Ben-Arieh (Eds.). *From child welfare to child well-being, children's well-being: Indicators and research* (pp. 339-368). Dordrecht, Netherlands: Springer Science+Business Media B.V.

Governors Highway Safety Association. (2005). *Survey of the states: Speeding.* http://www. ghsa.org/html/publications/survey/speeding2005.html [June 2012].

Green, K.M., Doherty, E.E., Zebrak, K.A., and Ensminger, M.E. (2011). Association between adolescent drinking and adult violence: Evidence from a longitudinal study of urban African Americans. *Journal of Studies on Alcohol and Drugs, 72*(5), 701-710.

Greene, W.R., Oyetunji, T.A., Bowers, U., Haider, A.H., Mellman, T.A., Cornwell, E.E., Siram, S.M., and Chang, D.C. (2010). Insurance status is a potent predictor of outcomes in both blunt and penetrating trauma. *American Journal of Surgery, 199*(4), 554-557.

Greenlund, K.J., Kiefe, C.I., Giles, W.H., and Liu, K. (2010). Associations of job strain and occupation with subclinical atherosclerosis: The CARDIA study. *Annals of Epidemiology, 20*(5), 323-331.

Grier, S.A., and Kumanyika, S.K. (2008). The context for choice: Health implications of targeted food and beverage marketing to African Americans. *American Journal of Public Health, 98*(9), 1,616-1,629.

Griffin, S.O., Jones, J.A., Brunson, D., Griffin, P.M., and Bailey, W.D. (2012). Burden of oral disease among older adults and implications for public health priorities. *American Journal of Public Health, 102*(3), 411-418.

Grossman, D.C., Kim, A., Macdonald, S.C., Klein, P., Copass, M.K., and Maier, R.V. (1997). Urban-rural differences in prehospital care of major trauma. *Journal of Trauma and Acute Care Surgery, 42*(4), 723-729.

Grumbach, K., Chattopadhyay, A., and Bindman, A. (2009). *Fewer and more specialized: A new assessment of physician supply in California.* Oakland: California Healthcare Foundation.

Guyer, B., Ma, S., Grason, H., Frick, K.D., Perry, D.F., Sharkey, A., and McIntosh, J. (2009). Early childhood health promotion and its life course health consequences. *Academic Pediatrics, 9*(3), 142-149.

Hadley, J. (2003). Sicker and poorer—the consequences of being uninsured: A review of the research on the relationship between health insurance, medical care use, health, work, and income. *Medical Care Research and Review, 60*(Suppl. 2), S3-S75.

Haider, A.H., Ong'uti, S., Efron, D.T., Oyetunji, T.A., Crandall, M.L., Scott, V.K., Haut, E.R., Schneider, E.B., Powe, N.R., Cooper, L.A., and Cornwell, E.E., III. (2012). Association between hospitals caring for a disproportionately high percentage of minority trauma patients and increased mortality: A nationwide analysis of 434 hospitals. *Archives of Surgery, 147*(1), 63-70.

Hales, C.N., Barker, D.J., Clark, P.M., Cox, L.J., Fall, C., Osmond, C., and Winter, P.D. (1991). Fetal and infant growth and impaired glucose tolerance at age 64. *British Medical Journal, 303*(6,809), 1,019-1,022.

Halfon, N., and Hochstein, M. (2002). Life course health development: An integrated framework for developing health, policy, and research. *Milbank Quarterly, 80*(3), 433-479.

Halfon, N., Larson, K., and Shirley, R. (2010). Why social determinants? *Healthcare Quarterly, 14*(Suppl.), 8-20.

Hall, P.A., and M. Lamont, M. (Eds.). (2009). *Successful societies: And how institutions and culture affect health.* Cambridge, UK: Cambridge University Press.

Hall, R.E., and Jones, C.I. (2007). The value of life and the rise in health spending. *Quarterly Journal of Economics, 122*(1), 39-72.

Hallal, P.C., Andersen, L.B., Bull, F.C., Guthold, R., Haskell, W., and Ekelund, U. for the Lancet Physical Activity Series Working Group. (2012). Global physical activity levels: Surveillance progress, pitfalls, and prospects. *Lancet, 380*(9,838), 247-257.

Han, H. (2011). *Does obesity lead to poor school performance? Estimates from propensity score matching.* Princeton University papers. Available: http://paa2012.princeton.edu/papers/120431 [December 2012].

Hanratty, M.J. (1996). Canadian national health insurance and infant health. *American Economic Review, 86*(1), 276-284.

Hanratty, M.J., and Blank, R.M. (1990). *Down and out in North America: Recent trends in poverty rates in the U.S. and Canada.* (NBER Working Paper No. 3462). Cambridge, MA: National Bureau of Economic Research.

Hans, B. (2009). A critical reflection on the role of social democracy in reducing socioeconomic inequalities in health: A commentary on Sekine, Chandola, Martikainen, Marmot, and Kagamimori. *Social Science and Medicine, 69*(10), 1,426-1,428.

Hanushek, E.A., Jamison, D.T., Jamison, E.A., and Woessmann, L. (2008). Education and economic growth: It's not just going to school, but learning something while there that matters. *Education Next, 8*(2), 62-70.

Harrell, C.J.P., Burford, T.I., Cage, B.N., Nelson, T.M., Shearon, S., Thompson, A., and Green, S. (2011). Multiple pathways linking racism to health outcomes. *Du Bois Review: Social Science Research on Race, 8*(1), 143-157.

Harris, A.R., Fisher, G.A., and Thomas, S.H. (2012). Homicide as a medical outcome: Racial disparity in deaths from assault in U.S. level I and II trauma centers. *Journal of Trauma: Injury, Infection, and Critical Care, 72*(3), 773-782.

Harris, J.L., Pomeranz, J.L., Lobstein, T., and Brownell, K.D. (2009). A crisis in the marketplace: How food marketing contributes to childhood obesity and what can be done. *Annual Review of Public Health, 30*, 211-225.

Harsha, B., and Hedlund, J. (2007). Changing America's culture of speed on the roads. In J. Hedlund (Ed.), *Improving traffic safety culture in the U.S.: The journey forward* (pp. 257-272). Ithaca, NY: AAA Foundation for Traffic Safety. Available: http://www.aaafoundation. org/pdf/HarshaHedlund.pdf [June 2012].

Hatcher, R., Trussell, J., and Nelson, A. (2007). *Contraceptive technology* (19th revised ed.). New York: Ardent Media.

Health Behaviour in School-Aged Children. (2002). *Overview.* Available: http://www.hbsc. org/overview.html [April 2012].

Heath, G.W. (2009). The role of the public health sector in promoting physical activity: National, state, and local applications. *Journal of Physical Activity and Health, 6*(Suppl. 2), S159-S167.

Heckman, J.J. (2007). Economics of health and mortality special feature: The economics, technology, and neuroscience of human capability formation. *Proceedings of the National Academy of Sciences, 104*(33), 13,250-13,255.

Heinen, L. (2006). Obesity in (corporate) America: Large employer concerns and strategies of response. *North Carolina Medical Journal, 67*(4), 307-309.

Hemenway, D. (2010). Why we don't spend enough on public health. *New England Journal of Medicine, 362*(18), 1,657-1,658.

Hemenway, D., and Miller, M. (2000). Firearm availability and homicide rates across 26 high-income countries. *Journal of Trauma, 49*(6), 985-988.

Hemenway, D., Vriniotis, M., Johnson, R.M., Miller, M., and Azrael, D. (2011). Gun carrying by high school students in Boston, MA: Does overestimation of peer gun carrying matter? *Journal of Adolescence, 34*(5), 997-1,003.

Hepburn, L.M., and Hemenway, D. (2004). Firearm availability and homicide: A review of the literature. *Aggression and Violent Behavior, 9*(4), 417-440.

Hepburn, L., Miller, M., Azrael, D., and Hemenway, D. (2007). The U.S. gun stock: Results from the 2004 national firearms survey. *Injury Prevention, 13*(1), 15-19.

Hernán, M.A., and Robins, J.M. (2006). Instruments for causal inference: An epidemiologist's dream? *Epidemiology, 17*(4), 360-372.

Hertz, T. (2006). *Understanding mobility in America.* Washington, DC: Center for American Progress.

Hertzman, C. (2000). The biological embedding of early experience and its effects on health in adulthood. *Annals of the New York Academy of Sciences, 896*(1), 85-95.

Heuveline, P., Timberlake, J.M., and Furstenberg, F.F. (2003). Shifting childrearing to single mothers: Results from 17 Western countries. *Population and Development Review, 29*(1), 47-71.

Hillis, S.D., Anda, R.F., Dube, S.R., Felitti, V.J., Marchbanks, P.A., and Marks, J.S. (2004). The association between adverse childhood experiences and adolescent pregnancy, long-term psychosocial consequences, and fetal death. *Pediatrics, 113*(2), 320-327.

Himmelstein, D.U., Thorne, D., Warren, E., and Woolhandler, S. (2009). Medical bankruptcy in the United States, 2007: Results of a national study. *American Journal of Medicine, 122*(8), 741-746.

Hirschfeld, S., Kramer, B., and Guttmacher, A. (2010). Current status of the National Children's Study. *Epidemiology, 21*(5), 605-606.

Ho, J.Y., and Preston, S.H. (2010). U.S. mortality in an international context: Age variations. *Population and Development Review, 36*(4), 749-773.

Ho, J.Y., and Preston, S.H. (2011). *International comparisons of U.S. mortality.* Unpublished data analyses for the NAS/IOM Panel on Understanding Cross-National Health Differences Among High-Income Countries. Population Studies Center, University of Pennsylvania.

Hoehner, C.M., Barlow, C.E., Allen, P., and Schootman, M. (2012). Commuting distance, cardiorespiratory fitness, and metabolic risk. *American Journal of Preventive Medicine, 42*(6), 571-578.

Hoffman, S., and Maynard, R. (2008). The costs of adolescent childbearing. In S. Hoffman and R.A. Maynard (Eds.), *Kids having kids: The economic costs and social consequences of teen pregnancy.* Washington, DC: The Urban Institute Press.

Hoffmeyer-Zlotnik, J.H.P., and Harkness, J.A. (Eds.). (2005). *Methodological aspects in cross-national research.* Available: http://www.gesis.org/fileadmin/upload/forschung/publikatio nen/zeitschriften/zuma_nachrichten_spezial/znspezial11.pdf [June 2012].

Hofstede, G. (1998). Comparative studies of sexual behavior: Sex as achievement or as relationship. In G. Hofstede (Ed.), *Masculinity and femininity: The taboo dimension of national cultures* (pp. 153-178). Thousand Oaks, CA: Sage.

Horwitz, A.V., Widom, C.S., McLaughlin, J., and White, H.R. (2001). The impact of childhood abuse and neglect on adult mental health: A prospective study. *Journal of Health and Social Behavior, 42*(2), 184-201.

Howard, D.H., Richardson, L.C., and Thorpe, K.E. (2009). Cancer screening and age in the United States and Europe. *Health Affairs, 28*(6), 1,838-1,847.

Howell, G.M., Peitzman, A.B., Nirula, R., Rosengart, M.R., Alarcon, L.H., Billiar, T.R., and Sperry, J.L. (2010). Delay to therapeutic interventional radiology postinjury: Time is of the essence. *Journal of Trauma and Acute Care Surgery, 68*(6), 1,296-1,300.

Hruschka, D.J., Brewis, A.A., Wutich, A., and Morin, B. (2011). Shared norms and their explanation for the social clustering of obesity. *American Journal of Public Health, 101*(Suppl. 1), S295-S300.

Huijts, T., and Eikemo, T.A. (2009). Causality, social selectivity or artifacts? Why socioeconomic inequalities in health are not smallest in the Nordic countries. *European Journal of Public Health, 19*(5), 452-453.

Huisman, M., Kunst, A.E., and Mackenbach, J.P. (2005). Inequalities in the prevalence of smoking in the European Union: Comparing education and income. *Preventive Medicine, 40*(6), 756-764.

Human Mortality Database. (2012). University of California, Berkeley (USA) and Max Planck Institute for Demographic Research (Germany). Available: http://www.mortality.org or http://www.humanmortality.de.

Hunink, M.G.M., Goldman, L., Tosteson, A.N.A., Mittleman, M.A., Goldman, P.A., Williams, L.W., Tsevat, J., and Weinstein, M.C. (1997). The recent decline in mortality from coronary heart disease, 1980-1990. *Journal of the American Medical Association, 277*(7), 535-542.

Hurrelmann, K., Rathmann, K., and Richter, M. (2011). Health inequalities and welfare state regimes. A research note. *Journal of Public Health, 19*(1), 3-13.

Huynh, P.T., Schoen, C., Osborn, R., and Holmgren, A.L. (2006). *The U.S. health care divide: Disparities in primary care experiences by income.* New York: Commonwealth Fund.

Inglehart, R., and Welzel, C. (2005). *Modernization, cultural change, and democracy: The human development sequence.* New York: Cambridge University Press.

Inoue, S., Murase, N., Shimomitsu, T., Ohya, Y., Odagiri, Y., Takamiya, T., Ishii, K., Katsumura, T., and Sallis, J.F. (2009). Association of physical activity and neighborhood environment among Japanese adults. *Preventive Medicine, 48*(4), 321-325.

Institute for Social Research. (2011). The Panel Study of Income Dynamics. Available: http// psidonline.isr.umich.edu/ [April 2012].

Institute of Medicine. (1988). *The future of public health.* Committee for the Study of the Future of Public Health. Washington, DC: National Academy Press.

Institute of Medicine. (1996). *Primary care: America's health in a new era.* M.S. Donaldson, K.D. Yordy, K.N. Lohr, and N.A. Vanselow (Eds). Committee on the Future of Primary Care. Washington, DC: National Academy Press.

Institute of Medicine. (2000). *To err is human: Building a safer health system.* L.T. Kohn, J.M. Corrigan, and M.S. Donaldson (Eds.). Committee on Quality of Health Care in America. Washington, DC: National Academy Press.

Institute of Medicine. (2001). *Crossing the quality chasm: A new health system for the 21st century.* Committee on Quality of Health Care in America. Washington, DC: National Academy Press.

Institute of Medicine. (2003a). *The future of the public's health in the 21st century.* Committee on Assuring the Health of the Public in the 21st Century. Washington, DC: The National Academies Press.

Institute of Medicine. (2003b). *Hidden costs, value lost: Uninsurance in America.* Committee on the Consequences of Uninsurance. Washington, DC: The National Academies Press.

Institute of Medicine. (2003c). *Priority areas for national action: Transforming health care quality.* K. Adams and J.M. Corrigan (Eds.). Committee on Identifying Priority Areas for Quality Improvement. Washington, DC: The National Academies Press.

Institute of Medicine. (2003d). *Unequal treatment: Confronting racial and ethnic disparities in health care.* B.D. Smedley, A.Y. Stith, and A.R. Nelson (Eds.). Committee on Understanding and Eliminating Racial and Ethnic Disparities in Health Care. Washington, DC: The National Academies Press.

Institute of Medicine. (2006a). *Food marketing to children and youth: Threat or opportunity?* J.M. McGinnis, J. Appleton Gootman, and V.I. Kraak (Eds.). Committee on Food Marketing and the Diets of Children and Youth. Washington, DC: The National Academies Press.

Institute of Medicine. (2006b). *Genes, behavior, and the social environment: Moving beyond the nature/nurture debate.* L.M. Hernandez and D.G. Blazer (Eds.). Committee on Assessing Interactions Among Social, Behavioral, and Genetic Factors in Health. Washington, DC: The National Academies Press.

Institute of Medicine. (2007a). *Future of emergency care: Dissemination workshop summaries.* M. McHugh and P. Slavin (Rapporteurs). The Future of Emergency Care Workshop Planning Group. Washington, DC: The National Academies Press.

Institute of Medicine. (2007b). *Preventing medication errors; Quality chasm series.* P. Aspden, J. Wolcott, J.L. Bootman, and L.R. Cronenwett (Eds.). Committee on Identifying and Preventing Medication Errors. Washington, DC: The National Academies Press.

Institute of Medicine. (2007c). *Rewarding provider performance: Aligning incentives in Medicare (pathways to quality health care series).* Committee on Redesigning Health Insurance Performance Measures, Payment, and Performance Improvement Programs. Washington, DC: The National Academies Press.

Institute of Medicine. (2009a). *America's uninsured crisis: Consequences for health and health care.* Committee on Health Insurance Status and Its Consequences. Washington, DC: The National Academies Press.

Institute of Medicine. (2009b). *Local government actions to prevent childhood obesity*. L. Parker, A.C. Burns, and E. Sanchez (Eds.). Committee on Childhood Obesity Prevention Actions for Local Governments. Washington, DC: The National Academies Press.

Institute of Medicine. (2009c). *School meals: Building blocks for healthy children*. V.A. Stallings, C. West Suitor, and C.L. Taylor (Eds.). Committee on Nutrition Standards for National School Lunch and Breakfast Programs. Washington, DC: The National Academies Press.

Institute of Medicine. (2009d). *State of the USA health indicators: Letter report*. Committee on the State of the USA Health Indicators. Washington, DC: The National Academies Press.

Institute of Medicine. (2010). *The healthcare imperative: Lowering costs and improving outcomes: Workshop series summary*. P.L. Yong and L.A. Olsen (Rapporteurs). Roundtable on Evidence-Based Medicine. Washington, DC: The National Academies Press.

Institute of Medicine. (2011a). *Advancing oral health in America*. Committee on an Oral Health Initiative. Washington, DC: The National Academies Press.

Institute of Medicine. (2011b). *Digital infrastructure for the learning health system: The foundation for continuous improvement in health and health care: Workshop series summary*. C. Grossman, B. Powers, and J.M. McGinnis (Rapporteurs and Editors). Roundtable on Value & Science-Driven Health Care. Washington, DC: The National Academies Press.

Institute of Medicine. (2011c). *Early childhood obesity prevention policies*. L.L. Birch, L. Parker, and A.C. Burns (Eds.). Committee on Obesity Prevention Policies for Young Children. Washington, DC: The National Academies Press.

Institute of Medicine. (2011d). *For the public's health: Revitalizing law and policy to meet new challenges*. Committee on Public Health Strategies to Improve Health. Washington, DC: The National Academies Press.

Institute of Medicine. (2011e). *For the public's health: The role of measurement in action and accountability*. Committee on Public Health Strategies to Improve Health. Washington, DC: The National Academies Press.

Institute of Medicine. (2011f). *Hunger and obesity: Understanding a food insecurity paradigm: Workshop summary*. L.M. Troy, E.A. Miller, and S. Olson (Rapporteurs). Washington, DC: The National Academies Press.

Institute of Medicine. (2011g). *Leading health indicators for Healthy People 2020: Letter report*. Committee on Leading Health Indicators for Healthy People 2020. Washington, DC: The National Academies Press.

Institute of Medicine. (2012). *For the public's health: Investing in a healthier future*. Committee on Public Health Strategies to Improve Health. Washington, DC: The National Academies Press.

Institute of Medicine and Committee on the Future of Emergency Care in the U.S. Health System. (2006). The future of emergency care in the United States health system. *Annals of Emergency Medicine, 48*(2), 115-120.

Institute of Medicine and National Research Council. (2011). *Child and adolescent health and health care quality: Measuring what matters*. Committee on Pediatric Health and Health Care Quality Measures. Washington, DC: The National Academies Press.

International Association for the Study of Obesity. (2011). *World map of obesity: Global prevalence of adult obesity*. Available: http://www.iaso.org/resources/world-map-obesity/ [August 2012].

International Labour Office and OECD. (2011). *Toward national social protection floors*. A policy note for the G20 meeting of Labour and Employment Ministers, Paris, September 26-26, 2011. Geneva, Switzerland: International Labour Office and Paris: OECD.

International Transport Forum. (2012). *Road safety annual report 2011*. Available: http://www.internationaltransportforum.org/irtadpublic/pdf/11IrtadReport.pdf [August 2012].

International Union for Health Promotion and Education. (2009). *Making a difference to global health*. Available: http://www.iuhpe.org/index.html?page=5&lang=en [April 2012].

Isaacs, J., Sawhill, I.V., and Haskins, R. (2008). *Getting ahead or losing ground: Economic mobility in America*. Washington, DC: The Brookings Institution.

Isaacs, J., Toran, K., Hahn, H., Fortuny, K., and Steuerle, C.E. (2012). *Kids'share: 2012 report on federal expenditures on children through 2011*. Washington, DC: Urban Institute. Available: http://www.urban.org/UploadedPDF/412600-Kids-Share-2012.pdf [August 2012].

Jamison, D.T., and Sandbu, M.E. (2001). WHO ranking of health system performance. *Science, 293*(5535), 1595-1596.

Janssen, F., Peeters, A., Mackenbach, J.P., and Kunst, A.E. (2005a). Relation between trends in late middle age mortality and trends in old age mortality—Is there evidence for mortality selection? *Journal of Epidemiology and Community Health, 59*(9), 775-781.

Janssen, F., Kunst, A., and Mackenbach, J. (2007). Variations in the pace of old-age mortality decline in seven European countries, 1950-1999: The role of smoking and other factors earlier in life. *European Journal of Population (Revue européenne de Démographie), 23*(2), 171-188.

Janssen, F., Kunst, A.E., and Mackenbach, J.P. (2008). Variations in the pace of mortality decline in elderly in 7 northwestern European countries between 1950-1990: The impact of smoking. *Variaties in het tempo van sterftedaling onder ouderen in 7 Noordwest-Europese landen tussen 1950-1999: De rol van roken, 152*(26), 1,478-1,484.

Janssen, I., and LeBlanc, A.G. (2010). Systematic review of the health benefits of physical activity and fitness in school-aged children and youth. *International Journal of Behavioral Nutrition and Physical Activity, 7*(1), 40.

Janssen, I., Katzmarzyk, P.T., Boyce, W.F., Vereecken, C., Mulvihill, C., Roberts, C., Currie, C., Pickett, W., Nemeth, Á., Ojala, K., Ravens-Sieberer, U., Todd, J., and Woynarowska, B. (2005b). Comparison of overweight and obesity prevalence in school-aged youth from 34 countries and their relationships with physical activity and dietary patterns. *Obesity Reviews, 6*(2), 123-132.

Jäntti, M., and Danziger, S. (1994). Child poverty in Sweden and the United States: The effect of social transfers and parental labor force participation. *Industrial and Labor Relations Review, 48*(1), 48-64.

Jäntti, M., Bratsberg, B., Røed, K., Raaum, O., Naylor, R., Österbacka, E., Björklund, A., and Eriksson, T. (2005). *American exceptionalism in a new light: A comparison of intergenerational earnings mobility in the Nordic countries, the United Kingdom and the United States*. (No. 781, Warwick Economic Research Papers). Warwick, UK: Department of Economics, University of Warwick.

Jemal, A., Thun, M.J., Ward, E.E., Henley, S.J., Cokkinides, V.E., and Murray, T.E. (2008a). Mortality from leading causes by education and race in the United States, 2001. *American Journal of Preventive Medicine, 34*(1), 1-8.

Jemal, A., Ward, E., Anderson, R.N., Murray, T., and Thun, M.J. (2008b). Widening of socioeconomic inequalities in U.S. death rates, 1993-2001. *PLoS ONE, 3*(5).

Jenkins, C.D., and Runyan, D.K. (2005). What's killing Americans in the prime of life? *International Journal of Health Services, 35*(2), 291-311.

Johnson, R.M., Barber, C., Azrael, D., Clark, D.E., and Hemenway, D. (2010). Who are the owners of firearms used in adolescent suicides? *Suicide and Life-Threatening Behavior, 40*(6), 609-611.

Kaczynski, A.T., and Henderson, K.A. (2008). Parks and recreation settings and active living: A review of associations with physical activity function and intensity. *Journal of Physical Activity and Health, 5*(4), 619-632.

Kahn, J.R., and Anderson, K.E. (1992). Intergenerational patterns of teenage fertility. *Demography, 29*(1), 39-57.

Kalaydjian, A., Swendsen, J., Chiu, W.T., Dierker, L., Degenhardt, L., Glantz, M., Merikangas, K.R., Sampson, N., and Kessler, R. (2009). Sociodemographic predictors of transitions across stages of alcohol use, disorders, and remission in the national comorbidity survey replication. *Comprehensive Psychiatry, 50*(4), 299-306.

Karaca-Mandic, P., Jena, A.B., Joyce, G.F., and Goldman, D.P. (2012). Out-of-pocket medication costs and use of medications and health care services among children with asthma. *Journal of the American Medical Association, 307*(12), 1,284-1,291.

Karch, D.L., Logan, J., and Patel, N. (2011). Surveillance for violent deaths—National Violent Death Reporting System, 16 states, 2008. *Morbidity and Mortality Weekly Report Surveillance Summaries, 60*(SS10), 1-54.

Karim, S.A., Eikemo, T.A., and Bambra, C. (2010). Welfare state regimes and population health: Integrating the East Asian welfare states. *Health Policy, 94*(1), 45-53.

Karoly, L.A., Kilburn, M.R., and Cannon, J., S. (2005). *Early childhood interventions: Proven results, future promise.* Santa Monica, CA: RAND Corporation.

Kasza, G. (2002). The illusion of welfare regimes. *Journal of Social Policy, 31,* 271-287.

Katz, D.L. (2009). School-based interventions for health promotion and weight control: Not just waiting on the world to change. *Annual Review of Public Health, 30,* 253-272.

Kaul, P., Armstrong, P.W., Chang, W.-C., Naylor, C.D., Granger, C.B., Lee, K.L., Peterson, E.D., Califf, R.M., Topol, E.J., and Mark, D.B. (2004). Long-term mortality of patients with acute myocardial infarction in the United States and Canada. *Circulation, 110*(13), 1,754-1,760.

Kawachi, I. (1999). Social capital and community effects on population and individual health. *Annals of New York Academy of Sciences, 896*(1), 120-130.

Kawachi, I., and Berkman, L. (2000). Social cohesion, social capital, and health. In L.F. Berkman and I. Kawachi (Eds.). *Social epidemiology* (pp. 174-190). New York: Oxford University Press.

Kawachi, I., and Subramanian, S.V. (2007). Neighbourhood influences on health. *Journal of Epidemiology and Community Health, 61*(1), 3-4.

Kawachi, I., Kennedy, B.P., Lochner, K., and Prothrow-Stith, D. (1997). Social capital, income inequality, and mortality. *American Journal of Public Health, 87*(9), 1,491-1,498.

Kawachi, I., Daniels, N., and Robinson, D.E. (2005). Health disparities by race and class: Why both matter. *Health Affairs, 24*(2), 343-352.

Kawachi, I., Adler, N.E., and Dow, W.H. (2010). Money, schooling, and health: Mechanisms and causal evidence. *Annals of the New York Academy of Sciences, 1,186*(1), 56-68.

Keating, D.P., and Hertzman, C. (1999). *Developmental health and the wealth of nations: Social, biological, and educational dynamics.* New York: Guilford Press.

Kelly, M.P., Bonnefoy, J., Morgan, A., and Florenzano, F. (2006). *The development of the evidence base about the social determinants of health.* Geneva, Switzerland: WHO, Commission on Social Determinants of Health, Measurement and Evidence Knowledge Network.

Kenworthy, L. (1998). *Luxembourg Income Study, Working Paper No. 188. Do social-welfare policies reduce poverty?: A cross-national assessment.* Syracuse, NY: Maxwell School of Citizenship and Public Affairs, Syracuse University.

Kessler, D. (2001). *A question of intent: a great American battle with a deadly industry.* New York: PublicAffairs.

Kessler, R.C. (2007). The global burden of anxiety and mood disorders: Putting the European Study of the Epidemiology of Mental (ESEMeD) findings into perspective. *Journal of Clinical Psychiatry, 68*(Suppl. 2), 10-19.

Kessler, R.C., Berglund, P., Demler, O., Jin, R., Merikangas, K.R., and Walters, E.E. (2005). Lifetime prevalence and age-of-onset distributions of DSM-IV disorders in the National Comorbidity Survey Replication. *Archives of General Psychiatry, 62,* 593-602.

Kessler, R.C., Amminger, G.P., Aguilar-Gaxiola, S., Alonso, J., Lee, S., and Ustun, T.B. (2007). Age of onset of mental disorders: A review of recent literature. *Current Opinion in Psychiatry, 20,* 359-364.

Keyfitz, N., and Flieger, W. (1990). *World population growth and aging: Demographic trends in the late twentieth century.* Chicago: University of Chicago Press.

Khush, K.K., Rapaport, E., and Waters, D. (2005). The history of the coronary care unit. *Canadian Journal of Cardiology, 21*(12), 1,041-1,045.

Killias, M. (1993). International correlations between gun ownership and rates of homicide and suicide. *Canadian Medical Association Journal, 148*(10), 1,721-1,725.

Kim, Y.J. (2011). The impact of time from ED arrival to surgery on mortality and hospital length of stay in patients with traumatic brain injury. *Journal of Emergency Nursing, 37*(4), 328-333.

Kindig, D., Peppard, P., Booske, B. (2010). How healthy could a state be? *Public Health Reports, 125*(2), 160-167.

Knudsen, E.I., Heckman, J.J., Cameron, J.L., and Shonkoff, J.P. (2006). Economic, neurobiological, and behavioral perspectives on building America's future workforce. *Proceedings of the National Academy of Sciences, 103*(27), 10,155-10,162.

Kociol, R.D., Lopes, R.D., Clare, R., Thomas, L., Mehta, R.H., Kaul, P., Pieper, K.S., Hochman, J.S., Weaver, W.D., Armstrong, P.W., Granger, C.B., and Patel, M.R. (2012). International variation in and factors associated with hospital readmission after myocardial infarction. *Journal of the American Medical Association, 307*(1), 66-74.

Korpi, W., and Palme, J. (1998). The paradox of redistribution and the strategy of equality: Welfare state institutions, inequality and poverty in the Western countries. *American Sociological Review, 63,* 662-687.

Krieger, N., Rehkopf, D.H., Chen, J.T., Waterman, P.D., Marcelli, E., and Kennedy, M. (2008). The fall and rise of U.S. inequities in premature mortality: 1960-2002. *PLoS Medicine, 5*(2), e46.

Krieger, N., Kosheleva, A., Waterman, P.D., Chen, J.T., and Koenen, K. (2011). Racial discrimination, psychological distress, and self-rated health among U.S.-born and foreign-born black Americans. *American Journal of Public Health, 101*(9), 1,704-1,713.

Krug, E., Powell, K., and Dahlberg, L. (1998). Firearm-related deaths in the United States and 35 other high- and upper-middle-income countries. *International Journal of Epidemiology, 27*(2), 214-221.

Kuehn, B.M. (2012). Poisonings top crashes for injury-related deaths. *Journal of the American Medical Association, 307*(3), 242.

Kuh, D., and Ben-Shlomo, Y. (2004). *A life course approach to chronic disease epidemiology* (2nd ed.). New York: Oxford University Press.

Kuh, D., Power, C., Blane, D., and Bartley, M. (1997). Social pathways between childhood and adult health. In D. Kuh and Y. Ben-Shlomo (Eds.), *A life course approach to chronic disease epidemiology* (pp. 169-198). New York: Oxford University Press.

Kuh, D., Hardy, R., Langenberg, C., Richards, M., and Wadsworth, M. (2002). Mortality in adults aged 26-54 years related to socioeconomic conditions in childhood and adulthood: Post war birth cohort study. *British Medical Journal, 325*(7,372), 1,076-1,080.

Kulkarni, S., Levin-Rector, A., Ezzati, M., and Murray, C. (2011). Falling behind: Life expectancy in U.S. counties from 2000 to 2007 in an international context. *Population Health Metrics, 9*(1), 16.

Kunitz, S.J., and Pesis-Katz, I. (2005). Mortality of white Americans, African Americans, and Canadians: The causes and consequences for health of welfare state institutions and policies. *Milbank Quarterly, 83*(1), 5-39.

Kunst, A.E., Groenhof, F., and Mackenbach, J.P. (1998). Mortality by occupational class among men 30-64 years in 11 European countries. EU Working Group on Socioeconomic Inequalities in Health. *Social Science and Medicine, 46*(11), 1,459-1,476.

Kwate, N.O., and Meyer, I.H. (2009). Association between residential exposure to outdoor alcohol advertising and problem drinking among African American women in New York City. *American Journal of Public Health, 99*(2), 228-230.

Laatikainen, T., Critchley, J., Vartiainen, E., Salomaa, V., Ketonen, M., and Capewell, S. (2005). Explaining the decline in coronary heart disease mortality in Finland between 1982 and 1997. *American Journal of Epidemiology, 162*(8), 764-773.

Lahelma, E., and Lundberg, O. (2009). Health inequalities in European welfare states. *European Journal of Public Health, 19*(5), 445-446.

Lanphear, B.P., Kahn, R.S., Berger, O., Auinger, P., Bortnick, S.M., and Nahhas, R.W. (2001). Contribution of residential exposures to asthma in U.S. children and adolescents. *Pediatrics, 107*(6), e98.

Lantz, P.M., Golberstein, E., House, J.S., and Morenoff, J. (2010). Socioeconomic and behavioral risk factors for mortality in a national 19-year prospective study of U.S. adults. *Social Science and Medicine, 70*(10), 1,558-1,566.

Larson, N.I., Story, M.T., and Nelson, M.C. (2009). Neighborhood environments: Disparities in access to healthy foods in the U.S. *American Journal of Preventive Medicine, 36*(1), 74-81.

Larsson, S., Lawyer, P., Garellick, G., Lindahl, B., and Lundström, M. (2012). Use of 13 disease registries in 5 countries demonstrates the potential to use outcome data to improve health care's value. *Health Affairs, 31*(1), 220-227.

Laumbach, R.J., and Kipen, H.M. (2012). Respiratory health effects of air pollution: Update on biomass smoke and traffic pollution. *Journal of Allergy and Clinical Immunology, 129*(1), 3-11.

LaVeist, T., Gaskin, D., and Richard, P. (2009) *The economic burden of health inequalities in the United States.* Joint Center for Political and Economic Studies. Available: http://www.jointcenter.org/hpi/sites/all/files/Burden_Of_Health_FINAL_0.pdf [August 2012].

LaVeist, T., Pollack, K., Thorpe, R., Jr., Fesahazion, R., and Gaskin, D. (2011). Place, not race: Disparities dissipate in southwest Baltimore when blacks and whites live under similar conditions. *Health Affairs, 30*(10), 1,880-1,887.

Lawlor, D.A., Batty, G.D., Morton, S.M.B., Clark, H., Macintyre, S., and Leon, D.A. (2005). Childhood socioeconomic position, educational attainment, and adult cardiovascular risk factors: The Aberdeen children of the 1950s cohort study. *American Journal of Public Health, 95*(7), 1,245-1,251.

Lee, I.-M., Shiroma, E.J., Lobelo, F., Puska, P., Blair, S.N., and Katzmarzyk, P.T. for the Lancet Physical Activity Series Working Group. (2012). Effect of physical inactivity on major non-communicable diseases worldwide: An analysis of burden of disease and life expectancy. *Lancet, 380*(9,838), 219-229.

Lee, K.S., Park, S.C., Khoshnood, B., Hsieh, H.L., and Mittendorf, R. (1997). Human development index as a predictor of infant and maternal mortality rates. *Journal of Pediatrics, 131*(3), 430-433.

Leibfried, S. (1992). Towards a European welfare state. In Z. Ferge and J.E. Kolberg (Eds.), *Social policy in a changing Europe.* Frankfurt, Germany: Campus-Verlag.

Levin, R., Brown, M.J., Kashtock, M.E., Jacobs, D.E., Whelan, E.A., Rodman, J., Schock, M.R., Padilla, A., and Sinks, T. (2008). Lead exposures in U.S. children, 2008: Implications for prevention. *Environmental Health Perspectives, 116*(10), 1,285-1,293.

Link, B.G., and Phelan, J. (1995). Social conditions as fundamental causes of disease. *Journal of Health and Social Behavior*, Spec. No. 80-94.

Liss, D.T., Chubak, J., Anderson, M.L., Saunders, K.W., Tuzzio, L., and Reid, R.J. (2011). Patient-reported care coordination: Associations with primary care continuity and specialty care use. *Annals of Family Medicine, 9*(4), 323-329.

Lleras-Muney, A. (2005). The relationship between education and adult mortality in the United States. *Review of Economic Studies, 72*(1), 189-221.

Lock, K., Pomerleau, J., Causer, L., and McKee, M. (2004). Low fruit and vegetable consumption. In M. Ezzati, A. Lopez, A. Rodgers, and C. Murray (Eds.), *Comparative quantification of health risks: Global and regional burden of disease attributable to selected major risk factors*. Geneva, Switzerland: World Health Organization.

Lopez, A.D., Collishaw, N.E., and Piha, T. (1994). A descriptive model of the cigarette epidemic in developed countries. *Tobacco Control, 3*(3), 242-247.

Lopez, A.D., Mathers, C., Ezzati, M., Jamison, D., and Murray, C.J.L. (Eds.). (2006). *Global burden of disease and risk factors*. Washington, DC: World Bank.

Lovasi, G.S., Hutson, M.A., Guerra, M., and Neckerman, K.M. (2009). Built environments and obesity in disadvantaged populations. *Epidemiologic Reviews, 31*(1), 7-20.

Lovato, C., Linn, G., Stead, L.F., and Best, A. (2003). Impact of tobacco advertising and promotion on increasing adolescent smoking behaviours. *Cochrane Database of Systematic Review* (4), CD003439.

Low, C.A., Thurston, R.C., and Matthews, K.A. (2010). Psychosocial factors in the development of heart disease in women: Current research and future directions. *Psychosomatic Medicine, 72*(9), 842-854.

Lu, C.Y., and Roughead, E. (2011). Determinants of patient-reported medication errors: A comparison among seven countries. *International Journal of Clinical Practice, 65*(7), 733-740.

Ludwig, J., Sanbonmatsu, L., Gennetian, L., Adam, E., Duncan, G.J., Katz, L.F., Kessler, R.C., Kling, J.R., Lindau, S.T., Whitaker, R.C., and McDade, T.W. (2011). Neighborhoods, obesity, and diabetes—A randomized social experiment. *New England Journal of Medicine, 365*(16), 1,509-1,519.

Lundberg, O., Yngwe, M.Å., Stjärne, M.K., Elstad, J.I., Ferrarini, T., Kangas, O., Norström, T., Palme, J., and Fritzell, J. (2008). The role of welfare state principles and generosity in social policy programmes for public health: An international comparative study. *Lancet, 372*(9,650), 1,633-1,640.

Luxembourg Income Study. (2012). *Cross-national data center: Key figures*. Available: http://www.lisdatacenter.org/data-access/key-figures/download-key-figures/ [April 2012].

Lynch, J., and Davey Smith, G. (2005). A life course approach to chronic disease epidemiology. *Annual Review of Public Health, 26*, 1-35.

Lynch, J., Davey Smith, G., Hillemeier, M., Shaw, M., Raghunathan, T., and Kaplan, G. (2001). Income inequality, the psychosocial environment, and health: Comparisons of wealthy nations. *Lancet, 358*(9,277), 194-200.

Lynch, J., Davey Smith, G., Harper, S., and Hillemeier, M. (2004a). Is income inequality a determinant of population health? Part 2. U.S. national and regional trends in income inequality and age- and cause-specific mortality. *Milbank Quarterly, 82*(2), 355-400.

Lynch, J., Davey Smith, G., Harper, S., Hillemeier, M., Ross, N., Kaplan, G.A., and Wolfson, M. (2004b). Is income inequality a determinant of population health? Part 1. A systematic review. *Milbank Quarterly, 82*(1), 5-99.

MacDorman, M., and Mathews, T.J. (2009). Behind international rankings of infant mortality: How the United States compares with Europe. *NCHS Data Brief* (23), 1-8.

Macinko, J., Starfield, B., and Shi, L. (2003). The contribution of primary care systems to health outcomes within organization for economic cooperation and development (OECD) countries, 1970-1998. *Health Services Research, 38*(3), 831-865.

Macinko, J., Starfield, B., and Shi, L. (2007). Quantifying the health benefits of primary care physician supply in the United States. *International Journal of Health Services, 37*(1), 111-126.

Mackenbach, J. (2012). The persistence of health inequalities in modern welfare states: The explanation of a paradox. *Social Science & Medicine, 75*(4), 761-769.

Mackenbach, J.P., Kunst, A.E., Cavelaars, A.E., Groenhof, F., and Geurts, J.J. (1997). Socioeconomic inequalities in morbidity and mortality in Western Europe. The EU Working Group on Socioeconomic Inequalities in Health. *Lancet, 349*(9,066), 1,655-1,659.

Mackenbach, J.P., Cavelaars, A.E.J.M., Kunst, A.E., Groenhof, F., Andersen, O., Borgan, J.K., Costa, G., Crialesi, R., Desplanques, G., Faggiano, F., Geurts, J.J.M., Grötvedt, L., Harding, S., Helmert, U., Junker, C.H.R., Lahelma, E., Lundberg, O., Matikainen, P., Matheson, J., Mielk, A., Minder, C., Mizrahi, A.R., Mizrahi, A., Nolan, B., Rasmussen, N., Regidor, E., Giraldes, M.D.R., Spuhler, T., Vâgerö, D., and Valkonen, T. (2000). Socioeconomic inequalities in cardiovascular disease mortality. An international study. *European Heart Journal, 21*(14), 1,141-1,151.

Mackenbach, J.P., Martikainen, P., Looman, C.W.N., Dalstra, J.A.A., Kunst, A.E., Lahelma, E., Breeze, E., Cambois, E., Grundy, E., Lunde, E., van Oyen, H., and Rasmussen, N. (2005). The shape of the relationship between income and self-assessed health: An international study. *International Journal of Epidemiology, 34*(2), 286-293.

Mackenbach, J.P., Stirbu, I., Roskam, A.-J.R., Schaap, M.M., Menvielle, G., Leinsalu, M., and Kunst, A.E. (2008). Socioeconomic inequalities in health in 22 European countries. *New England Journal of Medicine, 358*(23), 2,468-2,481.

MacKenzie, E.J., Rivara, F.P., Jurkovich, G.J., Nathens, A.B., Frey, K.P., Egleston, B.L., Salkever, D.S., and Scharfstein, D.O. (2006). A national evaluation of the effect of trauma-center care on mortality. *New England Journal of Medicine, 354*(4), 366-378.

Madsen, M., Andersen, A.-M.N., Christensen, K., Andersen, P.K., and Osler, M. (2010). Does educational status impact adult mortality in Denmark? A twin approach. *American Journal of Epidemiology, 172*(2), 225-234.

Mair, C.F., Roux, A.V.D., and Galea, S. (2008). Are neighborhood characteristics associated with depressive symptoms? A critical review. *Journal of Epidemiology and Community Health, 62*(11), 940-946.

Mann, T.E., and Ornstein, N.J. (2008). *The broken branch: How Congress is failing America and how to get it back on track.* New York: Oxford University Press.

Manton, K.G., and Vaupel, J.W. (1995). Survival after the age of 80 in the United States, Sweden, France, England, and Japan. *New England Journal of Medicine, 333*(18), 1,232-1,235.

Markides, K.S., and Eschbach, K. (2005). Aging, migration, and mortality: Current status of research on the Hispanic paradox. *Journals of Gerontology Series B: Psychological Sciences and Social Sciences, 60*(Special Issue 2), S68-S75.

Marmor, T., Freeman, R., and Okma, K. (2009). Comparative policy analysis and health care: An introduction. In T. Marmor, R. Freeman, and K. Okma (Eds.), *Comparative studies and the politics of modern medical care* (pp. 1-23). New Haven, CT: Yale University Press.

Marmot, M.G. (2005). Social determinants of health inequalities. *Lancet, 365*(9,464), 1,099-1,104.

Marmot, M.G. (2006). Status syndrome: A challenge to medicine. *Journal of the American Medical Association, 295*(11), 1,304-1,307.

Marmot, M.G. (2010). *Fair society, healthy lives: The Marmot review; strategic review of health inequalities in England post-2010.* Available: http://www.marmotreview.org/Asset Library/pdfs/Reports/FairSocietyHealthyLives.pdf [April 2012].

Marmot, M.G., Davey Smith, G., Stansfeld, S., Patel, C., North, F., Head, J., White, I., Brunner, E., and Feeney, A. (1991). Health inequalities among British civil servants: The Whitehall II study. *Lancet, 337*, 1,387-1,393.

Marmot, M.G., Bosma, H., Hemingway, H., Brunner, E., and Stansfeld, S. (1997). Contribution of job control and other risk factors to social variations in coronary heart disease incidence. *Lancet, 350*, 235-239.

Martin, A., Lassman, D., Washington, B., Catlin, A., and National Health Expenditure Accounts Team. (2012). Growth in U.S. health spending remained slow in 2010; health share of gross domestic product was unchanged from 2009. *Health Affairs, 31*(1), 208-219.

Martin, J.A., Hamilton, B.E., Sutton, P.D., Ventura, S.J., Mathews, T.J., and Osterman, M. (2010). Births: Final data for 2008. *National Vital Statistics Reports, 59*(1), 3-71.

Martinson, M.L. (2012). Income inequality in health at all ages: A comparison of the United States and England. *American Journal of Public Health, 102*(11), 2,049-2,056.

Martinson, M.L., Teitler, J.O., and Reichman, N.E. (2011a). Health across the life span in the United States and England. *American Journal of Epidemiology, 173*(8), 858-865.

Martinson, M.L., Teitler, J.O., and Reichman, N.E. (2011b). Martinson et al. respond to "search for explanations of the American health disadvantage." *American Journal of Epidemiology, 173*(8), 870.

Maternal and Child Health Life Course Research Network. (2012). *Maternal and Child Health Life Course Research Network*. Available: http://www.healthychild.ucla.edu/LCRN.asp [June 2012].

Mathers, C.D., Lopez, A., and Murray, C. (2006). The burden of disease and mortality by condition: Data, methods, and results for 2001. In A. Lopez, C. Mathers, M. Ezzati, C. Murray, and D. Jamison (Eds.), *Global burden of disease and risk factors* (pp. 45-240). New York: Oxford University Press.

Mathews, T.J., and MacDorman, M. (2007). *Infant mortality statistics from the 2004 period linked birth/infant death data set*. Hyattsville, MD: National Center for Health Statistics.

Matthews, K.A., and Gallo, L.C. (2011). Psychological perspectives on pathways linking socioeconomic status and physical health. *Annual Review of Psychology, 62*, 501-530.

Matthews, K.A., Gallo, L.C., and Taylor, S.E. (2010). Are psychosocial factors mediators of socioeconomic status and health connections? *Annals of the New York Academy of Sciences, 1,186*(1), 146-173.

Maty, S.C., Lynch, J.W., Raghunathan, T.E., and Kaplan, G.A. (2008). Childhood socioeconomic position, gender, adult body mass index, and incidence of type 2 diabetes mellitus over 34 years in the Alameda county study. *American Journal of Public Health, 98*(8), 1,486-1,494.

Mayer, K.U. (2009). New directions in life course research. *Annual Review of Sociology, 35*(1), 413-433.

Mays, G.P., and Smith, S.A. (2011). Evidence links increases in public health spending to declines in preventable deaths. *Health Affairs, 30*(8), 1,585-1,593.

McAlister, A., Sandström, P., Puska, P., Veijo, A., Chereches, R., and Heidmets, L.T. (2001). Attitudes towards war, killing, and punishment of children among young people in Estonia, Finland, Romania, the Russian Federation, and the USA. *Bulletin of the World Health Organization, 79*(5), 382-387.

McCann-Mortimer, P., Augoustinos, M., and LeCouteur, A. (2004). "Race" and the human genome project: Constructions of scientific legitimacy. *Discourse and Society, 15*(4), 409-432.

McCormack, G.R., and Shiell, A. (2011). In search of causality: A systematic review of the relationship between the built environment and physical activity among adults. *International Journal of Behavior Nutrition and Physical Activity 8*(1), 125.

McDade, T.W., Chyu, L., Duncan, G.J., Hoyt, L.T., Doane, L.D., and Adam, E.K. (2011). Adolescents' expectations for the future predict health behaviors in early adulthood. *Social Science and Medicine, 73*(3), 391-398.

McDonald, S.D., Han, Z., Mulla, S., Beyene, J. (2010). Overweight and obesity in mothers and risk of pre-term birth and low birth weight infants: Systematic review and meta analyses. *British Medical Journal 341*, c3428.

McEwen, B.S. (2000). The neurobiology of stress: From serendipity to clinical relevance. *Brain Research, 886*(1-2), 172-189.

McEwen, B.S., and Gianaros, P.J. (2010). Central role of the brain in stress and adaptation: Links to socioeconomic status, health, and disease. *Annals of the New York Academy of Sciences, 1,186*(1), 190-222.

McGinnis, J.M., and Foege, W.H. (1993). Actual causes of death in the United States. *Journal of the American Medical Association, 270*(18), 2,207-2,212.

McGinnis, J.M., Williams-Russo, P., and Knickman, J.R. (2002). The case for more active policy attention to health promotion. *Health Affairs, 21*(2), 78-93.

McGlynn, E.A., Asch, S.M., Adams, J., Keesey, J., Hicks, J., DeCristofaro, A., and Kerr, E.A. (2003). The quality of health care delivered to adults in the United States. *New England Journal of Medicine, 348*(26), 2,635-2,645.

McGue, M., Osler, M., and Christensen, K. (2010). Causal inference and observational research. *Perspectives on Psychological Science, 5*(5), 546-556.

McKee, M. (2002). Values, beliefs, and implications. In M. Marinker (Ed.), *Health targets in Europe: Polity, progress and promise* (p. 181). London, UK: BMJ Books.

McKeown, T. (1976). *The role of medicine: Dream, mirage, or nemesis?* London, UK: Nuffield Provincial Hospitals Trust.

McKinlay, J.B., and McKinlay, S.M. (1977). The questionable contribution of medical measures to the decline of mortality in the United States in the twentieth century. *Milbank Memorial Fund Quarterly Health and Society, 55*(3), 405-428.

McLeod, C.B., Lavis, J.N., Macnab, Y.C., and Hertzman, C. (2012). Unemployment and mortality: A comparative study of Germany and the United States. *American Journal of Public Health, 102*(8):1,542-1,550.

McNally, R.J. (2011). *What is mental illness?* Cambridge, MA: Belknap Press of Harvard University Press.

McQueen, D.V. (2009). The evaluation of health promotion practice: 21st century debates on evidence and effectiveness. In J. Douglas, S. Earle, S. Handsley, C. Lloyd and S.M. Spurr (Eds.), *A reader in promoting public health: Challenge and controversy* (2nd ed.). London, UK: Open University/Sage.

Meara, E.R., Richards, S., and Cutler, D.M. (2008). The gap gets bigger: Changes in mortality and life expectancy, by education, 1981-2000. *Health Affairs, 27*(2), 350-360.

Measure of America. (2012). *Mapping the measure of America.* Brooklyn, NY: Social Science Research Council. Available: http://www.measureofamerica.org/maps/ [June 2012].

Melchior, M., Moffitt, T.E., Milne, B.J., Poulton, R., and Caspi, A. (2007). Why do children from socioeconomically disadvantaged families suffer from poor health when they reach adulthood? A life-course study. *American Journal of Epidemiology, 166*(8), 966-974.

Melton, S.M., McGwin, G.J., Abernathy, J.H.I., MacLennan, P., Cross, J.M., and Rue, L.W.I. (2003). Motor vehicle crash-related mortality is associated with prehospital and hospital-based resource availability. *Journal of Trauma and Acute Care Surgery, 54*(2), 273-279.

Merikangas, K.R., He, J.-p., Burstein, M., Swanson, S.A., Avenevoli, S., Cui, L., Benjet, C., Georgiades, K., and Swendsen, J. (2010). Lifetime prevalence of mental disorders in U.S. adolescents: Results from the National Comorbidity Survey Replication-Adolescent Supplement (NCS-A). *Journal of the American Academy of Child and Adolescent Psychiatry, 49*(10), 980-989.

Merkin, S.S., Basurto-Dávila, R., Karlamangla, A., Bird, C.E., Lurie, N., Escarce, J., and Seeman, T. (2009). Neighborhoods and cumulative biological risk profiles by race/ethnicity in a national sample of U.S. Adults: NHANES III. *Annals of Epidemiology, 19*(3), 194-201.

Meslé, F., and Vallin, J. (2006). Diverging trends in female old-age mortality: The United States and the Netherlands versus France and Japan. *Population and Development Review, 32*(1), 123-145.

Michaud, P.-C., Van Soest, A., and Andreyeva, T. (2007). *Cross-country variation in obesity patterns among older Americans and Europeans.* Santa Monica, CA: RAND Corporation.

Michaud, P.-C., Goldman, D., Lakdawalla, D., Gailey, A., and Zheng, Y. (2011). Differences in health between Americans and western Europeans: Effects on longevity and public finance. *Social Science and Medicine, 73*(2), 254-263.

Miller, G., Roehrig, C., Hughes-Cromwick, P., and Lake, C. (2008). Quantifying national spending on wellness and prevention. *Advances in Health Economics and Health Services Research, 19*, 1-24.

Miller, G., Roehrig, C., Hughes-Cromwick, P., and Turner, A. (2012). What is currently spent on prevention as compared to treatment? In H.S. Faust and P.T. Menzel (Eds.), *Prevention vs treatment: What's the right balance?* New York: Oxford University Press.

Miller, M., and Hemenway, D. (2008). Guns and suicide in the United States. *New England Journal of Medicine, 359*(10), 989-991.

Miller, M., Azrael, D., and Barber, C. (2011a). Suicide mortality in the United States: The importance of attending to method in understanding population-level disparities in the burden of suicide. *Annual Review of Public Health, 33*(ePub), 393-408.

Miller, W.D., Pollack, C.E., and Williams, D.R. (2011b). Healthy homes and communities: Putting the pieces together. *American Journal of Preventive Medicine, 40*(1, Suppl. 1), S48-S57.

Miller, W.D., Sadegh-Nobari, T., and Lillie-Blanton, M. (2011c). Healthy starts for all: Policy prescriptions. *American Journal of Preventive Medicine, 40*(1, Suppl.1), S19-S37.

Milstein, B., Homer, J., Briss, P., Burton, D., and Pechacek, T. (2011). Why behavioral and environmental interventions are needed to improve health at lower cost. *Health Affairs, 30*(5), 823-832.

Moffitt, T.E., Caspi, A., Belsky, J., and Silva, P.A. (1992). Childhood experience and the onset of menarche: A test of a sociobiological model. *Child Development, 63*(1), 47-58.

Mohai, P., Lantz, P.M., Morenoff, J., House, J.S., and Mero, R.P. (2009). Racial and socioeconomic disparities in residential proximity to polluting industrial facilities: Evidence from the Americans' changing lives study. *American Journal of Public Health, 99*(Suppl. 3), S649-S656.

Moise, P., Jacobzone, S., and the ARD-IHD Experts Group. (2003). *OECD study of cross-national differences in the treatment, costs and outcomes of ischaemic heart disease.* (OECD Health Working Papers, No. 3). Paris: OECD.

Mokdad, A.H., Marks, J.S., Stroup, D.F., and Gerberding, J.L. (2004). Actual causes of death in the United States, 2000. *Journal of the American Medical Association, 291*(10), 1,238-1,245.

Mokdad, A.H., Marks, J.S., Stroup, D.F., and Gerberding, J.L. (2005). Correction: Actual causes of death in the United States, 2000. *Journal of the American Medical Association, 293*, 293-294.

Moon, L., Moïse, P., and Jacobzone, S. (2003). *Stroke care in OECD countries: Outcomes in 17 countries.* Paris: OECD.

Moore, L.V., and Diez Roux, A.V. (2006). Associations of neighborhood characteristics with the location and type of food stores. *American Journal of Public Health, 96*(2), 325-331.

Moore, L.V., Diez Roux, A.V., Nettleton, J.A., and Jacobs, D.R. (2008). Associations of the local food environment with diet quality—A comparison of assessments based on surveys and geographic information systems. *American Journal of Epidemiology, 167*(8), 917-924.

Morello-Frosch, R., Zuk, M., Jerrett, M., Shamasunder, B., and Kyle, A.D. (2011). Understanding the cumulative impacts of inequalities in environmental health: Implications for policy. *Health Affairs, 30*(5), 879-887.

Morenoff, J.D. (2003). Neighborhood mechanisms and the spatial dynamics of birth weight. *American Journal of Sociology, 108*(5), 976-1,017.

Morenoff, J.D., Sampson, R.J., and Raudenbush., S.W. (2001). Neighborhood inequality, collective efficacy, and the spatial dynamics of urban violence. *Criminology, 39*(3), 517-558.

Morgan, S., and Kennedy, J. (2010). Prescription drug accessibility and affordability in the United States and abroad. *Issue Brief (The Commonwealth Fund), 89*, 1-12.

Morland, K., Wing, S., and Diez Roux, A. (2002). The contextual effect of the local food environment on residents' diets: The atherosclerosis risk in communities study. *American Journal of Public Health, 92*(11), 1,761-1,767.

Morris, J.A., Jr., MacKenzie, E.J., and Edelstein, S.L. (1990). The effect of preexisting conditions on mortality in trauma patients. *Journal of the American Medical Association, 263*(14), 1,942-1,946.

Mosher, J.F. (2011). Joe camel in a bottle: Diageo, the Smirnoff brand, and the transformation of the youth alcohol market. *American Journal of Public Health, 102*(1), 56-63.

Mosher, W., and Jones, J. (2010). *Use of contraception in the United States: 1982-2008.* Hyattsville, MD: National Center for Health Statistics.

Mosher, W.D., Martinez, G.M., Chandra, A., Abma, J.C., and Willson, S.J. (2004). *Use of contraception and use of family planning services in the United States: 1982-2002.* Atlanta, GA: Centers for Disease Control and Prevention.

Muennig, P., Fiscella, K., Tancredi, D., and Franks, P. (2010). The relative health burden of selected social and behavioral risk factors in the United States: Implications for policy. *American Journal of Public Health, 100*(9), 1,758-1,764.

Muntaner, C., Borrell, C., Ng, E., Chung, H., Espelt, A., Rodriguez-Sanz, M., Benach, J., and O'Campo, P. (2011). Politics, welfare regimes, and population health: Controversies and evidence. *Sociology of Health and Illness, 33*(6), 946-964.

Murray, C.J.L., Kulkarni, S.C., Michaud, C., Tomijima, N., Bulzacchelli, M.T., Iandiorio, T.J., and Ezzati, M. (2006). Eight Americas: Investigating mortality disparities across races, counties, and race-counties in the United States. *PLoS Medicine, 3*(9), e260.

Murray, E.T., Mishra, G.D., Kuh, D., Guralnik, J., Black, S., and Hardy, R. (2011). Life course models of socioeconomic position and cardiovascular risk factors: 1946 birth cohort. *Annals of Epidemiology, 21*(8), 589-597.

Mustafić, H., Jabre, P., Caussin, C., Murad, M.H., Escolano, S., Tafflet, M., Périer, M.-C., Marijon, E., Vernerey, D., Empana, J.-P., and Jouven, X. (2012). Main air pollutants and myocardial infarction. *Journal of the American Medical Association, 307*(7), 713-721.

Nathanson, C. (2009). *Disease prevention as social change: The state, society, and public health in the United States, France, Great Britain, and Canada.* New York: Russell Sage Foundation.

National Academy of Sciences, National Academy of Engineering, and Institute of Medicine. (2007). *Rising above the gathering storm: Energizing and employing America for a brighter economic future.* Committee on Prospering in the Global Economy of the 21st Century: An Agenda for American Science and Technology. Washington, DC: The National Academies Press.

National Association of County and City Health Officials. (2011). *2010 national profile of local health departments*. Washington, DC: National Association of County and City Health Officials.

National Center for Education Statistics. (2009). *Digest of educational statistics. Table 413. Number of bachelor's degree recipients per 100 persons of the typical age of graduation, by sex and country: 2002 through 2006*. Available: http://nces.ed.gov/programs/digest/d09/tables/dt09_413.asp [August 2012].

National Center for Education Statistics. (2012a). *Early childhood longitudinal program. Kindergarten class of 1998-99 (ECLS-K)*. Available: http://nces.ed.gov/ecls/kindergarten.asp [April 2012].

National Center for Education Statistics. (2012b). *International data table library*. Education Outcomes, Education Attainment. Table B.3.04. OECD: Trends in tertiary graduation rates, by program type and country: 1995 and 2000 to 2008. Available: http://nces.ed.gov/surveys/international/table-library.asp [April 2012].

National Center for Education Statistics. (2012c). *International data table library*. Education Outcomes, Education Attainment. Table B.3.06. Number of bachelor's degree recipients per 100 persons of the typical age of graduation, by sex and country: 2002 through 2006. Available: http://nces.ed.gov/surveys/international/table-library.asp [April 2012].

National Center for Health Statistics. (2007). *Health, United States 2007: With chartbook on trends in the health of Americans*. Hyattsville, MD: National Center for Health Statistics.

National Center for Health Statistics. (2011). *Health, United States, 2010: With special feature on death and dying*. Hyattsville, MD: National Center for Health Statistics.

National Center for Health Statistics. (2012). *Health, United States, 2011: With special feature on socioeconomic status and health*. Hyattsville, MD: National Center for Health Statistics.

National Highway Traffic Safety Administration. (2007). *Traffic safety facts: 2006 Traffic safety annual assessment—alcohol-related fatalities*. Washington, DC: National Highway Traffic Safety Administration.

National Highway Traffic Safety Administration. (2009). *Traffic safety facts: 2008. A compilation of motor vehicle crash data from the fatality analysis reporting system and the general estimates system*. Washington, DC: National Highway Traffic Safety Administration.

National Highway Traffic Safety Administration. (2010). *Drug involvement of fatally injured drivers*. Washington, DC: National Highway Traffic Safety Administration.

National Institute on Aging. (2012). *Publicly available databases for aging-related secondary analyses in the behavioral and social sciences*. Unpublished table prepared for the NAS/IOM Panel on Understanding Cross-National Health Differences Among High-Income Countries. Washington, DC: U.S. Department of Health and Human Services.

National Prevention Council. (2011). *National prevention strategy*. Washington, DC: U.S. Department of Health and Human Services, Office of the Surgeon General.

National Research Council. (2010). *International differences in mortality at older ages: Dimensions and sources*. E.M. Crimmins, S.H. Preston, and B. Cohen (Eds.), Panel on Understanding Divergent Trends in Longevity in High-Income Countries. Committee on Population. Division of Behavioral and Social Sciences and Education. Washington, DC: The National Academies Press.

National Research Council. (2011). *Explaining divergent levels of longevity in high-income countries*. E.M. Crimmins, S.H. Preston, and B. Cohen (Eds.), Panel on Understanding Divergent Trends in Longevity in High-Income Countries. Committee on Population, Division of Behavioral and Social Sciences and Education. Washington, DC: The National Academies Press.

National Research Council and Institute of Medicine. (2000). *From neurons to neighborhoods: The science of early child development*. J.P. Shonkoff and D.A. Phillips (Eds.), Committee on Adolescent Health Care Services and Models of Care for Treatment, Prevention, and Healthy Development. Washington, DC: National Academy Press.

National Research Council and Institute of Medicine. (2004). *Children's health, the nation's wealth: Assessing and improving child health*. Committee on Evaluation of Children's Health. Washington, DC: The National Academies Press.

National Research Council and Institute of Medicine. (2008). *Adolescent health services: Missing opportunities*. Committee on Adolescent Health Care Services and Models of Care for Treatment, Prevention, and Healthy Development. Washington, DC: The National Academies Press.

Navarro, V. (2002). The world health report 2000: Can health care systems be compared using a single measure of performance? *American Journal of Public Health, 92*(31), 33-34.

Navarro, V., and Shi, L. (2001). The political context of social inequalities and health. *Social Science and Medicine, 52*, 481-491.

Navarro, V., Borrell, C., Benach, J., Muntaner, C., Quiroga, A., Rodriguez-Sanz, M., Verges, N., Guma, J., and Pasarin, M.I. (2003). The importance of the political and the social in explaining mortality differentials among the countries of the OECD, 1950-1998. *International Journal of Health Services, 33*(3), 419-494.

Navarro, V., Muntaner, C., Borrell, C., Benach, J., Quiroga, Á., Rodríguez-Sanz, M., Vergés, N., and Pasarín, M.I. (2006). Politics and health outcomes. *Lancet, 368*(9,540), 1,033-1,037.

Nestle, M. (2002). *Food politics: How the food industry influences nutrition and health*. Berkeley: University of California Press.

Newcomer, J.W. (2007). Metabolic syndrome and mental illness. *American Journal of Managed Care, 13*(7 Suppl.), S170-S177. Erratum: *14*(2):76.

Nickell, S., Nunziata, L., and Ochel, W. (2005). Unemployment in the OECD since the 1960s. What do we know? *Economic Journal, 115*(500), 1-27.

Nolte, E., and McKee, M. (2003). Measuring the health of nations: Analysis of mortality amenable to health care. *British Medical Journal, 327*(7,424), 1,129.

Nolte, E., and McKee, M. (2011). Variations in amenable mortality—Trends in 16 high-income nations. *Health Policy, 103*(1), 47-52.

Nolte, E., Bain, C., and McKee, M. (2006). Diabetes as a tracer condition in international benchmarking of health systems. *Diabetes Care, 29*(5), 1,007-1,011.

Northridge, M.E. (1995). Public health methods—Attributable risk as a link between causality and public health action. *American Journal of Public Health, 85*(9), 1,202-1,204.

Nuru-Jeter, A., Dominguez, T., Hammond, W., Leu, J., Skaff, M., Egerter, S., Jones, C., and Braveman, P. (2009). "It's the skin you're in": African-American women talk about their experiences of racism. An exploratory study to develop measures of racism for birth outcome studies. *Maternal and Child Health Journal, 13*(1), 29-39.

OECD. (2003). *Stroke care in OECD countries: A comparison of treatment, costs, and outcomes in 17 countries*. Paris: OECD.

OECD. (2006). *Starting strong II: Early childhood education and care*. Paris: OECD.

OECD. (2007). *Health care quality indicators project 2006: Data collection update report*. Paris: OECD.

OECD. (2008a). *Going for growth: 2008*. Paris: OECD.

OECD. (2008b). *OECD key environmental indicators*. Paris: OECD.

OECD. (2009a). *Doing better for children*. Paris: OECD.

OECD. (2009b). *Education at a glance, 2009*. Paris: OECD.

OECD. (2009c). *Health at a glance 2009: OECD indicators*. Available: http://www.oecd.org/health/healthpoliciesanddata/44117530.pdf [June 2012].

OECD. (2010a). *Economic policy reforms 2010: Going for growth*. Paris: OECD.

OECD. (2010b). *Health care systems: Getting more value for money*. Paris: OECD.

OECD. (2010c). *Social capital, human capital, and health. What is the evidence?* OECD and Centre for Educational Research and Innovation. Available: www.oecd.org/data oecd/40/24/45760738.pdf [June 2012].

OECD. (2011a). *Education at a glance: OECD indicators 2011*. Paris: OECD.

OECD. (2011b). *Health at a glance 2011: OECD indicators*. Paris: OECD.

OECD. (2011c). *OECD health data 2011, frequently requested data*. Available: http://www. oecd.org/document/16/0,3746,en_2649_37407_2085200_1_1_1_37407,00.html [April 2012].

OECD. (2011d). *OECD regions at a glance 2011*. Paris: OECD. Available: http://www. oecd-ilibrary.org/urban-rural-and-regional-development/oecd-regions-at-a-glance-2011_ reg_glance-2011-en [June 2012].

OECD. (2011e). *Society at a glance 2011: OECD social indicators*. Paris: OECD.

OECD. (2012a). *OECD better life index: United States*. Available: http://www.oecdbetterlife index.org/countries/united-states/ [June 2012].

OECD. (2012b). *OECD environmental outlook to 2050: Consequences of inaction*. Paris: OECD.

OECD. (2012c). *OECD factbook 2011-2012: Economic, environmental and social statistics*. Paris: OECD.

OECD. (2012d). *OECD family database*. Paris: OECD. Available: http://www.oecd.org/social/ family/database [June 2012].

OECD. (2012e). *OECD family database: CO2.2: Child poverty*. Paris: OECD. Available: http://www.oecd.org/els/familiesandchildren/41929552.pdf [June 2012].

OECD. (2012f). *OECD family database: CO3.1: Educational attainment by gender and average years spent in formal education*. Paris: OECD. Available: http://www.oecd.org/els/ familiesandchildren/37863998.pdf [June 2012].

OECD. (2012g). *OECD family database: LMF2.6: Time spent travelling to and from work*. Paris: OECD. Available: http://www.oecd.org/els/familiesandchildren/43199696.pdf [June 2012].

OECD. (2012h). *OECD family database: PF1.2: Public spending on education*. Paris: OECD. Available: http://www.oecd.org/els/familiesandchildren/37864432.pdf [June 2012].

OECD. (2012i). *OECD family database: PF3.1: Public spending on childcare and early education*. Paris: OECD. Available: http://www.oecd.org/els/familiesandchildren/37864512. pdf [June 2012].

OECD. (2012j). *OECD family database: PF3.2: Enrollment in childcare and pre-schools*. Paris: OECD. Available: http://www.oecd.org/els/familiesandchildren/37864698.pdf [June 2012].

OECD. (2012k). *OECD family database: SF3.4: Family violence*. Paris: OECD. Available: http://www.oecd.org/els/familiesandchildren/45583188.pdf [June 2012].

OECD. (2012l). *OECD.stat extracts*. Paris: OECD. Available: http://http://www.oecd-ilibrary. org/economics/data/oecd-stat_data-00285-en [June 2012].

OECD. (2012m). *Online OECD employment database*. Available: http://www.oecd.org/ employment/database [April 2012].

OECD. (2012n). *The PISA 2009 profiles by country/economy*. Available: http://stats.oecd.org/ PISA2009Profiles/# [April 2012].

Office of Management and Budget. (2012). *Outlays for mandatory and related programs, 1962-2017*. H. Tables. Washington, DC: Office of Management and Budget. Available: http://www.whitehouse.gov/omb/budget/Historicals [May 2012].

Ogden, C.L., Carroll, M.D., Kit, B.K., and Flegal, K.M. (2012a). Prevalence of obesity and trends in body mass index among U.S. children and adolescents, 1999-2010. *Journal of the American Medical Association, 307*(5), 483-490.

Ogden, L.L., Richards, C.L., and Shenson, D. (2012b). Clinical preventive services for older adults: The interface between personal health care and public health services. *American Journal of Public Health, 102*(3), 419-425.

Ogle, L.T. (2003). *International comparisons in fourth-grade reading literacy: Findings from the Progress in International Reading Literacy Study (PIRLS) of 2001.* Washington, DC: U.S. Department of Education, Institute of Education Sciences, National Center for Education Statistics.

Oh, D., Heck, J., Dresler, C., Allwright, S., Haglund, M., Del Mazo, S., Kralikova, E., Stucker, I., Tamang, E., Gritz, E., and Hashibe, M. (2010). Determinants of smoking initiation among women in five European countries: A cross-sectional survey. *BMC Public Health, 10*(1), 74.

Oksuzyan, A., Crimmins, E., Saito, Y., O'Rand, A., Il, J.W., and Christensen, K. (2010). Cross-national comparison of sex differences in health and mortality in Denmark, Japan, and the U.S. *European Journal of Epidemiology, 25*(7), 471-480.

Olshansky, S.J., and Ault, A.B. (1986). The fourth stage of the epidemiologic transition: The age of delayed degenerative diseases. *Milbank Quarterly, 64*(3), 355-391.

Olshansky, S.J., Antonucci, T., Berkman, L., Binstock, R.H., Boersch-Supan, A., Cacioppo, J.T., Carnes, B.A., Carstensen, L.L., Fried, L.P., Goldman, D.P., Jackson, J., Kohli, M., Rother, J., Zheng, Y., and Rowe, J. (2012). Differences in life expectancy due to race and educational differences are widening, and many may not catch up. *Health Affairs, 31*(8), 1,803-1,813.

Or, Z., Wang, J., and Jamison, D. (2005). International differences in the impact of doctors on health: A multilevel analysis of OECD countries. *Journal of Health Economics, 24*(3), 531-560.

Ozbay, F., Johnson, D., Dimoulas, E., Morgan, C., Charney, D., and Southwick, S. (2007). Social support and resilience to stress: From neurobiology to clinical practice. *Psychiatry (Edgmont), 4*(5), 35-40.

Ozbay, F., Fitterling, H., Charney, D., and Southwick, S. (2008). Social support and resilience to stress across the life span: A neurobiologic framework. *Current Psychiatry Reports, 10*(4), 304-310.

Paczkowski, M.M., and Galea, S. (2010). Sociodemographic characteristics of the neighborhood and depressive symptoms. *Current Opinion in Psychiatry, 23*(4), 337-341.

Palloni, A. (2006). Reproducing inequalities: Luck, wallets, and the enduring effects of childhood health. *Demography, 43*(4), 587-615.

Palloni, A., and Yonker, J. (2012). *Health in the U.S. at young ages: Preliminary findings.* (CDE Working Paper 2012-04). Madison: Center for Demography and Ecology, University of Wisconsin.

Palloni, A., Milesi, C., White, R.G., and Turner, A. (2009). Early childhood health, reproduction of economic inequalities and the persistence of health and mortality differentials. *Social Science and Medicine, 68*(9), 1,574-1,582.

Pampel, F.C. (2002). Cigarette use and the narrowing sex differential in mortality. *Population and Development Review, 28*(1), 77-104.

Pampel, F.C. (2010). Divergent patterns of smoking across high-income nations. In National Research Council, *International differences in mortality at older ages: Dimensions and sources* (pp. 132-163). E.M. Crimmins, S.H. Preston, and B. Cohen (Eds.), Panel on Understanding Divergent Trends in Longevity in High-Income Countries. Committee on Population. Division of Behavioral and Social Sciences and Education. Washington, DC: The National Academies Press.

Pampel, F.C., and Denney, J.T. (2011). Cross-national sources of health inequality: Education and tobacco use in the world health survey. *Demography, 48*(2), 653-674.

Pampel, F.C., Krueger, P.M., and Denney, J.T. (2010). Socioeconomic disparities in health behaviors. *Annual Review of Sociology, 36*, 349-370.

Panchaud, C., Singh, S., Feivelson, D., and Darroch, J. (2000). Sexually transmitted diseases among adolescents in developed countries. *Family Planning Perspectives, 32*(1), 24-32, 45.

Papas, M.A., Alberg, A.J., Ewing, R., Helzlsouer, K.J., Gary, T.L., and Klassen, A.C. (2007). The built environment and obesity. *Epidemiologic Reviews, 29*(1), 129-143.

Pappas, G., Queen, S., Hadden, W., and Fisher, G. (1993). The increasing disparity in mortality between socioeconomic groups in the United States, 1960 and 1986. *New England Journal of Medicine, 329*(2), 103-109.

Parker, J.D., Rich, D.Q., Glinianaia, S.V., Leem, J.H., Wartenberg, D., Bell, M.L., Bonzini, M., Brauer, M., Darrow, L., Gehring, U., Gouveia, N., Grillo, P., Ha, E., van den Hooven, E.H., Jalaludin, B., Jesdale, B.M., Lepeule, J., Morello-Frosch, R., Morgan, G.G., Slama, R., Pierik, F.H., Pesatori, A.C., Sathyanarayana, S., Seo, J., Strickland, M., Tamburic, L., and Woodruff, T.J. (2011a). The international collaboration on air pollution and pregnancy outcomes: Initial results. *Environmental Health Perspectives, 119*(7), 1,023-1,028.

Parker, R.N., Williams, K.R., McCaffree, K.J., Acensio, E.K., Browne, A., Strom, K.J., and Barrick, K. (2011b). Alcohol availability and youth homicide in the 91 largest U.S. cities, 1984-2006. *Drug and Alcohol Review, 30*(5), 505-514.

Parks, D.J., Svedsen, P. Singer, and Foti, M.E. (Eds.). (2006). *Morbidity and mortality in people with serious mental illness.* Alexandria, VA: National Association of State Mental Health Program Directors Medical Directors Council. Available: http://www.nasmhpd. org/general_files/publications/med_directors_pubs/Technical%20Report%20on%20 Morbidity%20and%20Mortaility%20-%20Final%2011-06.pdf [June 2012].

Pastor, M., Morello-Frosch, R., and Sadd, J.L. (2005). The air is always cleaner on the other side: Race, space, and air toxics exposures in California. *Journal of Urban Affairs, 27*(2), 127-148.

Patel, V., Flisher, A.J., Hetrick, S., and McGorry, P. (2007). Mental health of young people: A global public-health challenge. *Lancet, 369*(9,569), 1,302-1,313.

Patton, G.C., Coffey, C., Sawyer, S.M., Viner, R.M., Haller, D.M., Bose, K., Vos, T., Ferguson, J., and Mathers, C.D. (2009). Global patterns of mortality in young people: A systematic analysis of population health data. *Lancet, 374*(9,693), 881-892.

Paxson, C., and Schady, N. (2005). Child health and economic crisis in Peru. *World Bank Economic Review, 19*, 203-223.

Paxton, A.E., Strycker, L.A., Toobert, D.J., Ammerman, A.S., and Glasgow, R.E. (2011). Starting the conversation: Performance of a brief dietary assessment and intervention tool for health professionals. *American Journal of Preventive Medicine, 40*(1), 67-71.

Pearce, J., and Maddison, R. (2011). Do enhancements to the urban built environment improve physical activity levels among socially disadvantaged populations? *International Journal for Equity in Health, 10*(1), 28.

Pearce, J., Witten, K., Hiscock, R., and Blakely, T. (2007). Are socially disadvantaged neighbourhoods deprived of health-related community resources? *International Journal of Epidemiology, 36*(2), 348-355.

Pearce, J., Hiscock, R., Blakely, T., and Witten, K. (2008). The contextual effects of neighbourhood access to supermarkets and convenience stores on individual fruit and vegetable consumption. *Journal of Epidemiology and Community Health, 62*(3), 198-201.

Peto, R., Lopez, A.D., Boreham, J., Thun, M., and Heath, C., Jr. (1992). Mortality from tobacco in developed countries: Indirect estimation from national vital statistics. *Lancet, 339*(8,804), 1,268-1,278.

Petticrew, M., and Roberts, H. (2003). Evidence, hierarchies, and typologies: Horses for courses. *Journal of Epidemiology and Community Health, 57*(7), 527-529.

Pettit, B., and Hook, J.L. (2009). *Gendered tradeoffs: Family, social policy, and economic inequality in twenty-one countries*. New York: Russell Sage Foundation.

Pew Center on the States. (2008). *One in 100: Behind bars in America 2008*. Available: http://www.pewstates.org/uploadedFiles/PCS_Assets/2008/one%20in%20100.pdf [June 2012].

Pew Research Center. (2009). *Few see U.S. health care as "best in the world."* Washington, DC: Pew Research Center.

Phillips, R.L. (2012). International learning on increasing the value and effectiveness of primary care (I LIVE PC). *Journal of the American Board of Family Medicine, 25*(Suppl. 1), S2-S5.

Phillips, R.L., and Bazemore, A.W. (2010). Primary care and why it matters for U.S. health system reform. *Health Affairs, 29*(5), 806-810.

Pickett, K.E., and Pearl, M. (2001). Multilevel analyses of neighbourhood socioeconomic context and health outcomes: A critical review. *Journal of Epidemiology and Community Health, 55*(2), 111-122.

Pickett, K.E., and Wilkinson, R.G. (2009). Greater equality and better health. *British Medical Journal, 339*, b4320.

Pickett, K.E., and Wilkinson, R.G. (2010). Inequality: An underacknowledged source of mental illness and distress. *The British Journal of Psychiatry, 197*(6), 426-428.

Pickett, W., Craig, W., Harel, Y., Cunningham, J., Simpson, K., Molcho, M., Mazur, J., Dostaler, S., Overpeck, M.D., Currie, C.E., and on behalf of the HBSC Violence Injuries Writing Group. (2005). Cross-national study of fighting and weapon carrying as determinants of adolescent injury. *Pediatrics, 116*(6), e855-e863.

Pinto, S., and Beckfield J. (2011). Organized labor in an evolving Europe. *Research in the Sociology of Work, 22*(2), 153-179.

Plug, I., Hoffmann, R., Artnik, B., Bopp, M., Borrell, C., Costa, G., Deboosere, P., Esnaola, S., Kalediene, R., Leinsalu, M., Lundberg, O., Martikainen, P., Regidor, E., Rychtarikova, J., Strand, B.H., Wojtyniak, B., and Mackenbach, J. (2012). Socioeconomic inequalities in mortality from conditions amenable to medical interventions: Do they reflect inequalities in access or quality of health care? *BMC Public Health, 12*(1), 346.

Pollack, C.E., Chideya, S., Cubbin, C., Williams, B., Dekker, M., and Braveman, P. (2007). Should health studies measure wealth? A systematic review. *American Journal of Preventive Medicine, 33*(3), 250-264.

Pollitt, R., Rose, K., and Kaufman, J. (2005). Evaluating the evidence for models of life course socioeconomic factors and cardiovascular outcomes: A systematic review. *BMC Public Health, 5*(1), 7.

Popova, S., Giesbrecht, N., Bekmuradov, D., and Patra, J. (2009). Hours and days of sale and density of alcohol outlets: Impacts on alcohol consumption and damage: A systematic review. *Alcohol and Alcoholism, 44*(5), 500-516.

Poundstone, K.E., Strathdee, S.A., and Celentano, D.D. (2004). The social epidemiology of human immunodeficiency virus/acquired immunodeficiency syndrome. *Epidemiologic Reviews, 26*(1), 22-35.

Power, C., and Hertzman, C. (1997). Social and biological pathways linking early life and adult disease. *British Medical Bulletin, 53*(1), 210-221.

Power, C., Manor, O., and Fox, J. (1991). *Health and class: The early years*. New York: Chapman & Hall.

Preston, S. (1984). Children and the elderly: Divergent paths for America's dependents. *Demography 21*(4), 435-457.

Preston, S.H., and Haines, M. (1991). *Fatal years: Child mortality in late nineteenth-century America*. Princeton, NJ: Princeton University Press.

Preston, S.H., and Ho, J.Y. (2009). Low life expectancy in the United States: Is the health care system at fault? In National Research Council, *International differences in mortality at older ages: Dimensions and sources* (pp. 259-298). E.M. Crimmins, S.H. Preston, and B. Cohen (Eds.), Panel on Understanding Divergent Trends in Longevity in High-Income Countries. Committee on Population. Division of Behavioral and Social Sciences and Education. Washington, DC: The National Academies Press.

Preston, S.H., and Stokes, A. (2011). Contribution of obesity to international differences in life expectancy. *American Journal of Public Health, 101*(11), 2,137-2,143.

Preston, S.H., and Wang, H. (2006). Sex mortality differences in the United States: The role of cohort smoking patterns. *Demography, 43*(4), 631-646.

Preston, S.H., Glei, D.A., and Wilmoth, J.R. (2010a). A new method for estimating smoking-attributable mortality in high-income countries. *International Journal of Epidemiology, 39*(2), 430-438.

Preston, S.H., Glei, D.A., and Wilmoth, J.R. (2010b). Contributions of smoking to international differences in life expectancy. In National Research Council, *International differences in mortality at older ages: Dimensions and sources* (pp. 105-131). E.M. Crimmins, S.H. Preston, and B. Cohen (Eds.), Panel on Understanding Divergent Trends in Longevity in High-Income Countries. Committee on Population. Division of Behavioral and Social Sciences and Education. Washington, DC: The National Academies Press.

Primack, B., Bost, J., Land, S., and Fine, M. (2007). Volume of tobacco advertising in African American markets: Systematic review and meta-analysis. *Public Health Reports, 122*(5), 607-615.

Pucher, J., and Dijkstra, L. (2003). Promoting safe walking and cycling to improve public health: lessons from the Netherlands and Germany. *American Journal of Public Health, 93*(9), 1,509-1,516.

Pucher, J., Buehler, R., Bassett, D.R., and Dannenberg, A.L. (2010a). Walking and cycling to health: A comparative analysis of city, state, and international data. *American Journal of Public Health, 100*(10), 1,986-1,992.

Pucher, J., Dill, J., and Handy, S. (2010b). Infrastructure, programs, and policies to increase bicycling: An international review. *Preventive Medicine, 50*(Suppl. 1), S106-S125.

Pudrovska, T., Schieman, S., Pearlin, L.I., and Nguyen, K. (2005). The sense of mastery as a mediator and moderator in the association between economic hardship and health in late life. *Journal of Aging and Health, 17*(5), 634-660.

Purnell, J.Q., Peppone, L.J., Alcaraz, K., McQueen, A., Guido, J.J., Carroll, J.K., Shacham, E., and Morrow, G.R. (2012). Perceived discrimination, psychological distress, and current smoking status: Results from the Behavioral Risk Factor Surveillance System reactions to race module, 2004-2008. *American Journal of Public Health, 102*(5), 844-851.

Putnam, R. (1993). *Making democracy work.* Princeton, NJ: Princeton University Press.

Putnam, R. (2000). *Bowling alone: The collapse and revival of American community.* New York: Simon & Schuster.

Quinn, M.J. (2003). Cancer trends in the United States—A view from Europe. *Journal of the National Cancer Institute, 95*(17), 1,258-1,261.

Rässler, S. (2002). *Statistical matching: A frequentist theory, practical applications, and alternative Bayesian approaches.* New York: Springer.

Rau, R., Soroko, E., Jasilionis, D., and Vaupel, J.W. (2008). Continued reductions in mortality at advanced ages. *Population and Development Review, 34*(4), 747-768.

Ravelli, A.C.J., van der Meulen, J.H.P., Michels, R.P.J., Osmond, C., Barker, D.J.P., Hales, C.N., and Bleker, O.P. (1998). Glucose tolerance in adults after prenatal exposure to famine. *Lancet, 351*(9,097), 173-177.

Reardon, S.F., and Bischoff, K. (2011). Income inequality and income segregation. *American Journal of Sociology, 116*(4), 1,092-1,153.

Regidor, E., Pascual, C., Martínez, D., Calle, M., Ortega, P., and Astasio, P. (2011). The role of political and welfare state characteristics in infant mortality: A comparative study in wealthy countries since the late 19th century. *International Journal of Epidemiology, 40*(5), 1,187-1,195.

Rehm, J., Mathers, C., Popova, S., Thavorncharoensap, M., Teerawattananon, Y., and Patra, J. (2009). Global burden of disease and injury and economic cost attributed to alcohol use and alcohol-use disorders. *Lancet, 373,* 2,223-2,233.

Reinhardt, U.E., Hussey, P.S., and Anderson, G.F. (2004). U.S. health care spending in an international context. *Health Affairs, 23*(3), 10-25.

Retherford, R.D. (1975). *The changing sex differential in mortality.* Westport, CT: Greenwood Press.

Reynolds, S.L., Hagedorn, A., Yeom, J., Saito, Y., Yokoyama, E., and Crimmins, E.M. (2008). A tale of two countries—the United States and Japan: Are differences in health due to differences in overweight? *Journal of Epidemiology, 18*(6), 280-290.

Richards, T.J., and Patterson, P.M. (2003). *A bilateral comparison of fruit and vegetable consumption: United States and Canada.* Montreal, Canada: Agricultural and Applied Economics Association.

Richardson, E.G., and Hemenway, D. (2011). Homicide, suicide, and unintentional firearm fatality: Comparing the United States with other high-income countries, 2003. *Journal of Trauma and Acute Care Surgery, 70*(1), 238-243.

Richardson, H.W., and Bae, C.-H.C. (2004). *Urban sprawl in Western Europe and the United States.* Burlington, VT: Ashgate.

Richter, M., Erhart, M., Vereecken, C.A., Zambon, A., Boyce, W., and Gabhainn, S.N. (2009). The role of behavioural factors in explaining socio-economic differences in adolescent health: A multilevel study in 33 countries. *Social Science and Medicine, 69*(3), 396-403.

Riley, J.C. (2001). *Rising life expectancy: A global history.* New York: Cambridge University Press.

Robert, S.A., and Booske, B.C. (2011). U.S. opinions on health determinants and social policy as health policy. *American Journal of Public Health, 101*(9), 1,655-1,663.

Robert Wood Johnson Foundation Commission to Build a Healthier America. (2009). *Beyond health care: New directions to a healthier America.* Recommendations from the Robert Wood Johnson Foundation Commission to Build a Healthier America. Princeton, NJ: RWJF Commission to Build a Healthier America.

Roberts, C., Freeman, J., Samdal, O., Schnohr, C., De Looze, M., Nic Gabhainn, S., Iannotti, R., and Rasmussen, M. (2009). The health behaviour in school-aged children (HBSC) study: Methodological developments and current tensions. *International Journal of Public Health, 54*(Suppl. 2), 140-150.

Robinson, T.N., Borzekowski, D.L.G., Matheson, D.M., and Kraemer, H.C. (2007). Effects of fast food branding on young children's taste preferences. *Archives of Pediatrics and Adolescent Medicine, 161*(8), 792-797.

Rolnick, A., and Grunewald, R. (2003). *Early childhood development: Economic development with a high public return.* Minneapolis, MN: Federal Reserve Bank of Minneapolis.

Ross, C.E., and Wu, C.-L. (1995). The links between education and health. *American Sociological Review, 60*(5), 719-745.

Ross, C.E., Danziger, S., and Smolensky, E. (1987). The level and trend of poverty in the United States, 1939-1979. *Demography, 24*(4), 587-600.

Russell, L. (2011). *Reducing disparities in life expectancy: What factors matter?* Background paper prepared for the workshop on Reducing Disparities in Life Expectancy held by the Roundtable on the Promotion of Health Equity and the Elimination of Health Disparities of the Institute of Medicine. Available: http://www.iom.edu/~/media/Files/Activity%20 Files/SelectPops/HealthDisparities/2011-FEB-24/Commissioned%20Paper%20by%20 Lesley%20Russell.pdf [August 2012].

Rutstein, D.D., Berenberg, W., Chalmers, T.C., Child, C.G., Fishman, A.P., Perrin, E.B., Feldman, J.J., Leaverton, P.E., Lane, J.M., Sencer, D.J., and Evans, C.C. (1976). Measuring the quality of medical care. *New England Journal of Medicine, 294*(11), 582-588.

Rychetnik, L., Frommer, M., Hawe, P., and Shiell, A. (2002). Criteria for evaluating evidence on public health interventions. *Journal of Epidemiology and Community Health, 56*(2), 119-127.

Sabia, J.J. (2007). The effect of body weight on adolescent academic performance. *Southern Economic Journal, 73*(4), 871-900.

Sallis, J.F., and Glanz, K. (2006). The role of built environments in physical activity, eating, and obesity in childhood. *Future of Children, 16*(1), 89-108.

Sallis, J.F., and Glanz, K. (2009). Physical activity and food environments: Solutions to the obesity epidemic. *Milbank Quarterly, 87*(1), 123-154.

Sallis, J.F., and Saelens, B.E. (2000). Assessment of physical activity by self-report: Status, limitations, and future directions. *Research Quarterly for Exercise and Sport, 71*(Suppl. 2), S1-S14.

Sampson, R.J., Raudenbush, S.W., and Earls, F. (1997). Neighborhoods and violent crime: A multilevel study of collective efficacy. *Science, 277*(5,328), 918-924.

Sandoval, J., and Esteller, M. (2012). Cancer epigenomics: Beyond genomics. *Current Opinion in Genetics & Development, 22*(1), 50-55.

Sandy, L.G., Bodenheimer, T., Pawlson, L.G., and Starfield, B. (2009). The political economy of U.S. primary care. *Health Affairs, 28*(4), 1,136-1,145.

Santelli, J.S., and Schalet, A.T. (2009). *A new vision for adolescent sexual and reproductive health.* Ithaca, NY: ACT for Youth Center of Excellence.

Santelli, J., Sandfort, T., and Orr, M. (2008). Transnational comparisons of adolescent contraceptive use: What can we learn from these comparisons? *Archives of Pediatrics and Adolescent Medicine, 162*(1), 92-94.

Santelli, J., Sandfort, T.G., and Orr, M. (2009). U.S./European differences in condom use. *Journal of Adolescent Health, 44*(3), 306.

Sarlio-Lahteenkorva, S., Silventoinen, K., and Lahelma, E. (2004). Relative weight and income at different levels of socioeconomic status. *American Journal of Public Health, 94*(3), 468-472.

Satcher, D., Fryer, G.E., McCann, J., Troutman, A., Woolf, S.H., and Rust, G. (2005). What if we were equal? A comparison of the black-white mortality gap in 1960 and 2000. *Health Affairs, 24*(2), 459-464.

Schalet, A.T. (2000). Raging hormones, regulated love: Adolescent sexuality and the constitution of the modern individual in the United States and the Netherlands. *Body & Society, 6*(1), 75-105.

Schilling, E.A., Aseltine, R. H., and Gore, S. (2007). Adverse childhood experiences and mental health in young adults: A longitudinal survey. *BMC Public Health, 7*(1), 30.

Schoen, C., Osborn, R., Huynh, P.T., Doty, M., Davis, K., Zapert, K., and Peugh, J. (2004). Primary care and health system performance: Adults experiences in five countries. *Health Affairs* (Suppl. Web Exclusive), w4-487-503.

Schoen, C., Osborn, R., Doty, M.M., Squires, D., Peugh, J., and Applebaum, S. (2009a). A survey of primary care physicians in eleven countries, 2009: Perspectives on care, costs, and experiences. *Health Affairs, 28*(6), w1171-w1183.

Schoen, C., Osborn, R., How, S.K.H., Doty, M.M., and Peugh, J. (2009b). In chronic condition: Experiences of patients with complex health care needs, in eight countries, 2008. *Health Affairs, 28*(1), w1-w16.

Schoen, C., Osborn, R., Squires, D., Doty, M.M., Pierson, R., and Applebaum, S. (2010). How health insurance design affects access to care and costs, by income, in eleven countries. *Health Affairs, 29*(12), 2,323-2,334.

Schoen, C., Osborn, R., Squires, D., Doty, M., Pierson, R., and Applebaum, S. (2011). New 2011 survey of patients with complex care needs in eleven countries finds that care is often poorly coordinated. *Health Affairs, 30*(12), 2,437-2,448.

Schoeni, R.F., Dow, W.H., Miller, W.D., and Pamuk, E.R. (2011). The economic value of improving the health of disadvantaged Americans. *American Journal of Preventive Medicine, 40*(1, Suppl. 1), S67-S72.

Scholz, R.D., and Maier, H. (2003). *German unification and the plasticity of mortality at older ages.* Rostock, Germany: Max Planck Institute for Demographic Research.

Schroeder, S.A., and Morris, C.D. (2010). Confronting a neglected epidemic: Tobacco cessation for persons with mental illnesses ad substance abuse problems. *Annual Review of Public Health, 31*, 297-314.

Schulte, P.A., Pandalai, S., Wulsin, V., and Chun, H. (2011). Interaction of occupational and personal risk factors in workforce health and safety. *American Journal of Public Health, 102*(3), 434-448.

Schultz, J., O'Brien, A.M., and Tadesse, B. (2008). Social capital and self-rated health: Results from the U.S. 2006 social capital survey of one community. *Social Science and Medicine, 67*(4), 606-617.

Schyns, P., and Koop, C. (2010). Political distrust and social capital in Europe and the USA. *Social Indicators Research, 96*(1), 145-167.

Seeman, T., Epel, E., Gruenewald, T., Karlamangla, A., and McEwen, B.S. (2010). Socioeconomic differentials in peripheral biology: Cumulative allostatic load. *Annals of the New York Academy of Sciences, 1,186*(1), 223-239.

Sellers, J.M. (1999). Public goods and the politics of segregation: An analysis and cross-national comparison. *Journal of Urban Affairs, 21*(2), 237-262.

Sexton, K., and Linder, S.H. (2011). Cumulative risk assessment for combined health effects from chemical and nonchemical stressors. *American Journal of Public Health, 101*(Suppl. 1), S81-S88.

Sharpe, A. (2011). *Estimates of relative and absolute poverty rates for the working population in developed countries.* Ottawa, Ontario: Centre for the Study of Living Standard.

Shaw, M. (2004). Housing and public health. *Annual Review of Public Health, 25*, 397-418.

Shkolnikov, V.M., Andreev, E.M., and Begun, A.Z. (2003). Gini coefficient as a life table function: Computation from discrete data, decomposition of differences and empirical examples. *Demographic Research,* (8), 305-358.

Shkolnikov, V.M., Andreev, E., Zhang, Z., Oeppen, J., and Vaupel, J. (2011). Losses of expected lifetime in the United States and other developed countries: Methods and empirical analyses. *Demography, 48*(1), 211-239.

Shonkoff, J.P., Boyce, W.T., and McEwen, B.S. (2009). Neuroscience, molecular biology, and the childhood roots of health disparities: Building a new framework for health promotion and disease prevention. *Journal of the American Medical Association, 301*(21), 2,252-2,259.

Shults, R.A., Elder, R.W., Sleet, D.A., Nichols, J.L., Alao, M.O., Carande-Kulis, V.G., Zaza, S., Sosin, D.M., and Thompson, R.S. (2001). Reviews of evidence regarding interventions to reduce alcohol-impaired driving. *American Journal of Preventive Medicine, 21*(4) (Suppl. 1), 66-88.

Siega-Riz, A.M., Siega-Riz, A.M., and Laraia, B. (2006). The implications of maternal overweight and obesity on the course of pregnancy and birth outcomes. *Maternal and Child Health Journal, 10*(Suppl. 5), S153-S156.

Sihler, K.C., and Hemmila, M.R. (2009). Injuries in nonurban areas are associated with increased disability at hospital discharge. *Journal of Trauma and Acute Care Surgery, 67*(5), 903-909.

Sims, M., Diez-Roux, A.V., Dudley, A., Gebreab, S., Wyatt, S.B., Bruce, M.A., James, S.A., Robinson, J.C., Williams, D.R., and Taylor, H.A. (2012). Perceived discrimination and hypertension among African Americans in the Jackson Heart Study. *American Journal of Public Health, 102*(Suppl. 2), S258-S265.

Singh-Manoux, A., Adler, N.E., and Marmot, M.G. (2003). Subjective social status: Its determinants and its association with measures of ill-health in the Whitehall II study. *Social Science and Medicine, 56*(6), 1,321-1,333.

Singh-Manoux, A., Marmot, M.G., and Adler, N.E. (2005). Does subjective social status predict health and change in health status better than objective status? *Psychosomatic Medicine, 67*(6), 855-861.

Sly, P.D., and Flack, F. (2008). Susceptibility of children to environmental pollutants. *Annals of the New York Academy of Sciences, 1,140*(1), 163-183.

Small Arms Survey. (2007). *Guns and the city.* Cambridge, UK: Cambridge University Press. Available: http://www.smallarmssurvey.org/publications/by-type/yearbook/small-arms-survey-2007.html [June 2012].

Smedley, B.D. (2012). The lived experience of race and its health consequences. *American Journal of Public Health, 102*(5), 933-935.

Smeeding, T. (2006). Poor people in rich nations: The United States in comparative perspective. *Journal of Economic Perspectives, 20*(1), 69-90.

Smith-Khuri, E., Iachan, R., Scheidt, P.C., Overpeck, M.D., Gabhainn, S.N., Pickett, W., and Harel, Y. (2004). A cross-national study of violence-related behaviors in adolescents. *Archives of Pediatrics and Adolescent Medicine, 158*(6), 539-544.

Smolensky, M.H., Di Milia, L., Ohayon, M.M., and Philip, P. (2011). Sleep disorders, medical conditions, and road accident risk. *Accident, Analysis & Prevention, 43*(2), 533-548.

Snyder, T.D., and Dillow, S.A. (2011). *Digest of education statistics 2011.* Washington, DC: National Center for Education Statistics, Institute of Education Sciences, U.S. Department of Education.

Solon, G. (2002). Cross-country differences in intergenerational earnings mobility. *Journal of Economic Perspectives, 16*(3), 59-66.

Spano, R., Pridemore, W.A., and Bolland, J. (2012). Specifying the role of exposure to violence and violent behavior on initiation of gun carrying. *Journal of Interpersonal Violence, 27*(1), 158-176.

Squires, D.A. (2011). *The U.S. health system in perspective: A comparison of twelve industrialized nations.* New York: The Commonwealth Fund.

Staetsky, L. (2009). Diverging trends in female old-age mortality: A reappraisal. *Demographic Research, 21*(30), 885-914.

Stafford, M., Martikainen, P., Lahelma, E., and Marmot, M. (2004). Neighbourhoods and self-rated health: A comparison of public sector employees in London and Helsinki. *Journal of Epidemiology and Community Health, 58*(9), 772-778.

Staley, K. (2001). *Voices, values and health: Involving the public in moral decisions.* London, UK: King's Fund.

Stanistreet, D., Bambra, C., and Scott-Samuel, A. (2005). Is patriarchy the source of men's higher mortality? *Journal of Epidemiology and Community Health, 59*(10), 873-876.

Stansfeld, S., and Candy, B. (2006). Psychosocial work environment and mental health: A meta-analytic review. *Scandinavian Journal of Work, Environment and Health, 32*(6), 443-462.

Stansfeld, S.A., Bosma, H., Hemingway, H., and Marmot, M.G. (1998). Psychosocial work characteristics and social support as predictors of SF-36 health functioning: The Whitehall II study. *Psychosomatic Medicine, 60*(3), 247-257.

Starfield, B. (1996). Is strong primary care good for health outcomes? In J. Griffin (Ed.), *The future of primary care* (pp. 18-29). Papers from a symposium September 13, 1995. London, UK: Office of Health Economics.

Starfield, B., Shi, L., and Macinko, J. (2005). Contribution of primary care to health systems and health. *Milbank Quarterly, 83*(3), 457-502.

Starr, P. (2011). *Remedy and reaction: The peculiar American struggle over health care reform.* New Haven, CT: Yale University Press.

Steptoe, A., and Wikman, A. (2010). The contribution of physical activity to divergent trends in longevity. In National Research Council, *International differences in mortality at older ages: Dimensions and sources* (pp. 193-216). E.M. Crimmins, S.H. Preston, and B. Cohen (Eds.), Panel on Understanding Divergent Trends in Longevity in High-Income Countries. Committee on Population. Division of Behavioral and Social Sciences and Education. Washington, DC: The National Academies Press.

Stewart, S.T., Cutler, D.M., and Rosen, A.B. (2009). Forecasting the effects of obesity and smoking on U.S. life expectancy. *New England Journal of Medicine, 361*(23), 2,252-2,260.

Stirbu, I., Kunst, A.E., Bopp, M., Leinsalu, M., Regidor, E., Esnaola, S., Costa, G., Martikainen, P., Borrell, C., Deboosere, P., Kalediene, R., Rychtarikova, J., Artnik, B., and Mackenbach, J.P. (2010). Educational inequalities in avoidable mortality in Europe. *Journal of Epidemiology and Community Health, 64*(10), 913-920.

Story, M., Kaphingst, K.M., Robinson-O'Brien, R., and Glanz, K. (2008). Creating healthy food and eating environments: Policy and environmental approaches. *Annual Review of Public Health, 29,* 253-272.

Strully, K.W. (2009). Job loss and health in the U.S. labor market. *Demography, 46*(2), 221-246.

Stuckler, D., Basu, S., and McKee, M. (2010). Budget crises, health, and social welfare programmes. *British Medical Journal, 340,* c3311.

Subramanian, S.V., and Kawachi, I. (2004). Income inequality and health: What have we learned so far? *Epidemiologic Reviews, 26*(1), 78-91.

Subramanian, S.V., Kim, D.J., and Kawachi, I. (2002). Social trust and self-rated health in U.S. communities: A multilevel analysis. *Journal of Urban Health, 79*(4 Suppl. 1), S21-S34.

Subramanian, S.V., Acevedo-Garcia, D., and Osypuk, T.L. (2005). Racial residential segregation and geographic heterogeneity in black/white disparity in poor self-rated health in the U.S.: A multilevel statistical analysis. *Social Science and Medicine, 60*(8), 1,667-1,679.

Sullivan, D., and vonWachter, T. (2009). Job displacement and mortality: An analysis using administrative data. *Quarterly Journal of Economics, 124,* 1,265-1,306.

Sullivan, H., and Skelcher, C. (2002). *Working across boundaries.* Basingstoke, UK: Palgrave Macmillan.

Sullivent, E.E., Faul, M., and Wald, M.M. (2011). Reduced mortality in injured adults transported by helicopter emergency medical services. *Prehospital Emergency Care, 15*(3), 295-302.

Sundquist, K., Eriksson, U., Kawakami, N., Skog, L., Ohlsson, H., and Arvidsson, D. (2011). Neighborhood walkability, physical activity, and walking behavior: The Swedish Neighborhood and Physical Activity (SNAP) study. *Social Science and Medicine, 72*(8), 1,266-1,273.

Tamers, S.L., Agurs-Collins, T., Dodd, K.W., and Nebeling, L. (2009). U.S. and France adult fruit and vegetable consumption patterns: An international comparison. *European Journal of Clinical Nutrition, 63*(1), 11-17.

Taylor, H. (2003). *While most Americans believe in God, only 36 percent attend a religious service once a month or more.* The Harris Poll® #59, October 15, 2003, Creators Syndacate, Inc. http://www.harrisinteractive.com/vault/Harris-Interactive-Poll-Research-While-Most-Americans-Believe-in-God-Only-36-pct-A-2003-10.pdf [June 2012].

Taylor, P., Kochhar, R., Fry, R., Velasco, G., and Motel, S. (2011). *Wealth gaps rise to record highs between whites, blacks and Hispanics.* Washington, DC: Pew Research Center.

Taylor, S.E., Lehman, B.J., Kiefe, C.I., and Seeman, T.E. (2006). Relationship of early life stress and psychological functioning to adult c-reactive protein in the coronary artery risk development in young adults study. *Biological Psychiatry, 60*(8), 819-824.

Thompson, M.P., Arias, I., Basile, K.C., and Desai, S. (2002). The association between childhood physical and sexual victimization and health problems in adulthood in a nationally representative sample of women. *Journal of Interpersonal Violence, 17*(10), 1,115-1,129.

Thornicroft, G. (2011). Physical health disparities and mental illness: The scandal of premature mortality. *British Journal of Psychiatry, 199*, 441-442.

Thorpe, K.E., Howard, D.H., and Galactionova, K. (2007). Differences in disease prevalence as a source of the U.S.-European health care spending gap. *Health Affairs, 26*(6), w678-w686.

Thun, M., Peto, R., Boreham, J., and Lopez, A.D. (2012). Stages of the cigarette epidemic on entering its second century. *Tobacco Control, 21*(2), 96-101.

Tirosh, A., Shai, I., Afek, A., Dubnov-Raz, G., Ayalon, N., Gordon, B., Derazne, E., Tzur, D., Shamis, A., Vinker, S., and Rudich, A. (2011). Adolescent BMI trajectory and risk of diabetes versus coronary disease. *New England Journal of Medicine, 364*(14), 1,315-1,325.

Toomey, T.L., Erickson, D.J., Carlin, B.P., Lenk, K.M., Quick, H.S., Jones, A.M., and Harwood, E.M. (2012). The association between density of alcohol establishments and violent crime within urban neighborhoods. *Alcohol Clinical Experimental Research, 36*(8), 1,468-1,473.

Toro, P.A., Tompsett, C.J., Lombardo, S., Philippot, P., Nachtergael, H., Galand, B., Schlienz, N., Stammel, N., Yabar, Y., Blume, M., MacKay, L., and Harvey, K. (2007). Homelessness in Europe and the United States: A comparison of prevalence and public opinion. *Journal of Social Issues, 63*(3), 505-524.

Transportation Research Board. (2005). *TRB special report 282: Does the built environment influence physical activity? Examining the evidence.* Washington, DC: Transportation Research Board.

Transportation Research Board. (2009). *TRB special report 298: Driving and the built environment. The effects of compact development on motorized travel, energy use, and CO_2 emissions.* Washington, DC: Transportation Research Board.

Transportation Research Board. (2011). *TRB special report 300: Achieving traffic safety goals in the United States: Lessons from other nations.* Washington, DC: Transportation Research Board.

Trasande, L., Cronk, C., Durkin, M., Weiss, M., Schoeller, D.A., Gall, E.A., Hewitt, J.B., Carrel, A.L., Landrigan, P.J., and Gillman, M.W. (2009). Environment and obesity in the National Children's Study. *Environmental Health Perspectives, 117*(2), 159-166.

Treuhaft, S., and Karpyn, A. (2010). *The grocery gap: Who has access to healthy food and why it matters.* Oakland, CA: PolicyLink.

Trzesniewski, K.H., Donnellan, M.B., Moffitt, T.E., Robins, R.W., Poulton, R., and Caspi, A. (2006). Low self-esteem during adolescence predicts poor health, criminal behavior, and limited economic prospects during adulthood. *Developmental Psychology, 42*(2), 381-390.

Turrell, G., Lynch, J.W., Leite, C., Raghunathan, T., and Kaplan, G.A. (2007). Socioeconomic disadvantage in childhood and across the life course and all-cause mortality and physical function in adulthood: Evidence from the Alameda county study. *Journal of Epidemiology and Community Health, 61*(8), 723-730.

Tzivian, L. (2011). Outdoor air pollution and asthma in children. *Journal of Asthma, 48*(5), 470-481.

Umberson, D., Liu, H., and Reczek, C. (2008). Stress and health behaviour over the life course. *Advances in Life Course Research, 13*, 19-44.

Unal, B., Critchley, J.A., and Capewell, S. (2005). Modelling the decline in coronary heart disease deaths in England and Wales, 1981-2000: Comparing contributions from primary prevention and secondary prevention. *British Medical Journal, 331*(7,517), 614.

UNICEF. (2007). *Child poverty in perspective: An overview of child well-being in rich countries, Innocenti report card, vol. 7*. Florence, Italy: UNICEF Innocenti Research Centre.

United Nations. (2012a). *A/RES/66/2. Political declaration of the high-level meeting of the General Assembly on the prevention and control of non-communicable diseases resolution adopted by the General Assembly [without reference to a main committee (A/66/L.1)].* Available: //www.who.int/nmh/events/un_ncd_summit2011/political_declaration_en.pdf [September 2012].

United Nations. (2012b). *Table 3. Population by sex, annual rate of population increase, surface area and density*. Available: http://unstats.un.org/unsd/demographic/products/dyb/dyb2006/Table03.pdf [August 2012].

United Nations Development Programme. (2011). *International human development indicators: Adolescent fertility rates*. Available: http://hdrstats.undp.org/en/indicators/36806.html [August 2012].

United Nations Office on Drugs and Crime. (2010). *World drug report 2010*. Vienna, Austria: United Nations Office on Drugs and Crime.

United Nations Office on Drugs and Crime. (2012). *Data*. Available: http://www.unodc.org/unodc/en/data-and-analysis/statistics/data.html [June 2012].

University of Minnesota. (2012a). *Integrated health interview series*. Available: http://www.ihis.us/ihis-action/faq#ques0 [April 2012].

University of Minnesota. (2012b). *Integrated public use microdata series international*. Available: https://international.ipums.org/international/overview.shtml [April 2012].

U.S. Bureau of Labor Statistics. (2012). *National longitudinal surveys*. Available: http://www.bls.gov/nls/#overview [April 2012].

U.S. Census Bureau. (1970). *Statistical abstract of the United States*. Washington, DC: U.S. Census Bureau.

U.S. Census Bureau. (1980). *Statistical abstract of the United States*. Washington, DC: U.S. Census Bureau.

U.S. Census Bureau. (1995). *Statistical abstract of the United States*. Washington, DC: U.S. Census Bureau.

U.S. Census Bureau. (2003). *Statistical abstract of the United States*. Washington, DC: U.S. Census Bureau.

U.S. Census Bureau. (2008). *Statistical abstract of the United States*. Washington, DC: U.S. Census Bureau.

U.S. Census Bureau. (2011). *Table 1336. Marriage and divorce rates by country*. Washington, DC: U.S. Census Bureau.

U.S. Census Bureau. (2012). *The 2012 statistical abstract, international statistics: Labor force*. Available: http://www.census.gov/compendia/statab/cats/international_statistics/labor_force.html [April 2012].

U.S. Department of Agriculture. (2003). *Agriculture fact book 2001-2002*. Washington, DC: U.S. Government Printing Office. Available: http://www.usda.gov/factbook/tables/ch2table26.jpg [June 2012].

U.S. Department of Agriculture and U.S. Department of Health and Human Services. (2010). *Dietary guidelines for Americans, 2010*. Washington, DC: U.S. Department of Agriculture and U.S. Department of Health and Human Services.

U.S. Department of Health and Human Services. (1964). *Smoking and health: Report of the advisory committee to the Surgeon General of the public health service: A report of the Surgeon General*. Washington, DC: U.S. Department of Education and Welfare, Public Health Service. Available: http://profiles.nlm.nih.gov/NN/B/B/M/Q/ [June 2012].

U.S. Department of Health and Human Services. (2000). *Reducing tobacco use: A report of the Surgeon General—executive summary.* Atlanta, GA: Centers for Disease Control and Prevention, National Center for Chronic Disease Prevention and Health Promotion, Office on Smoking and Health. Available: http://www.cdc.gov/tobacco/data_statistics/sgr/2000/index.htm [June 2012].

U.S. Department of Health and Human Services. (2001a). *Mental health: Culture, race, and ethnicity—A supplement to mental health: A report of the Surgeon General.* Rockville, MD: U.S. Department of Health and Human Services, Substance Abuse and Mental Health Services Administration, Center for Mental Health Services. Available: http://137.187.25.243/library/mentalhealth/cre/execsummary-1.html [April 2012].

U.S. Department of Health and Human Services. (2001b). *Youth violence: A report of the Surgeon General.* Rockville, MD: U.S. Department of Health and Human Services. Available: http://www.ncbi.nlm.nih.gov/books/NBK44294/ [April 2012].

U.S. Department of Health and Human Services. (2002). *Women and smoking: A report of the Surgeon General.* Rockville, MD: U.S. Department of Health and Human Services. Available: http://www.surgeongeneral.gov/library/reports/womenandtobacco/ [April 2012].

U.S. Department of Health and Human Services. (2008a). *2008 physical activity guidelines for Americans: Be active, healthy, and happy!* U.S. Department of Health and Human Services. Available: http://www.health.gov/paguidelines/pdf/paguide.pdf [April 2012].

U.S. Department of Health and Human Services. (2008b). *The secretary's advisory committee on national health promotion and disease prevention objectives for 2020.* Rockville, MD: U.S. Department of Health and Human Services.

U.S. Department of Health and Human Services. (2012a). *About healthy people.* Available: http://www.healthypeople.gov/2020/about/default.aspx [June 2012].

U.S. Department of Health and Human Services. (2012b). *Preventing tobacco use among youth and young adults: A report of the Surgeon General.* Atlanta, GA: U.S. Department of Health and Human Services, Centers for Disease Control and Prevention, National Center for Chronic Disease Prevention and Health Promotion, Office on Smoking and Health. Available: http://www.surgeongeneral.gov/library/reports/preventing-youth-tobacco-use/ [August 2012].

U.S. Department of Health and Human Services. (2012c). *Shortage designation: Health professional shortage areas & medically underserved areas/populations.* Available: http://bhpr.hrsa.gov/shortage/ [August 2012].

U.S. Department of Justice. (2007). *Adolescents, neighborhoods, and violence: Recent findings from the Project on Human Development in Chicago Neighborhoods.* Washington, DC: U.S. Department of Justice.

U.S. Environmental Protection Agency. (2007). *PM10 NAAQS implementation.* Available: http://www.epa.gov/ttn/naaqs/pm/pm10_index.html [August 2012].

U.S. Food and Drug Administration. (2011). *AERS patient outcomes by year.* Available: http://www.fda.gov/Drugs/GuidanceComplianceRegulatoryInformation/Surveillance/AdverseDrugEffects/ucm070461.htm [November 2011].

U.S. Preventive Services Task Force. (2012). *U.S. Preventive Services Task Force.* Available: http://www.uspreventiveservicestaskforce.org/index.html [June 2012].

Valkonen, T., and Van Poppel, F. (1997). The contribution of smoking to sex differences in life expectancy: Four Nordic countries and the Netherlands 1970-1989. *European Journal of Public Health, 7*(3), 302-310.

Van der Heyden, J.H., Schaap, M.M., Kunst, A.E., Esnaola, S., Borrell, C., Cox, B., Leinsalu, M., Stirbu, I., Kalediene, R., Deboosere, P., Mackenbach, J.P., and Van Oyen, H. (2009). Socioeconomic inequalities in lung cancer mortality in 16 European populations. *Lung Cancer, 63*(3), 322-330.

Van Dyck, D., Cardon, G., Deforche, B., Sallis, J.F., Owen, N., and De Bourdeaudhuij, I. (2010). Neighborhood SES and walkability are related to physical activity behavior in Belgian adults. *Preventive Medicine, 50*(Suppl.), S74-S79.

Van Lenthe, F.J., Borrell, L.N., Costa, G., Diez Roux, A.V., Kauppinen, T.M., Marinacci, C., Martikainen, P., Regidor, E., Stafford, M., and Valkonen, T. (2005). Neighbourhood unemployment and all cause mortality: A comparison of six countries. *Journal of Epidemiology and Community Health, 59*(3), 231-237.

Vaupel, J.W., Carey, J.R., and Christensen, K. (2003). It's never too late. *Science, 301*(5,640), 1,679-1,681.

Verdecchia, A., Francisci, S., Brenner, H., Gatta, G., Micheli, A., Mangone, L., Kunkler, I., and Eurocare Working Group. (2007). Recent cancer survival in Europe: A 2000-02 period analysis of EUROCARE-4 data. *Lancet Oncology, 8*(9), 784-796.

Verguet, S., and Jamison, D. (2011). *How many years behind is the U.S. in mortality rates?* Unpublished paper prepared for the NAS/IOM Panel on Understanding Cross-National Health Differences Among High-Income Countries. Department of Global Health, University of Washington.

Vermeiren, R., Schwab-Stone, M., Deboutte, D., Leckman, P.E., and Ruchkin, V. (2003). Violence exposure and substance use in adolescents: Findings from three countries. *Pediatrics, 111*(3), 535-540.

Vettukattil, A.S., Haider, A.H., Haut, E.R., Chang, D.C., Oyetunji, T., Cornwell, E.E., III, Stevens, K.A., and Efron, D.T. (2011). Do trauma safety-net hospitals deliver truly safe trauma care? A multilevel analysis of the national trauma data bank. *Journal of Trauma, 70*(4), 978-984.

Victora, C.G., Habicht, J.-P., and Bryce, J. (2004). Evidence-based public health: Moving beyond randomized trials. *American Journal of Public Health, 94*(3), 400-405.

Viner, R. (2012). *Adolescence and the development of health differences in the USA compared with other high income countries.* Unpublished paper prepared for the NAS/IOM Panel on Understanding Cross-National Health Differences Among High-Income Countries. UCL Institute of Child Health, University College London.

von Hinke Kessler Scholder, S., Davey Smith, G., Lawlor, D.A., Propper, C., and Windmeijer, F. (2012). The effect of fat mass on educational attainment: Examining the sensitivity to different identification strategies. *Economics and Human Biology, 10*(4), 405-418.

Wadhwa, P.D., Buss, C., Entringer, S., and Swanson, J.M. (2009). Developmental origins of health and disease: Brief history of the approach and current focus on epigenetic mechanisms. *Seminars in Reproductive Medicine, 27*(5), 358-368.

Wadman, M. (2012). Growing pains for Children's Study. *Nature, 482*(7,386), 448.

Wadsworth, M.E.J. (1991). *The imprint of time: Childhood, history, and adult life.* New York: Oxford University Press.

Wagner, E.H., Austin, B.T., and Von Korff, M. (1996). Organizing care for patients with chronic illness. *Milbank Quarterly, 74*(4), 511-544.

Wahlbeck, K., Westman, J., Nordentoft, M., Gissler, M., and Laursen, T.M. (2011). Outcomes of Nordic mental health systems: Life expectancy of patients with mental disorders. *British Journal of Psychiatry, 199*(6), 453-458.

Waidmann, T. (2009). *Estimating the cost of racial and ethnic health disparities.* Washington, DC: The Urban Institute. Available: http://www.urban.org/uploadedpdf/411962_health_disparities.pdf [August 2012].

Waldron, I. (1986). The contribution of smoking to sex differences in mortality. *Public Health Reports, 101*(2), 163-173.

Walker, A., and Wong, C. (2005). *East Asian welfare regimes in transition: From confucianism to globalisation.* Bristol, UK: Policy Press.

Wang, H., and Preston, S.H. (2009). Forecasting United States mortality using cohort smoking histories. *Proceedings of the National Academy of Sciences, 106*(2), 393-398.

Wang, Y.C., Coxson, P., Shen, Y.-M., Goldman, L., and Bibbins-Domingo, K. (2012). A penny-per-ounce tax on sugar-sweetened beverages would cut health and cost burdens of diabetes. *Health Affairs, 31*(1), 199-207.

Warner, D.F., and Hayward, M.D. (2006). Early-life origins of the race gap in men's mortality. *Journal of Health and Social Behavior, 47*(3), 209-226.

Warner, M., Chen, L., Makuc, D., Anderson, R., and Miniño, A. (2011). *Drug poisoning deaths in the United States, 1980-2008.* Hyattsville, MD: National Center for Health Statistics.

Wechsler, H., and Nelson, T.F. (2008). What we have learned from the Harvard School of Public Health College Alcohol Study: Focusing attention on college student alcohol consumption and the environmental conditions that promote it. *Journal of Studies on Alcohol and Drugs, 69*(4), 481-490.

Wendt, C., Mischke, M., Pfeifer, M., and Reibling, N. (2011). Cost barriers reduce confidence in receiving medical care when seriously ill. *International Journal of Clinical Practice, 65*(11), 1,115-1,117.

Whiteford, P., and Adema, W. (2007). *What works best in reducing child poverty: A benefit or work strategy?* Paris: OECD.

Wildeman, C., and Western, B. (2010). Incarceration in fragile families. *Future of Children, 20*(2), 157-177.

Wilkinson, R.G., and Pickett, K.E. (2007). The problems of relative deprivation: Why some societies do better than others. *Social Science and Medicine, 65*(9), 1,965-1,978.

Wilkinson, R.G., and Pickett, K. (2009). *The spirit level: Why greater equality makes societies stronger.* New York: Bloomsbury Press.

Willett, W. (1998). *Nutritional epidemiology* (2nd ed.). New York: Oxford University Press.

Williams, D.R. (1999). Race, socioeconomic status, and health. The added effects of racism and discrimination. *Annals of the New York Academy of Sciences, 896*(1), 173-188.

Williams, D.R., and Collins, C. (1995). U.S. socioeconomic and racial differences in health: Patterns and explanations. *Annual Review of Sociology, 21*, 349-386.

Williams, D.R., and Collins, C. (2001). Racial residential segregation: A fundamental cause of racial disparities in health. *Public Health Reports, 116*(5), 404-416.

Williams, D.R., and Mohammed, S.A. (2009). Discrimination and racial disparities in health: Evidence and needed research. *Journal of Behavioral Medicine, 32*(1), 20-47.

Wilmoth, J.R., and Horiuchi, S. (1999). Rectangularization revisited: Variability of age at death within human populations. *Demography, 36*(4), 475-495.

Wilper, A.P., Woolhandler, S., Lasser, K.E., McCormick, D., Bor, D.H., and Himmelstein, D.U. (2009). Health insurance and mortality in U.S. adults. *American Journal of Public Health, 99*(12), 2,289-2,295.

Winker, M.A. (2004). Measuring race and ethnicity: Why and how? *Journal of the American Medical Association, 292*(13), 1,612-1,614.

Wolf-Maier, K., Cooper, R.S., Banegas, J.R., Giampaoli, S., Hense, H.-W., Joffres, M., Kastarinen, M., Poulter, N., Primatesta, P., Rodríguez-Artalejo, F., Stegmayr, B., Thamm, M., Tuomilehto, J., Vanuzzo, D., and Vescio, F. (2003). Hypertension prevalence and blood pressure levels in 6 European countries, Canada, and the United States. *Journal of the American Medical Association, 289*(18), 2,363-2,369.

Woolf, S.H. (2011). Public health implications of government spending reductions. *Journal of the American Medical Association, 305*(18), 1,902-1,903.

Woolf, S.H., Johnson, R.E., Fryer, G.E., Rust, G., and Satcher, D. (2004). The health impact of resolving racial disparities: An analysis of U.S. mortality data. *American Journal of Public Health, 94*(12), 2,078-2,081.

Woolf, S.H., Glasgow, R.E., Krist, A., Bartz, C., Flocke, S.A., Holtrop, J.S., Rothemich, S.F., and Wald, E.R. (2005). Putting it together: Finding success in behavior change through integration of services. *Annals of Family Medicine, 3*(Suppl. 2), S20-S27.

Woolf, S.H., Johnson, R.E., Phillips, R.L., and Philipsen, M. (2007). Giving everyone the health of the educated: An examination of whether social change would save more lives than medical advances. *American Journal of Public Health, 97*(4), 679-683.

Woolf, S.H., Jones, R.M., Johnson, R.E., Phillips, Jr., R.L., Norman Oliver, M., Bazemore, A., and Vichare, A. (2010). Avertable deaths associated with household income in Virginia. *American Journal of Public Health, 100*(4), 750-755.

Woolf, S.H., Dekker, M.M., Byrne, F.R., and Miller, W.D. (2011). Citizen-centered health promotion: Building collaborations to facilitate healthy living. *American Journal of Preventive Medicine, 40*(1, Suppl. 1), S38-S47.

World Bank. (2007). *The world development report: Development and the next generation.* Washington, DC: World Bank.

World Bank. (2012a). *GDP per capita (current US$).* Available: http://data.worldbank.org/indicator/NY.GDP.PCAP.CD [April 2012].

World Bank. (2012b). *Indicators.* Available: http://data.worldbank.org/indicator [June 2012].

World Cancer Research Fund. (1997). *Food, nutrition and the prevention of cancer: A global perspective.* London, UK: World Cancer Research Fund.

World Economic Forum. (2011). *The global competitiveness report 2011-2012.* Geneva, Switzerland: World Economic Forum.

World Health Organization. (1948). *The constitution of the World Health Organization.* Available: http://apps.who.int/gb/bd/PDF/bd47/EN/constitution-en.pdf [June 2012].

World Health Organization. (2000a). *Health and health behaviour among young people.* Geneva, Switzerland: World Health Organization.

World Health Organization. (2000b). *The world health report 2000—Health systems: Improving performance.* Geneva, Switzerland: World Health Organization.

World Health Organization. (2002). *World report on violence and health.* Geneva, Switzerland: World Health Organization.

World Health Organization. (2003). *Investing in mental health.* Geneva, Switzerland: Department of Mental Health and Substance Dependence, Noncommunicable Diseases and Mental Health, World Health Organization.

World Health Organization. (2004a). *Global status report on alcohol 2004.* Geneva, Switzerland: World Health Organization.

World Health Organization. (2004b). *Global strategy on diet, physical activity and health.* Geneva, Switzerland: World Health Organization.

World Health Organization. (2008a). *The global burden of disease: 2004 update.* Geneva, Switzerland: World Health Organization.

World Health Organization. (2008b). *The world health report 2008: Primary health care: Now more than ever.* Geneva, Switzerland: World Health Organization.

World Health Organization. (2009). *Global status report on road safety: Time for action.* Geneva, Switzerland: World Health Organization.

World Health Organization. (2010). *World health statistics, 2010.* Available: http://www.who.int/whosis/whostat/EN_WHS10_Full.pdf [August 2012].

World Health Organization. (2011a). *Global health observatory data repository, Summary estimates of mortality for WHO member states for the year 2008.* Available: http://www.who.int/gho/mortality_burden_disease/global_burden_disease_death_estimates_sex_2008.xls [August 2012].

World Health Organization. (2011b). *Global information system on alcohol and health (GISAH).* Available: http://apps.who.int/ghodata/?theme=GISAH [June 2012].

World Health Organization. (2011c). *Rio political declaration on social determinants of health*. World Conference on Social Determinants of Health, Rio de Janeiro, Brazil, October 21, 2011. Available: http://www.who.int/sdhconference/declaration/Rio_political _declaration.pdf, page 7 [August 2012].

World Value Survey Association. (2012). *World values survey*. Available: http://www.world valuessurvey.org/ [June 2012].

Wright, R.J. (2006). Health effects of socially toxic neighborhoods: The violence and urban asthma paradigm. *Clinics in Chest Medicine, 27*(3), 413-421.

Wu, L.-T., Woody, G.E., Yang, C., Pan, J.-J., and Blazer, D.G. (2011). Racial/ethnic variations in substance-related disorders among adolescents in the United States. *Archives of General Psychiatry, 68*(11), 1,176-1,185.

Wunsch, H., Angus, D.C., Harrison, D.A., Collange, O., Fowler, R., Hoste, E.A., de Keizer, N.F., Kersten, A., Linde-Zwirble, W.T., Sandiumenge, A., and Rowan, K.M. (2008). Variation in critical care services across North America and Western Europe. *Critical Care Medicine, 36*(10), 2,787-2,793, e2,781-e2,789.

Young, F., Capewell, S., Ford, E.S., and Critchley, J.A. (2010). Coronary mortality declines in the U.S. between 1980 and 2000: Quantifying the contributions from primary and secondary prevention. *American Journal of Preventive Medicine, 39*(3), 228-234.

Zhang, T.Y., Chrétien, P., Meaney, M.J., and Gratton, A. (2005). Influence of naturally occurring variations in maternal care on prepulse inhibition of acoustic startle and the medial prefrontal cortical dopamine response to stress in adult rats. *Journal of Neuroscience, 25*(6), 1,493-1,502.

Ziersch, A.M., Gallaher, G., Baum, F., and Bentley, M. (2011). Responding to racism: Insights on how racism can damage health from an urban study of Australian aboriginal people. *Social Science and Medicine, 73*(7), 1,045-1,053.

Zito, J.M., Safer, D.J., Berg, L.T., Janhsen, K., Fegert, J.M., Gardner, J.F., Glaeske, G., and Valluri, S.C. (2008). A three-country comparison of psychotropic medication prevalence in youth. *Child and Adolescent Psychiatry and Mental Health, 2*(1), 1-8.

Zwerling, C., Peek-Asa, C., Whitten, P.S., Choi, S.W., Sprince, N.L., and Jones, M.P. (2005). Fatal motor vehicle crashes in rural and urban areas: Decomposing rates into contributing factors. *Injury Prevention, 11*(1), 24-28.

Appendixes

Appendix A

Recommendations of the National Prevention Council and Evidence Cited in Its Report

Recommendation	Supporting Evidence-Based Interventions
HEALTHY AND SAFE COMMUNITY ENVIRONMENTS	
Improve quality of air, land, and water.	• HP: Reduce exposure to selected environmental chemicals in the population, as measured by blood and urine concentrations of the substances or their metabolites. http://www.healthypeople.gov/2020/topicsobjectives2020/ objectiveslist.aspx?topicid=12 • HP: Improve quality, utility, awareness, and use of existing information systems for environmental health. http://www.healthypeople.gov/2020/topicsobjectives2020/ objectiveslist.aspx?topicid=12 • HP: Increase the number of states, territories, tribes, and the District of Columbia that monitor diseases or conditions that can be caused by exposure to environmental hazards. http://www.healthypeople.gov/2020/topicsobjectives2020/ objectiveslist.aspx?topicid=12
Design and promote affordable, accessible, safe, and healthy housing.	• HP: Reduce indoor allergen levels. http://www.healthypeople.gov/2020/topicsobjectives2020/ objectiveslist.aspx?topicid=12 • HP: Increase the number of homes with an operating radon mitigation system for persons living in homes at risk for radon exposure. http://www.healthypeople.gov/2020/topicsobjectives2020/ objectiveslist.aspx?topicid=12 • HP: Increase the percentage of new single family homes (SFHs) constructed with radon-reducing features, especially in high-radon-potential areas. http://www.healthypeople.gov/2020/topicsobjectives2020/ objectiveslist.aspx?topicid=12

continued

Recommendation	Supporting Evidence-Based Interventions
HEALTHY AND SAFE COMMUNITY ENVIRONMENTS Continued	
	• HP: Increase the percentage of new single family homes (SFHs) constructed with radon-reducing features, especially in high-radon-potential areas. http://www.healthypeople.gov/2020/topicsobjectives2020/objectiveslist.aspx?topicid=12 • HP: Reduce the number of U.S. homes that are found to have lead-based paint or related hazards. http://www.healthypeople.gov/2020/topicsobjectives2020/objectiveslist.aspx?topicid=12 • HP: Reduce the proportion of occupied housing units that have moderate or severe physical problems. http://www.healthypeople.gov/2020/topicsobjectives2020/objectiveslist.aspx?topicid=12
Strengthen state, tribal, local, and territorial public health departments to provide essential services.	• HP: Increase the proportion of tribal and state public health agencies that provide or assure comprehensive laboratory services to support essential public health services. http://www.healthypeople.gov/2020/topicsobjectives2020/objectiveslist.aspx?topicid=35 • HP: Increase the proportion of tribal, state, and local public health agencies that provide or assure comprehensive epidemiology services to support essential public health services. http://www.healthypeople.gov/2020/topicsobjectives2020/objectiveslist.aspx?topicid=35 • IOM: The committee finds that the core functions of public health agencies at all levels of government are assessment, policy development, and assurance. http://books.nap.edu/openbook.php?record_id=10548&page=411
Integrate health criteria into decision making, where appropriate, across multiple sectors.	• HP: Reduce the number of new schools sited within 500 feet of an interstate or federal or state highway. http://www.healthypeople.gov/2020/topicsobjectives2020/objectiveslist.aspx?topicid=12
Enhance cross-sector collaboration in community planning and design to promote health and safety.	• IOM: Private and public purchasers, health care organizations, clinicians, and patients should work together to redesign health care. http://www.nap.edu/openbook.php?record_id=10027&page=8
Expand and increase access to information technology and integrated data systems to promote cross-sector information exchange.	• HP: Increase the number of states that record vital events using the latest U.S. standard certificates and report. http://www.healthypeople.gov/2020/topicsobjectives2020/objectiveslist.aspx?topicid=35 • HP: Increase the proportion of quality, health-related websites. http://www.healthypeople.gov/2020/topicsobjectives2020/objectiveslist.aspx?topicid=18 • HP: Increase the proportion of online health information seekers who report easily accessing health information. http://www.healthypeople.gov/2020/topicsobjectives2020/objectiveslist.aspx?topicid=18

Recommendation	Supporting Evidence-Based Interventions
	• HP: Increase the proportion of medical practices that use electronic health records. http://www.healthypeople.gov/2020/topicsobjectives2020/objectiveslist.aspx?topicid=18
Identify and implement strategies that are proven to work and conduct research where evidence is lacking.	• IOM: Making evidence the foundation of decision making and the measure of success. http://books.nap.edu/openbook.php?record_id=10548&page=4
Maintain a skilled, cross-trained, and diverse prevention workforce.	• IOM: Greater emphasis in public health curricula should be placed on managerial and leadership skills, such as the ability to communicate important agency values to employees and enlist their commitment; to sense and deal with important changes in the environment; to plan, mobilize, and use resources effectively; and to relate the operation of the agency to its larger community role. http://books.nap.edu/openbook.php?record_id=10548&page=418
	• IOM: Schools of public health should strengthen their response to the needs for qualified personnel for important, but often neglected aspects of public health such as the health of minority groups and international health. http://books.nap.edu/openbook.php?record_id=10548&page=418
	• IOM: Schools of public health should encourage and assist other institutions to prepare appropriate, qualified public health personnel for positions in the field. When educational institutions other than schools of public health undertake to train personnel for work in the field, careful attention to the scope and capacity of the educational program is essential. http://books.nap.edu/openbook.php?record_id=10548&page=418

CLINICAL AND COMMUNITY PREVENTIVE SERVICES

Support the National Quality Strategy's focus on improving cardiovascular health.	• CG: Increasing Tobacco Use Cessation: Provider Reminders When Used Alone. http://www.thecommunityguide.org/tobacco/cessation/providerreminders.html
	• CG: Increasing Tobacco Use Cessation: Provider Reminders with Provider Education. http://www.thecommunityguide.org/tobacco/cessation/providerreminderedu.html
	• CG: Increasing Tobacco Use Cessation: Reducing Client Out-of-Pocket Costs for Cessation Therapies. http://www.thecommunityguide.org/tobacco/cessation/outofpocketcosts.html
	• USPSTF: Recommends that clinicians ask all adults about tobacco use and provide tobacco cessation interventions for those who use tobacco products. http://www.uspreventiveservicestaskforce.org/uspstf/uspstbac2.htm

continued

Recommendation	Supporting Evidence-Based Interventions

CLINICAL AND COMMUNITY PREVENTIVE SERVICES Continued

- USPSTF: Recommends that clinicians ask all pregnant women about tobacco use and provide augmented, pregnancy-tailored counseling for those who smoke.
  http://www.uspreventiveservicestaskforce.org/uspstf/uspstbac2.htm
- USPSTF: Recommends the use of aspirin for men aged 45 to 79 when the potential benefit due to a reduction in myocardial infarctions outweighs the potential harm due to an increase in gastrointestinal hemorrhage.
  http://www.uspreventiveservicestaskforce.org/uspstf/uspsasmi.htm
- USPSTF: Recommends the use of aspirin for women aged 55 to 79 when the potential benefit of a reduction in ischemic strokes outweighs the potential harm of an increase in gastrointestinal hemorrhage.
  http://www.uspreventiveservicestaskforce.org/uspstf/uspsasmi.htm
- USPSTF: Recommends screening for high blood pressure in adults age 18 and older.
  http://www.uspreventiveservicestaskforce.org/uspstf/uspshype.htm
- USPSTF: Strongly recommends screening men age 35 and older for lipid disorders.
  http://www.uspreventiveservicestaskforce.org/uspstf/uspschol.htm
- USPSTF: Recommends screening men aged 20 to 35 for lipid disorders if they are at increased risk for coronary heart disease.
  http://www.uspreventiveservicestaskforce.org/uspstf/uspschol.htm
- USPSTF: Strongly recommends screening women age 45 and older for lipid disorders if they are at increased risk for coronary heart disease.
  http://www.uspreventiveservicestaskforce.org/uspstf/uspschol.htm
- USPSTF: Recommends screening women aged 20 to 45 for lipid disorders if they are at increased risk for coronary heart disease.
  http://www.uspreventiveservicestaskforce.org/uspstf/uspschol.htm
- USPSTF: Recommends that clinicians ask all adults about tobacco use and provide tobacco cessation interventions for those who use tobacco products.
  http://www.uspreventiveservicestaskforce.org/uspstf/uspstbac2.htm
- USPSTF: Recommends that clinicians ask all pregnant women about tobacco use and provide augmented, pregnancy-tailored counseling for those who smoke.
  http://www.uspreventiveservicestaskforce.org/uspstf/uspstbac2.htm
- HP: Increase the proportion of adults who have had their blood pressure measured within the preceding 2 years and can state whether their blood pressure was normal or high.
  http://www.healthypeople.gov/2020/topicsobjectives2020/objectiveslist.aspx?topicid=21
- HP: Increase the proportion of adults who have had their blood cholesterol checked within the preceding 5 years.
  http://www.healthypeople.gov/2020/topicsobjectives2020/objectiveslist.aspx?topicid=21

Recommendation	Supporting Evidence-Based Interventions
	• HP: Increase the proportion of adults with hypertension who are taking the prescribed medications to lower their blood pressure. http://www.healthypeople.gov/2020/topicsobjectives2020/objectiveslist.aspx?topicid=21 • HP: Increase the proportion of adults with hypertension whose blood pressure is under control. http://www.healthypeople.gov/2020/topicsobjectives2020/objectiveslist.aspx?topicid=21 • HP: Increase smoking cessation attempts by adult smokers. http://www.healthypeople.gov/2020/topicsobjectives2020/objectiveslist.aspx?topicid=41 • HP: Increase recent smoking cessation success by adult smokers. http://www.healthypeople.gov/2020/topicsobjectives2020/objectiveslist.aspx?topicid=41 • HP: Increase tobacco cessation counseling in health care settings. http://www.healthypeople.gov/2020/topicsobjectives2020/objectiveslist.aspx?topicid=41
Use payment and reimbursement mechanisms to encourage delivery of clinical preventive services.	• IOM: That purchasers, regulators, health professions, educational institutions, and the Department of Health and Human Services create an environment that fosters and rewards improvement by (1) creating an infrastructure to support evidence-based practice, (2) facilitating the use of information technology, (3) aligning payment incentives, and (4) preparing the workforce to better serve patients in a world of expanding knowledge and rapid change. http://www.nap.edu/openbook.php?record_id=10027&page=5
Expand use of interoperable health information technology.	• HP: Increase the proportion of persons who use electronic personal health management tools. http://www.healthypeople.gov/2020/topicsobjectives2020/objectiveslist.aspx?topicid=18 • HP: Increase the proportion of quality, health-related websites. http://www.healthypeople.gov/2020/topicsobjectives2020/objectiveslist.aspx?topicid=18 • HP: Increase the proportion of medical practices that use electronic health records. http://www.healthypeople.gov/2020/topicsobjectives2020/objectiveslist.aspx?topicid=18
Support implementation of community-based preventive services and enhance linkages with clinical care.	• USPSTF: Integrating Evidence-Based Clinical and Community Strategies to Improve Health. http://www.uspreventiveservicestaskforce.org/uspstf07/methods/tfmethods.htm • IOM: Clinicians and patients, and the health care organizations that support care delivery, adopt a new set of principles to guide the redesign of care processes. http://www.nap.edu/openbook.php?record_id=10027&page=5

continued

Recommendation	Supporting Evidence-Based Interventions

CLINICAL AND COMMUNITY PREVENTIVE SERVICES Continued

Reduce barriers to accessing clinical and community preventive services, especially among populations at greatest risk.	• HP: Increase the proportion of persons with a usual primary care provider. http://www.healthypeople.gov/2020/topicsobjectives2020/objectiveslist.aspx?topicid=1 • HP: Increase the proportion of persons who have a specific source of ongoing care. http://www.healthypeople.gov/2020/topicsobjectives2020/objectiveslist.aspx?topicid=1 • HP: Reduce the proportion of individuals who are unable to obtain or delay in obtaining necessary medical care, dental care, or prescription medicines. http://www.healthypeople.gov/2020/topicsobjectives2020/objectiveslist.aspx?topicid=1
Enhance coordination and integration of clinical, behavioral, and complementary health strategies.	• HP: Increase the proportion of persons who use electronic personal health management tools. http://www.healthypeople.gov/2020/topicsobjectives2020/objectiveslist.aspx?topicid=18 • HP: Increase the proportion of quality, health-related websites. http://www.healthypeople.gov/2020/topicsobjectives2020/objectiveslist.aspx?topicid=18 • HP: Increase the proportion of medical practices that use electronic health records. http://www.healthypeople.gov/2020/topicsobjectives2020/objectiveslist.aspx?topicid=18 • IOM: All health care organizations, professional groups, and private and public purchasers should pursue six major aims; specifically, health care should be safe, effective, patient-centered, timely, efficient, and equitable. http://books.nap.edu/openbook.php?record_id=10027&page=6 • IOM: Private and public purchasers, health care organizations, clinicians, and patients should work together to redesign health care processes. http://books.nap.edu/openbook.php?record_id=10027&page=8

EMPOWERED PEOPLE

Provide people with tools and information to make healthy choices.	• HP: Increase the proportion of elementary, middle, and senior high schools that provide school health education to promote personal health and wellness in the following areas: hand washing or hand hygiene, oral health, growth and development, sun safety and skin cancer prevention, benefits of rest and sleep, ways to prevent vision and hearing loss, and the importance of health screenings and checkups. http://www.healthypeople.gov/2020/topicsobjectives2020/objectiveslist.aspx?topicid=11

Recommendation	Supporting Evidence-Based Interventions
	• HP: Increase the proportion of college and university students who receive information from their institution on each of the priority health risk behavior areas (all priority areas; unintentional injury; violence; suicide; tobacco use and addiction; alcohol and other drug use; unintended pregnancy, HIV/AIDS, and STD infection; unhealthy dietary patterns; and inadequate physical activity). http://www.healthypeople.gov/2020/topicsobjectives2020/objectiveslist.aspx?topicid=11
Provide people with tools and information to make healthy choices.	• IOM: Industry should make obesity prevention in children and youth a priority by developing and promoting products, opportunities, and information that will encourage healthful eating behaviors and regular physical activity. http://www.nap.edu/openbook.php?record_id=11015&page=8 • IOM: Nutrition labeling should be clear and useful so that parents and youth can make informed product comparisons and decisions to achieve and maintain energy balance at a healthy weight. http://www.nap.edu/openbook.php?record_id=11015&page=8
Promote positive social interactions and support healthy decision making.	• HP: Increase the proportion of the nation's elementary, middle, and high schools that have official school policies and engage in practices that promote a healthy and safe physical school environment. http://www.healthypeople.gov/2020/topicsobjectives2020/objectiveslist.aspx?topicid=12 • IOM: Schools should provide a consistent environment that is conducive to healthful eating behaviors and regular physical activity. http://www.nap.edu/openbook.php?record_id=11015&page=13 • IOM: Parents should promote healthful eating behaviors and regular physical activity for their children. http://www.nap.edu/openbook.php?record_id=11015&page=15 • IOM: Local governments, private developers, and community groups should expand opportunities for physical activity including recreational facilities, parks, playgrounds, sidewalks, bike paths, routes for walking or bicycling to school, and safe streets and neighborhoods, especially for populations at high risk of childhood obesity. http://www.nap.edu/openbook.php?record_id=11015&page=11
Engage and empower people and communities to plan and implement prevention policies and programs.	• HP: Increase the proportion of elementary, middle, and senior high schools that have health education goals or objectives that address the knowledge and skills articulated in the *National Health Education Standards* (high school, middle, elementary). http://www.healthypeople.gov/2020/topicsobjectives2020/objectiveslist.aspx?topicid=11

continued

Recommendation	Supporting Evidence-Based Interventions

EMPOWERED PEOPLE Continued

	• HP: Increase the proportion of the nation's public and private schools that require daily physical education for all students. http://healthypeople.gov/2020/topicsobjectives2020/objectiveslist.aspx?topicid=33 • IOM: Local governments, public health agencies, schools, and community organizations should collaboratively develop and promote programs that encourage healthful eating behaviors and regular physical activity, particularly for populations at high risk of childhood obesity. Community coalitions should be formed to facilitate and promote cross-cutting programs and community-wide efforts. http://www.nap.edu/openbook.php?record_id=11015&page=10
Engage and empower people and communities to plan and implement prevention policies and programs.	• IOM: Industry should develop and strictly adhere to marketing and advertising guidelines that minimize the risk of obesity in children and youth. http://www.nap.edu/openbook.php?record_id=11015&page=9
Improve education and employment opportunities.	• HP: Eliminate very low food security among children. http://www.healthypeople.gov/2020/topicsobjectives2020/objectiveslist.aspx?topicid=29 • HP: Reduce household food insecurity, and in doing so, reduce hunger. http://www.healthypeople.gov/2020/topicsobjectives2020/objectiveslist.aspx?topicid=29 • IOM: Health professions educational institutions (HPEIs) governing bodies should develop institutional objectives consistent with community benefit principles that support the goal of increasing health care workforce diversity including, but not limited to, efforts to ease financial and nonfinancial obstacles to URM participation, increase involvement of diverse local stakeholders in key decision-making processes, and undertake initiatives that are responsive to local, regional, and societal imperatives. http://www.nap.edu/openbook.php?record_id=10885&page=17

ELIMINATION OF HEALTH DISPARITIES

Ensure a strategic focus on communities at greatest risk.	• HP: Increase the number of community-based organizations (including local health departments, tribal health services, nongovernmental organizations, and state agencies) providing population-based primary prevention services in the following areas: injury, violence, mental illness, tobacco use, substance abuse, unintended pregnancy, chronic disease programs, nutrition, and physical activity. http://www.healthypeople.gov/2020/topicsobjectives2020/objectiveslist.aspx?topicid=11

Recommendation	Supporting Evidence-Based Interventions
	• IOM: Private and public (e.g., federal, state, and local governments) entities should convene major community benefit stakeholders (e.g., community advocates, academic institutions, health care providers), to inform them about community benefit standards and to build awareness that placing a priority on diversity and cultural competency programs is a societal expectation of all institutions that receive any form of public funding. http://www.nap.edu/openbook.php?record_id=10885&page=17
Reduce disparities in access to quality health care.	• USPSTF: To continue the improvement in the health of the people in the United States, we need to use the complete array of effective prevention tools at our disposal, increase their effectiveness and utilization by connecting them where possible, and systematically apply them at all levels of influence on behavior. http://www.uspreventiveservicestaskforce.org/uspstf07/methods/tfmethods.htm • HP: Increase individuals' access to the Internet. http://www.healthypeople.gov/2020/topicsobjectives2020/objectiveslist.aspx?topicid=18 • IOM: All health care organizations, professional groups, and private and public purchasers should pursue six major aims; specifically, health care should be safe, effective, patient-centered, timely, efficient, and equitable. http://books.nap.edu/openbook.php?record_id=10027&page=6 • IOM: HPEIs should be encouraged to affiliate with community-based health care facilities in order to attract and train a more diverse and culturally competent workforce and to increase access to health care. http://www.nap.edu/openbook.php?record_id=10885&page=15
Increase the capacity of the prevention workforce to identify and address disparities.	• IOM: Health professions education accreditation bodies should develop explicit policies articulating the value and importance of providing culturally competent health care and the role it sees for racial and ethnic diversity among health professionals in achieving this goal. http://www.nap.edu/openbook.php?record_id=10885&page=12 • IOM: Health professions education accreditation bodies should develop standards and criteria that more effectively encourage health professions schools to recruit URM students and faculty, to develop cultural competence curricula, and to develop an institutional climate that encourages and sustains the development of a critical mass of diversity. http://www.nap.edu/openbook.php?record_id=10885&page=12 • IOM: Private entities should be encouraged to collaborate through business partnerships and other entrepreneurial relationships with HPEIs to support the common goal of developing a more diverse health care workforce. http://www.nap.edu/openbook.php?record_id=10885&page=12

continued

Recommendation	Supporting Evidence-Based Interventions

ELIMINATION OF HEALTH DISPARITIES Continued

Support research to identify effective strategies to eliminate health disparities.	• IOM: Additional data collection and research are needed to more thoroughly characterize URM participation in the health professions and in health professions education and to further assess the benefits of diversity among health professionals, particularly with regard to the potential economic benefits of diversity. http://www.nap.edu/openbook.php?record_id=10885&page=18
Standardize and collect data to better identify and address disparities.	• IOM: Collect data on granular ethnicity using categories that are applicable to the populations it serves or studies. Categories should be selected from a national standard on the basis of health and health care quality issues, evidence or likelihood of disparities, or size of subgroups within the population. The selection of categories should also be informed by analysis of relevant data (e.g., Census data) on the service or study population. In addition, an open-ended option of "Other, please specify:—" should be provided for persons whose granular ethnicity is not listed as a response option. http://www.ahrq.gov/research/iomracereport/reldatasum.htm
.	• IOM: Pursue studies on different ways of framing the questions and related response categories for collecting race and ethnicity data at the level of the OMB categories, focusing on completeness and accuracy of response among all groups. http://www.ahrq.gov/research/iomracereport/reldatasum.htm

TOBACCO-FREE LIVING

Support comprehensive tobacco-free policies and other evidence-based tobacco control policies.	• CG: Reducing Exposure to Environmental Tobacco Smoke: Smoking Bans and Restrictions. http:// www.thecommunityguide.org/tobacco/environmental/ smokingbans.html • CG: Decreasing Tobacco Use Among Workers: Smoke-Free Policies to Reduce Tobacco Use. http://www.thecommunityguide.org/tobacco/worksite/smoke freepolicies.html • HP: Reduce the proportion of nonsmokers exposed to second-hand smoke. http://www.healthypeople.gov/2020/topicsobjectives2020/ objectiveslist.aspx?topicid=41 • HP: Increase the proportion of persons covered by indoor work-site policies that prohibit smoking. http://www.healthypeople.gov/2020/topicsobjectives2020/ objectiveslist.aspx?topicid=41 • HP: Increase tobacco-free environments in schools, including all school facilities, property, vehicles, and school events. http://www.healthypeople.gov/2020/topicsobjectives2020/ objectiveslist.aspx?topicid=41

Recommendation	Supporting Evidence-Based Interventions
Support full implementation of the 2009 Family Smoking Prevention and Tobacco Control Act (Tobacco Control Act).	• CG: Restricting Minors' Access to Tobacco Products: Community Mobilization with Additional Interventions. http://www.thecommunityguide.org/tobacco/restrictingaccess/communityinterventions.html • HP: Reduce tobacco use by adolescents. http://www.healthypeople.gov/2020/topicsobjectives2020/objectiveslist.aspx?topicid=41 • HP: Reduce the initiation of tobacco use among children, adolescents, and young adults. http://www.healthypeople.gov/2020/topicsobjectives2020/objectiveslist.aspx?topicid=41 • HP: Reduce the proportion of adolescents and young adults in grades 6-12 who are exposed to tobacco advertising and promotion. http://www.healthypeople.gov/2020/topicsobjectives2020/objectiveslist.aspx?topicid=41 • HP: Reduce the illegal sales rate to minors through enforcement of laws prohibiting the sale of tobacco products to minors. http://www.healthypeople.gov/2020/topicsobjectives2020/objectiveslist.aspx?topicid=41
Expand use of tobacco cessation services.	• CG: Decreasing Tobacco Use Among Workers: Incentives & Competitions When Combined with Additional Interventions. http://www.thecommunityguide.org/tobacco/worksite/incentives.html • CG: Increasing Tobacco Use Cessation: Provider Reminders When Used Alone. http://www.thecommunityguide.org/tobacco/cessation/providerreminders.html • CG: Increasing Tobacco Use Cessation: Provider Reminders with Provider Education. http://www.thecommunityguide.org/tobacco/cessation/providerreminderedu.html • CG: Increasing Tobacco Use Cessation: Reducing Client Out-of-Pocket Costs for Cessation Therapies. http://www.thecommunityguide.org/tobacco/cessation/outofpocketcosts.html • CG: Increasing Tobacco Use Cessation: Multicomponent Interventions That Include Telephone Support. http://www.thecommunityguide.org/tobacco/cessation/multicomponentinterventions.html • USPSTF: Clinicians ask all adults about tobacco use and provide tobacco cessation interventions for those who use tobacco products. http://www.uspreventiveservicestaskforce.org/uspstf/uspstbac2.htm • HP: Increase smoking cessation attempts by adult smokers. http://www.healthypeople.gov/2020/topicsobjectives2020/objectiveslist.aspx?topicid=41

continued

Recommendation	Supporting Evidence-Based Interventions

TOBACCO-FREE LIVING   Continued

| | • HP: Increase recent smoking cessation success by adult smokers. http://www.healthypeople.gov/2020/topicsobjectives2020/objectiveslist.aspx?topicid=41
• HP: Increase smoking cessation during pregnancy. http://www.healthypeople.gov/2020/topicsobjectives2020/objectiveslist.aspx?topicid=41
• HP: Increase smoking cessation attempts by adolescent smokers. http://www.healthypeople.gov/2020/topicsobjectives2020/objectiveslist.aspx?topicid=41
• HP: Increase tobacco screening in health care settings. http://www.healthypeople.gov/2020/topicsobjectives2020/objectiveslist.aspx?topicid=41
• HP: Increase tobacco cessation counseling in health care settings. http://www.healthypeople.gov/2020/topicsobjectives2020/objectiveslist.aspx?topicid=41 |
| Use media to educate and encourage people to live tobacco free. | • CG: Reducing Tobacco Use Initiation: Mass Media Campaigns When Combined with Other Interventions. http://www.thecommunityguide.org/tobacco/initiation/massmediaeducation.html
• HP: Reduce the proportion of adolescents and young adults in grades 6-12 who are exposed to tobacco advertising and promotion. http://www.healthypeople.gov/2020/topicsobjectives2020/objectiveslist.aspx?topicid=41
• IOM: A national, youth-oriented media campaign should be funded as a permanent component of the nation's strategy to reduce tobacco use. State and community tobacco control programs should supplement the national media campaign with co-ordinated youth prevention activities. The campaign should be implemented by an established public health organization with funds provided by the federal government, public-private partnerships, or the tobacco industry (voluntarily or under litigation settlement agreements or court orders) for media development, testing, and purchases of advertising time and space. Institute of Medicine. *Ending the Tobacco Problem: A Blueprint for the Nation.* http://books.nap.edu/catalog/11795.html |

PREVENTING DRUG ABUSE AND EXCESSIVE ALCOHOL USE

| Support state, tribal, local, and territorial implementation and enforcement of alcohol control policies. | • CG: Preventing Excessive Alcohol Consumption: Enhanced Enforcement of Laws Prohibiting Sales to Minors. http://www.thecommunityguide.org/alcohol/lawsprohibitingsales.html
• CG: Reducing Alcohol-Impaired Driving: Maintaining Current Minimum Legal Drinking Age (MLDA) Laws. http://www.thecommunityguide.org/mvoi/AID/mlda-laws.html |

Recommendation	Supporting Evidence-Based Interventions
	• CG: Reducing Alcohol-Impaired Driving: School-Based Programs. http://www.thecommunityguide.org/mvoi/AID/school-based.html
	• IOM: States should strengthen their compliance check programs in retail outlets using media campaigns and license revocation to increase deterrence. http://books.nap.edu/openbook.php?record_id=10729&page=6
	• IOM: States should require all sellers and servers of alcohol to complete state-approved training as a condition of employment. http://books.nap.edu/openbook.php?record_id=10729&page=7
	• IOM: States and localities should implement enforcement programs to deter adults from purchasing alcohol for minors. http://books.nap.edu/openbook.php?record_id=10729&page=7
	• IOM: States and communities should establish and implement a system requiring registration of beer kegs that records information on the identity of purchasers. http://books.nap.edu/openbook.php?record_id=10729&page=8
	• IOM: States should facilitate enforcement of zero-tolerance laws in order to increase their deterrent effect. http://books.nap.edu/openbook.php?record_id=10729&page=8
	• IOM: States and localities should routinely implement sobriety checkpoints. http://books.nap.edu/openbook.php?record_id=10729&page=8
	• IOM: Local police, working with community leaders, should adopt and announce policies for detecting and terminating underage drinking parties. http://books.nap.edu/openbook.php?record_id=10729&page=8
	• IOM: States should strengthen efforts to prevent and detect use of false identification by minors to make alcohol purchases. http://books.nap.edu/openbook.php?record_id=10729&page=8
	• IOM: States should establish administrative procedures and noncriminal penalties, such as fines or community service, for alcohol infractions by minors. http://books.nap.edu/openbook.php?record_id=10729&page=9
Create environments that empower young people not to drink or use other drugs.	• CG: Adolescent Health: Person-to-Person Interventions to Improve Caregivers' Parenting Skills. http://www.thecommunityguide.org/adolescenthealth/PersonToPerson.html
	• HP: Reduce the proportion of adolescents who report that they rode, during the past 30 days, with a driver who had been drinking alcohol. http://www.healthypeople.gov/2020/topicsobjectives2020/objectiveslist.aspx?topicid=40
	• HP: Increase the proportion of adolescents never using substances. http://www.healthypeople.gov/2020/topicsobjectives2020/objectiveslist.aspx?topicid=40
	• HP: Increase the proportion of adolescents who disapprove of substance abuse. http://www.healthypeople.gov/2020/topicsobjectives2020/objectiveslist.aspx?topicid=40

continued

Recommendation	Supporting Evidence-Based Interventions

PREVENTING DRUG ABUSE AND EXCESSIVE ALCOHOL USE Continued

	• HP: Increase the proportion of adolescents who perceive great risk associated with substance abuse. http://www.healthypeople.gov/2020/topicsobjectives2020/objectiveslist.aspx?topicid=40 • HP: Reduce past-month use of illicit substances. http://www.healthypeople.gov/2020/topicsobjectives2020/objectiveslist.aspx?topicid=40 • HP: Reduce the proportion of persons engaging in binge drinking of alcoholic beverages. http://www.healthypeople.gov/2020/topicsobjectives2020/objectiveslist.aspx?topicid=40 • HP: Reduce the proportion of adolescents who have been offered, sold, or given an illegal drug on school property. http://www.healthypeople.gov/2020/topicsobjectives2020/objectiveslist.aspx?topicid=2
Create environments that empower young people not to drink or use other drugs.	• IOM: Alcohol companies, advertising companies, and commercial media should refrain from marketing practices (including product design, advertising, and promotional techniques) that have substantial underage appeal and should take reasonable precautions in the time, place, and manner of placement and promotion to reduce youthful exposure to other alcohol advertising and marketing activity. http://books.nap.edu/openbook.php?record_id=10729&page=4 • IOM: The alcohol industry trade associations, as well as individual companies, should strengthen their advertising codes to preclude placement of commercial messages in venues where a significant proportion of the expected audience is underage, to prohibit the use of commercial messages that have substantial underage appeal, and to establish independent external review boards to investigate complaints and enforce the codes. http://books.nap.edu/openbook.php?record_id=10729&page=4 • IOM: The entertainment industries should use rating systems and marketing codes to reduce the likelihood that underage audiences will be exposed to movies, recordings, or television programs with unsuitable alcohol content, even if adults are expected to predominate in the viewing or listening audiences. http://books.nap.edu/openbook.php?record_id=10729&page=5 • IOM: The film rating board of the Motion Picture Association of America should consider alcohol content in rating films, avoiding G or PG ratings for films with unsuitable alcohol content, and assigning mature ratings for films that portray underage drinking in a favorable light. http://books.nap.edu/openbook.php?record_id=10729&page=5

Recommendation	Supporting Evidence-Based Interventions
	• IOM: The music recording industry should not market recordings that promote or glamorize alcohol use to young people; should include alcohol content in a comprehensive rating system, similar to those used by the television, film, and video game industries; and should establish an independent body to assign ratings and oversee the industry code. http://books.nap.edu/openbook.php?record_id=10729&page=5 • IOM: Television broadcasters and producers should take appropriate precautions to ensure that programs do not portray underage drinking in a favorable light, and that unsuitable alcohol content is included in the category of mature content for purposes of parental warnings. http://books.nap.edu/openbook.php?record_id=10729&page=5 • Cochrane: Social norms interventions to reduce alcohol misuse in university and college students. http://www2.cochrane.org/reviews/en/ab006748.html
Identify alcohol and other drug abuse disorders early and provide brief intervention, referral, and treatment.	• USPSTF: Recommends screening and behavioral counseling interventions to reduce alcohol misuse by adults, including pregnant women, in primary care settings. U.S. Preventive Services Task Force. Screening and Behavioral Counseling Interventions in Primary Care to Reduce Alcohol Misuse: Recommendation Statement. April 2004. http://www.uspreventiveservicestaskforce.org/3rduspstf/alcohol/alcomisrs.htm • HP: Increase the number of admissions to substance abuse treatment for injection drug use. http://www.healthypeople.gov/2020/topicsobjectives2020/objectiveslist.aspx?topicid=40 • HP: Increase the proportion of persons who need alcohol and/or illicit drug treatment and received specialty treatment for abuse or dependence in the past year. http://www.healthypeople.gov/2020/topicsobjectives2020/objectiveslist.aspx?topicid=40 • HP: Increase the number of Level I and Level II trauma centers and primary care settings that implement evidence-based alcohol Screening and Brief Intervention (SBI). http://www.healthypeople.gov/2020/topicsobjectives2020/objectiveslist.aspx?topicid=40 • IOM: Residential colleges and universities should adopt comprehensive prevention approaches, including evidence-based screening, brief intervention strategies, consistent policy enforcement, and environmental changes that limit underage access to alcohol. They should use universal education interventions, as well as selective and indicated approaches with relevant populations. http://books.nap.edu/openbook.php?record_id=10729&page=9
Reduce inappropriate access to and use of prescription drugs.	• HP: Reduce the past-year nonmedical use of prescription drugs. http://www.healthypeople.gov/2020/topicsobjectives2020/objectiveslist.aspx?topicid=40

continued

Recommendation	Supporting Evidence-Based Interventions
HEALTHY EATING	
Increase access to healthy and affordable foods in communities.	• HP: (Developmental) Increase the proportion of Americans who have access to a food retail outlet that sells a variety of foods that are encouraged by the Dietary Guidelines for Americans. http://www.healthypeople.gov/2020/topicsobjectives2020/objectiveslist.aspx?topicid=29 • HP: Increase the proportion of schools that offer nutritious foods and beverages outside of school meals. http://www.healthypeople.gov/2020/topicsobjectives2020/objectiveslist.aspx?topicid=29
Implement organizational and programmatic nutrition standards and policies.	• HP: Increase the proportion of schools that offer nutritious foods and beverages outside of school meals. http://www.healthypeople.gov/2020/topicsobjectives2020/objectiveslist.aspx?topicid=29
Improve nutritional quality of the food supply.	• HP: Increase the contribution of fruits to the diets of the population age 2 and older. http://www.healthypeople.gov/2020/topicsobjectives2020/objectiveslist.aspx?topicid=29 • HP: Increase the variety and contribution of vegetables to the diets of the population age 2 and older. http://www.healthypeople.gov/2020/topicsobjectives2020/objectiveslist.aspx?topicid=29 • HP: Increase the contribution of whole grains to the diets of the population age 2 and older. http://www.healthypeople.gov/2020/topicsobjectives2020/objectiveslist.aspx?topicid=29 • HP: Reduce consumption of calories from solid fats and added sugars in the population age 2 and older. http://www.healthypeople.gov/2020/topicsobjectives2020/objectiveslist.aspx?topicid=29 • HP: Reduce consumption of saturated fat in the population age 2 and older. http://www.healthypeople.gov/2020/topicsobjectives2020/objectiveslist.aspx?topicid=29 • HP: Reduce consumption of sodium in the population age 2 and older. http://www.healthypeople.gov/2020/topicsobjectives2020/objectiveslist.aspx?topicid=29 • HP: Increase consumption of calcium in the population age 2 and older. http://www.healthypeople.gov/2020/topicsobjectives2020/objectiveslist.aspx?topicid=29

Recommendation	Supporting Evidence-Based Interventions
Help people recognize and make healthy food and beverage choices.	• IOM: Food and beverage companies should use their creativity, resources, and full range of marketing practices to promote and support more healthful diets for children and youth. http://books.nap.edu/openbook.php?record_id=11514 &page=382 • IOM: Full-serve restaurant chains, family restaurants, and quick-serve restaurants should use their creativity, resources, and full range of marketing practices to promote healthful meals for children and youth. http://books.nap.edu/openbook.php?record_id=11514 &page=382
Support policies and programs that promote breastfeeding.	• HP: Increase the proportion of infants who are breastfed. http://www.healthypeople.gov/2020/topicsobjectives2020/ objectiveslist.aspx?topicid=26 • HP: Increase the proportion of employers that have work-site lactation support programs. http://www.healthypeople.gov/2020/topicsobjectives2020/ objectiveslist.aspx?topicid=26 • HP: Reduce the proportion of breastfed newborns who receive formula supplementation within the first 2 days of life. http://www.healthypeople.gov/2020/topicsobjectives2020/ objectiveslist.aspx?topicid=26 • HP: Increase the proportion of live births that occur in facilities that provide recommended care for lactating mothers and their babies. http://www.healthypeople.gov/2020/topicsobjectives2020/ objectiveslist.aspx?topicid=26 • Cochrane: Optimal duration of exclusive breastfeeding. http://onlinelibrary.wiley.com/o/cochrane/clsysrev/articles/ CD003517/frame.html
Enhance food safety.	• HP: Reduce infections caused by key pathogens transmitted commonly through food. http://www.healthypeople.gov/2020/topicsobjectives2020/ objectiveslist.aspx?topicid=14 • HP: Reduce the number of outbreak-associated infections due to Shiga toxin-producing *E. coli* O157, or Campylobacter, Listeria, or Salmonella species associated with food commodity groups. http://www.healthypeople.gov/2020/topicsobjectives2020/ objectiveslist.aspx?topicid=14 • HP: Prevent an increase in the proportion of nontyphoidal Salmonella and Campylobacter jejuni isolates from humans that are resistant to antimicrobial drugs. http://www.healthypeople.gov/2020/topicsobjectives2020/ objectiveslist.aspx?topicid=14

continued

Recommendation	Supporting Evidence-Based Interventions

HEALTHY EATING Continued

| | • HP: Reduce severe allergic reactions to food among adults with a food allergy diagnosis. http://www.healthypeople.gov/2020/topicsobjectives2020/objectiveslist.aspx?topicid=14
 • HP: Increase the proportion of consumers who follow key food safety practices. http://www.healthypeople.gov/2020/topicsobjectives2020/objectiveslist.aspx?topicid=14
 • IOM: Integrating Food Safety Programs and Educating the Public. http://www.iom.edu/Reports/2010/Enhancing-Food-Safety-The-Role-of-the-Food-and-Drug-Administration.aspx
 • IOM: Enhancing the Efficiency of Inspections. http://www.iom.edu/Reports/2010/Enhancing-Food-Safety-The-Role-of-the-Food-and-Drug-Administration.aspx |

ACTIVE LIVING

| Encourage community design and development that supports physical activity. | • CG: Environmental and Policy Approaches to Increase Physical Activity: Community-Scale Urban Design Land Use Policies. http://www.thecommunityguide.org/pa/environmental-policy/communitypolicies.html
 • CG: Environmental and Policy Approaches to Increase Physical Activity: Street-Scale Urban Design Land Use Policies. http://www.thecommunityguide.org/pa/environmental-policy/streetscale.html
 • CG: (Expanding Evidence) Environmental and Policy Approaches to Increase Physical Activity: Transportation and Travel Policies and Practices. http://www.thecommunityguide.org/pa/environmental-policy/travelpolicies.html
 • CG: (Expanding Evidence) The available studies do not provide sufficient evidence to determine if the intervention is, or is not, effective. This lack of evidence does NOT mean that the intervention does not work, but that additional research is needed to determine whether the intervention is effective. http://www.thecommunityguide.org/about/methods.html
 • HP: (Developmental) Increase legislative policies for the built environment that enhance access to and availability of physical activity opportunities. http://www.healthypeople.gov/2020/topicsobjectives2020/objectiveslist.aspx?topicid=33 |

Recommendation	Supporting Evidence-Based Interventions
Promote and strengthen school and early learning policies and programs that increase physical activity.	• CG: Behavioral and Social Approaches to Increase Physical Activity: Enhanced School-Based Physical Education. http://www.thecommunityguide.org/pa/behavioral-social/school based-pe.html • HP: Increase the proportion of the nation's public and private schools that require daily physical education for all students. http://www.healthypeople.gov/2020/topicsobjectives2020/ objectiveslist.aspx?topicid=33 • HP: Increase the proportion of adolescents who participate in daily school physical education. http://www.healthypeople.gov/2020/topicsobjectives2020/ objectiveslist.aspx?topicid=33 • HP: Increase regularly scheduled elementary school recess in the United States. http://www.healthypeople.gov/2020/topicsobjectives2020/ objectiveslist.aspx?topicid=33 • HP: Increase the proportion of school districts that require or recommend elementary school recess for an appropriate period of time. http://www.healthypeople.gov/2020/topicsobjectives2020/ objectiveslist.aspx?topicid=33 • HP: Increase the number of states with licensing regulations for physical activity provided in child care. http://www.healthypeople.gov/2020/topicsobjectives2020/ objectiveslist.aspx?topicid=33
Facilitate access to safe, accessible, and affordable places for physical activity.	• CG: Environmental and Policy Approaches to Increase Physical Activity: Creation of or Enhanced Access to Places for Physical Activity Combined with Informational Outreach Activities. http://www.thecommunityguide.org/pa/environmental-policy/ improvingaccess.html • HP: Reduce the proportion of adults who engage in no leisure-time physical activity. http://www.healthypeople.gov/2020/topicsobjectives2020/ objectiveslist.aspx?topicid=33 • HP: Increase the proportion of the nation's public and private schools that provide access to their physical activity spaces and facilities for all persons outside of normal school hours (that is, before and after the school day, on weekends, and during summer and other vacations). http://www.healthypeople.gov/2020/topicsobjectives2020/ objectiveslist.aspx?topicid=33 • IOM: Those responsible for modifications or additions to the built environment should facilitate access to, enhance the attractiveness of, and ensure the safety and security of places where people can be physically active. http://books.nap.edu/openbook.php?record_id=11203&page=14

continued

Recommendation	Supporting Evidence-Based Interventions

ACTIVE LIVING Continued

Support workplace policies and programs that increase physical activity.	• CG: Environmental and Policy Approaches to Increase Physical Activity: Point-of-Decision Prompts to Encourage Use of Stairs. http://www.thecommunityguide.org/pa/environmental-policy/podp.html • CG: Behavioral and Social Approaches to Increase Physical Activity: Social Support Interventions in Community Settings. http://www.thecommunityguide.org/pa/behavioral-social/community.html
Assess physical activity levels and provide education, counseling, and referrals.	• CG: Behavioral and Social Approaches to Increase Physical Activity: Individually-Adapted Health Behavior Change Programs. http://www.thecommunityguide.org/pa/behavioral-social/individuallyadapted.html • HP: Increase the proportion of physician office visits that include counseling or education related to physical activity. http://www.healthypeople.gov/2020/topicsobjectives2020/objectiveslist.aspx?topicid=33 • Cochrane: Interventions for promoting physical activity. http://www2.cochrane.org/reviews/en/ab003180.html

INJURY AND VIOLENCE-FREE LIVING

Implement and strengthen policies and programs to enhance transportation safety.	• CG: Use of Child Safety Seats: Community-Wide Information and Enhanced Enforcement Campaigns. http://www.thecommunityguide.org/mvoi/childsafetyseats/community.html • CG: Use of Child Safety Seats: Distribution and Education Programs. http://www.thecommunityguide.org/mvoi/childsafetyseats/distribution.html • CG: Use of Child Safety Seats: Incentive and Education Programs. http://www.thecommunityguide.org/mvoi/childsafetyseats/incentives.html • CG: Use of Safety Belts: Primary (vs. Secondary) Enforcement Laws. http://www.thecommunityguide.org/mvoi/safetybelts/enforcementlaws.html • CG: Use of Safety Belts: Enhanced Enforcement Programs. http://www.thecommunityguide.org/mvoi/safetybelts/enforcementprograms.html • CG: Reducing Alcohol-Impaired Driving: Maintaining Current Minimum Legal Drinking Age (MLDA) Laws. http://www.thecommunityguide.org/mvoi/AID/lowerbaclaws.html • CG: Reducing Alcohol-Impaired Driving: Sobriety Checkpoints. http://www.thecommunityguide.org/mvoi/AID/sobrietyckpts.html • CG: Reducing Alcohol-Impaired Driving: Mass Media Campaigns. http://www.thecommunityguide.org/mvoi/AID/massmedia.html

Recommendation	Supporting Evidence-Based Interventions
	• CG: Reducing Alcohol-Impaired Driving: Multicomponent Interventions with Community Mobilization. http://www.thecommunityguide.org/mvoi/AID/multicomponent.html
	• CG: Reducing Alcohol-Impaired Driving: Ignition Interlocks. http://www.thecommunityguide.org/mvoi/AID/ignitioninterlocks.html
	• CG: Reducing Alcohol-Impaired Driving: School-Based Instructional Programs. http://www.thecommunityguide.org/mvoi/AID/school-based.html
	• HP: Increase use of safety belts. http://www.healthypeople.gov/2020/topicsobjectives2020/objectiveslist.aspx?topicid=24
	• HP: Increase age-appropriate vehicle restraint system use in children. http://www.healthypeople.gov/2020/topicsobjectives2020/objectiveslist.aspx?topicid=24
	• HP: Increase the proportion of motorcycle operators and passengers using helmets. http://www.healthypeople.gov/2020/topicsobjectives2020/objectiveslist.aspx?topicid=24
Support community and streetscape design that promotes safety and prevents injuries.	• CG: Environmental and Policy Approaches to Increase Physical Activity: Street-Scale Urban Design Land Use Policies. http://www.thecommunityguide.org/pa/environmental-policy/streetscale.html
	• CG: Environmental and Policy Approaches to Increase Physical Activity: Community-Scale Urban Design Land Use Policies. http://www.thecommunityguide.org/pa/environmental-policy/communitypolicies.html
	• Cochrane: Interventions for increasing pedestrian and cyclist visibility for the prevention of death and injuries. http://onlinelibrary.wiley.com/o/cochrane/clsysrev/articles/CD003438/frame.html
Promote and strengthen policies and programs to prevent falls, especially among older adults.	• HP: Prevent an increase in the rate of fall-related deaths. http://www.healthypeople.gov/2020/topicsobjectives2020/objectiveslist.aspx?topicid=24
	• HP: Reduce the rate of emergency department visits due to falls among older adults. http://www.healthypeople.gov/2020/topicsobjectives2020/objectiveslist.aspx?topicid=31
	• Cochrane: Population-based interventions for the prevention of fall-related injuries in older people. http://onlinelibrary.wiley.com/o/cochrane/clsysrev/articles/CD004441/frame.html

continued

Recommendation	Supporting Evidence-Based Interventions

INJURY AND VIOLENCE-FREE LIVING Continued

Recommendation	Supporting Evidence-Based Interventions
Promote and enhance policies and programs to increase safety and prevent injury in the workplace.	• IOM: Develop and Implement Risk-Based Conformity Assessment Processes for Non-Respirator PPT. http://www.iom.edu/Reports/2010/Certifying-Personal-Protective-Technologies-Improving-Worker-Safety.aspx • IOM: Enhance Research, Standards Development, and Communication. http://www.iom.edu/Reports/2010/Certifying-Personal-Protective-Technologies-Improving-Worker-Safety.aspx • IOM: Establish a PPT and Occupational Safety and Health Surveillance System. http://www.iom.edu/Reports/2010/Certifying-Personal-Protective-Technologies-Improving-Worker-Safety.aspx
Strengthen policies and programs to prevent violence.	• CG: Early Childhood Home Visitation to Prevent Child Maltreatment. http://www.thecommunityguide.org/violence/home/home visitation.html • CG: Youth Violence Prevention: School-Based Programs to Reduce Violence. http://www.thecommunityguide.org/violence/schoolbased programs.html • CG: Therapeutic Foster Care to Reduce Violence for Chronically Delinquent Juveniles. http://www.thecommunityguide.org/violence/therapeutic fostercare/index.html
Provide individuals and families with the knowledge, skills, and tools to make safe choices that prevent violence and injuries.	• HP: Increase the proportion of adolescents who are connected to a parent or other positive adult caregiver. http://www.healthypeople.gov/2020/topicsobjectives2020/objectiveslist.aspx?topicid=2 • HP: Reduce bullying among adolescents. http://www.healthypeople.gov/2020/topicsobjectives2020/objectiveslist.aspx?topicid=24 • HP: Reduce children's exposure to violence. http://www.healthypeople.gov/2020/topicsobjectives2020/objectiveslist.aspx?topicid=24 • Cochrane: School-based secondary prevention programs for preventing violence. http://onlinelibrary.wiley.com/o/cochrane/clsysrev/articles/CD004606/frame.html • Cochrane: Safety education of pedestrians for injury prevention. http://onlinelibrary.wiley.com/o/cochrane/clsysrev/articles/CD001531/frame.html

Recommendation	Supporting Evidence-Based Interventions

REPRODUCTIVE AND SEXUAL HEALTH

Increase utilization of preconception and prenatal care.	• USPSTF: Recommends that all women planning or capable of pregnancy take a daily supplement containing 0.4 to 0.8 mg (400 to 800 mg) of folic acid. http://www.uspreventiveservicestaskforce.org/uspstf09/folicacid/folicacidrs.htm • USPSTF: Recommends that clinicians screen all pregnant women for syphilis infection. http://www.uspreventiveservicestaskforce.org/uspstf/uspssyphpg.htm • HP: Increase the proportion of pregnant women who receive early and adequate prenatal care. http://healthypeople.gov/2020/topicsobjectives2020/objectiveslist.aspx?topicid=26 • HP: Increase abstinence from alcohol, cigarettes, and illicit drugs among pregnant women. http://healthypeople.gov/2020/topicsobjectives2020/objectiveslist.aspx?topicid=26 • HP: Increase the proportion of women of childbearing potential with intake of at least 400 mg of folic acid from fortified foods or dietary supplements. http://healthypeople.gov/2020/topicsobjectives2020/objectiveslist.aspx?topicid=26 • HP: Reduce the proportion of women of childbearing potential who have low red blood cell folate concentrations. http://healthypeople.gov/2020/topicsobjectives2020/objectiveslist.aspx?topicid=26 • HP: Increase the proportion of women delivering a live birth who received preconception care services and practiced key recommended preconception health behaviors. http://healthypeople.gov/2020/topicsobjectives2020/objectiveslist.aspx?topicid=26 • CG: Prevention of Birth Defects: Community-Wide Campaigns to Promote the Use of Folic Acid Supplements. http://www.thecommunityguide.org/birthdefects/community.html • CG: Interventions to fortify food products with folic acid. http://www.thecommunityguide.org/birthdefects/index.html • Cochrane: Smoking cessation interventions in pregnancy reduce the proportion of women who continue to smoke in late pregnancy, and reduce low birthweight and preterm birth. Smoking cessation interventions in pregnancy need to be implemented in all maternity care settings. Given the difficulty many pregnant women addicted to tobacco have quitting during pregnancy, population-based measures to reduce smoking and social inequalities should be supported. http://onlinelibrary.wiley.com/o/cochrane/clsysrev/articles/CD001055/frame.html

continued

Recommendation	Supporting Evidence-Based Interventions

REPRODUCTIVE AND SEXUAL HEALTH Continued

Support reproductive and sexual health services and support services for pregnant and parenting women.	• CG: Prevention of HIV/AIDS, Other STIs and Pregnancy: Interventions to Reduce Sexual Risk Behaviors or Increase Protective Behaviors to Prevent Acquisition of HIV in Men Who Have Sex with Men (MSM). http://www.thecommunityguide.org/hiv/msm.html • USPSTF: Recommends high-intensity behavioral counseling to prevent STIs for all sexually active adolescents and for adults at increased risk for STIs. http://www.uspreventiveservicestaskforce.org/uspstf/uspsstds.htm • HP: Increase the proportion of sexually active persons who received reproductive health services. http://healthypeople.gov/2020/topicsobjectives2020/objectiveslist.aspx?topicid=13 • HP: Increase the proportion of sexually active persons aged 15 to 19 who use condoms to both effectively prevent pregnancy and provide barrier protection against disease. http://healthypeople.gov/2020/topicsobjectives2020/objectiveslist.aspx?topicid=13 • HP: Increase the proportion of sexually active persons aged 15 to 19 who use condoms and hormonal or intrauterine contraception to both effectively prevent pregnancy and provide barrier protection against disease. http://healthypeople.gov/2020/topicsobjectives2020/objectiveslist.aspx?topicid=13 • HP: Increase the proportion of females in need of publicly supported contraceptive services and supplies who receive those services and supplies. http://healthypeople.gov/2020/topicsobjectives2020/objectiveslist.aspx?topicid=13 • HP: Increase the proportion of sexually active persons who use condoms. http://healthypeople.gov/2020/topicsobjectives2020/objectiveslist.aspx?topicid=22

Recommendation	Supporting Evidence-Based Interventions
Provide effective sexual health education, especially for adolescents.	• CG: Prevention of HIV/AIDS, Other STIs and Pregnancy: Group-Based Comprehensive Risk Reduction Interventions for Adolescents. http://www.thecommunityguide.org/hiv/riskreduction.html • CG: Youth Development Behavioral Interventions Coordinated with Community Service to Reduce Sexual Risk Behaviors in Adolescents. http://www.thecommunityguide.org/hiv/youthdev-community.html • HP: Increase the proportion of adolescents who received formal instruction on reproductive health topics before they were 18 years old. http://healthypeople.gov/2020/topicsobjectives2020/objectives list.aspx?topicid=13 • HP: Increase the proportion of adolescents who talked to a parent or guardian about reproductive health topics before they were 18 years old. http://healthypeople.gov/2020/topicsobjectives2020/objectives list.aspx?topicid=13 • HP: Increase the proportion of substance abuse treatment facilities that offer HIV/AIDS education, counseling, and support. http://healthypeople.gov/2020/topicsobjectives2020/objectives list.aspx?topicid=22
Enhance early detection of HIV, viral hepatitis and other STIs, and improve linkage to care.	• CG: Interventions to Identify HIV-Positive People Through Partner Counseling and Referral Services. http://www.thecommunityguide.org/hiv/partnercounseling.html • USPSTF: Recommends screening for hepatitis B virus (HBV) infection in pregnant women at their first prenatal visit. http://www.uspreventiveservicestaskforce.org/uspstf/uspshepbpg.htm • USPSTF: Strongly recommends that clinicians screen persons at increased risk for syphilis infection. http://www.uspreventiveservicestaskforce.org/uspstf/uspssyph.htm • HP: Increase the proportion of sexually active females age 24 and under enrolled in Medicaid plans who are screened for genital Chlamydia infections during the measurement year. http://www.healthypeople.gov/2020/topicsobjectives2020/objectiveslist.aspx?topicid=37 • HP: Increase the proportion of sexually active females age 24 and under enrolled in commercial health insurance plans who are screened for genital Chlamydia infections during the measurement year. http://www.healthypeople.gov/2020/topicsobjectives2020/objectiveslist.aspx?topicid=37

continued

Recommendation	Supporting Evidence-Based Interventions

REPRODUCTIVE AND SEXUAL HEALTH Continued

	• HP: Increase the proportion of people living with HIV who know their serostatus. http://healthypeople.gov/2020/topicsobjectives2020/objectives list.aspx?topicid=22 • HP: Increase the proportion of adolescents and adults who have been tested for HIV in the past 12 months. http://healthypeople.gov/2020/topicsobjectives2020/objectives list.aspx?topicid=22 • HP: Increase the proportion of adults with tuberculosis (TB) who have been tested for HIV. http://healthypeople.gov/2020/topicsobjectives2020/objectives list.aspx?topicid=22 • IOM: *Hepatitis and Liver Cancer: A National Strategy for Prevention and Control of Hepatitis B and C.* http://www.iom.edu/Reports/2010/Hepatitis-and-Liver-Cancer-A-National-Strategy-for-Prevention-and-Control-of-Hepatitis-B-and-C.aspx

MENTAL AND EMOTIONAL WELL-BEING

Promote positive early childhood development, including positive parenting and violence-free homes.	• CG: Early Childhood Development Programs: Comprehensive, Center-Based Programs for Children of Low-Income Families. http://www.thecommunityguide.org/social/centerbasedprograms.html • CG: Violence Prevention Focused on Children and Youth: Early Childhood Home Visitation. http://www.thecommunityguide.org/violence/home/index.html • CG: Violence Prevention Focused on Children and Youth: Reducing Psychological Harm from Traumatic Events. http://www.thecommunityguide.org/violence/traumaticevents/index.html • CG: Violence Prevention Focused on Children and Youth: Therapeutic Foster Care. http://www.thecommunityguide.org/violence/therapeuticfostercare/index.html • HP: Increase the proportion of parents who use positive parenting and communicate with their doctors or other health care professionals about positive parenting. http://www.healthypeople.gov/2020/topicsobjectives2020/objectiveslist.aspx?topicid=10 • HP: Increase the proportion of children with disabilities, birth through age 2, who receive early intervention services in home or community-based settings. http://www.healthypeople.gov/2020/topicsobjectives2020/objectiveslist.aspx?topicid=9

Recommendation	Supporting Evidence-Based Interventions
Facilitate social connectedness and community engagement across the life span.	• CG: School-Based Programs to Reduce Violence. http://www.thecommunityguide.org/violence/schoolbased programs.html • HP: Increase the proportion of children and youth with disabilities who spend at least 80 percent of their time in regular education programs. http://www.healthypeople.gov/2020/topicsobjectives2020/objectiveslist.aspx?topicid=9 • HP: Increase the number of community-based organizations (including local health departments, tribal health services, nongovernmental organizations, and state agencies) providing population-based primary prevention services in the following areas: injury, violence, mental illness, tobacco use, substance abuse, unintended pregnancy, chronic disease programs, nutrition, and physical activity. http://www.healthypeople.gov/2020/topicsobjectives2020/objectiveslist.aspx?topicid=11
Provide individuals and families with the support necessary to maintain positive mental well-being.	• CG: Adolescent Health: Person-to-Person Interventions to Improve Caregivers' Parenting Skills. http://www.thecommunityguide.org/adolescenthealth/PersonToPerson.html • HP: Increase the proportion of students in grades 9-12 who get sufficient sleep. http://www.healthypeople.gov/2020/topicsobjectives2020/objectiveslist.aspx?topicid=38 • HP: Increase the proportion of adults who get sufficient sleep. http://www.healthypeople.gov/2020/topicsobjectives2020/objectiveslist.aspx?topicid=38 • HP: Increase the proportion of elementary, middle, and senior high schools that have health education goals or objectives that address the knowledge and skills articulated in the *National Health Education Standards* (high school, middle, elementary). http://www.healthypeople.gov/2020/topicsobjectives2020/objectiveslist.aspx?topicid=11 • HP: Increase the proportion of college and university students who receive information from their institution on each of the priority health risk behavior areas (all priority areas; unintentional injury; violence; suicide; tobacco use and addiction; alcohol and other drug use; unintended pregnancy, HIV/AIDS, and STD infection; unhealthy dietary patterns; and inadequate physical activity). http://www.healthypeople.gov/2020/topicsobjectives2020/objectiveslist.aspx?topicid=11

continued

Recommendation	Supporting Evidence-Based Interventions

MENTAL AND EMOTIONAL WELL-BEING Continued

	• IOM: States and communities should develop networked systems to apply resources to the promotion of mental health and prevention of mental, emotional, and behavioral disorders among their young people. These systems should involve individuals, families, schools, justice systems, health care systems, and relevant community-based programs. Such approaches should build on available evidence-based programs and involve local evaluators to assess the implementation process of individual programs or policies and to measure community-wide outcomes. http://books.nap.edu/openbook.php?record_id=12480&page=6
Promote early identification of mental health needs and access to quality services.	• CG: Collaborative Care for the Management of Depressive Disorders. http://www.thecommunityguide.org/mentalhealth/collab-care.html • CG: Interventions to Reduce Depression Among Older Adults: Clinic-Based Depression Care Management. http://www.thecommunityguide.org/mentalhealth/depression-clinic.html • CG: Interventions to Reduce Depression Among Older Adults: Home-Based Depression Care Management. http://www.thecommunityguide.org/mentalhealth/depression-home.html • USPSTF: Recommends screening of adolescents (ages 12-18) for MDD when systems are in place to ensure accurate diagnosis, psychotherapy (e.g., cognitive-behavioral, interpersonal), and follow-up. In 2002, the USPSTF concluded that there was insufficient evidence to recommend for or against routine screening of children or adolescents for MDD (I recommendation). http://www.uspreventiveservicestaskforce.org/uspstf09/depression/chdeprrs.htm • HP: Increase depression screening by primary care providers. http://www.healthypeople.gov/2020/topicsobjectives2020/objectiveslist.aspx?topicid=28 • HP: Increase the proportion of homeless adults with mental health problems who receive mental health services. http://www.healthypeople.gov/2020/topicsobjectives2020/objectiveslist.aspx?topicid=28 • Cochrane: Prompts to encourage appointment attendance for people with serious mental illness. http://onlinelibrary.wiley.com/o/cochrane/clsysrev/articles/CD002085/frame.html

SOURCE: Adapted from Appendix 5, National Prevention Council (2011).

Appendix B

Biographical Sketches of Panel Members and Staff

Steven H. Woolf (*Chair*) is director of the Center on Human Needs and professor of family medicine, both at Virginia Commonwealth University. He is board certified in family medicine and in preventive medicine and public health. His work has focused on promoting effective health care services and on highlighting the importance of behavioral and social determinants of health, particularly with regard to the role of poverty, education, and racial and ethnic disparities in determining the health of Americans. In addition to his work as a researcher, he has also been involved with health policy issues. He has served as science adviser, member, and senior adviser to the U.S. Preventive Services Task Force. He is a member of the Institute of Medicine. He has an M.D. from Emory University and an M.P.H. from Johns Hopkins University.

Laudan (Laudy) Aron (*Study Director*) is a senior program officer with the Division of Behavioral and Social Sciences and Education at the National Research Council. Previously, she worked as a senior research associate with the Urban Institute and as director of policy research at the National Alliance on Mental Illness (NAMI). She has conducted and managed research and policy analysis on many issues that affect vulnerable populations, including health, behavioral health, and disability; education, special education, and alternative education; child welfare and at-risk youth; housing and homelessness; and family violence and human trafficking. She has coauthored books on homelessness and publicly funded programs for children with disabilities. She holds a B.Sc. in mathematics from McGill University and an M.A. in demography from the University of Pennsylvania.

375

Paula A. Braveman is professor of family and community medicine and director of the Center on Social Disparities in Health at the University of California, San Francisco. Her research has focused on measuring, documenting, and understanding socioeconomic and racial/ethnic disparities in health, particularly in maternal and infant health. She collaborates extensively with local, state, federal, and international health agencies to support the translation of research into policies and programs. She serves on the Advisory Council of the National Institute for Minority Health and Health Disparities of the National Institutes of Health and in an advisory capacity to several federal agencies regarding research on social inequalities in health. She is a member of the Institute of Medicine. She holds an M.D. from the University of California, San Francisco and an M.P.H. in epidemiology from the University of California, Berkeley.

Kaare Christensen is professor of epidemiology at the University of Southern Denmark and senior research scientist in the Department of Public Policy Studies at Duke University. His research is focused on genetic epidemiology, twin studies, aging, age-related diseases, and fetal programming. He is engaged in interdisciplinary aging research combining methods from epidemiology, genetics, and demography. His recent work has covered divergent life expectancy trends in Denmark and Sweden, with some potential explanations, and genetic factors and adult mortality. He holds a D.M.Sc. from the University of Southern Denmark and an M.D. and a Ph.D. from Odense University.

Eileen M. Crimmins is the AARP chair in gerontology at the University of Southern California (USC). She leads the Center on Biodemography and Population Health, a joint endeavor of USC and the University of California, Los Angeles, and codirects the Network on Biological Risk sponsored by the National Institute on Aging. Her research focuses on the connections between social and environmental factors and life expectancy and other health outcomes. Much of her work has been on trends in mortality, morbidity, and healthy life expectancy. She is a recipient of the Kleemeier Award for Research from the Gerontological Society of America. She is a member of the Institute of Medicine, and a fellow of the American Association for the Advancement of Science. She holds an M.A. and a Ph.D. in demography, both from the University of Pennsylvania.

Ana V. Diez Roux is a professor of epidemiology and chair of the Department of Epidemiology at the School of Public Health at the University of Michigan, where she also directs the Center for Social Epidemiology and Population Health. Her research areas include social epidemiology, environmental health effects, urban health, psychosocial factors in health, health disparities, and cardiovascular disease epidemiology. She has been deeply involved in the investigation of neighborhood and community health effects and the application of multilevel analysis in public health. Other areas of research include the integration of social and biologic factors in health research, complex systems approaches to population health, the impact of stress on cardiovascular disease, and air pollution effects on health. She is a member of the Institute of Medicine. She was recently awarded the Wade Hampton Frost Award of the American Public Health Association for her contributions to public health. She holds an M.D. from the University of Buenos Aires and an M.P.H. and a Ph.D. from Johns Hopkins University.

Dean T. Jamison is a professor of global health at the University of Washington. Previously, he was on the faculty of the University of California, Los Angeles, and he also spent many years at the World Bank where he was a senior economist in the research department; division chief for education policy; and division chief for population, health, and nutrition. He served as director of the World Development Report Office and as lead author for the World Bank's *World Development Report 1993: Investing in Health*. His research and publications are in the areas of economic theory, public health, and education. He is a member of the Institute of Medicine. He holds an M.S. in engineering sciences from Stanford University and a Ph.D. in economics from Harvard University.

Johan P. Mackenbach is a professor of public health and chair of the Department of Public Health at Erasmus University Medical Center at Erasmus University in Rotterdam. He is also an honorary professor at the London School of Hygiene and Tropical Medicine. His research interests are in social epidemiology, medical demography, and health services research. He has served as the editor-in-chief of the *European Journal of Public Health* and has coordinated a number of international comparative studies funded by the European Commission. His current research focuses on socioeconomic inequalities in health, on issues related to aging and the compression of morbidity, and on the effectiveness and quality of health services. He is actively engaged in exchanges between researchers and policy analysts, among others, as a member of the Health Council and the Council for Public Health and Health Care in the Netherlands. He holds an M.D. and a Ph.D. in public health, both from Erasmus University in Rotterdam.

David V. McQueen is a behavioral scientist, currently serving as an adjunct professor at the Rollins School of Public Health at Emory University. Previously, he held several positions at the Centers for Disease Control and Prevention, including associate director for Global Health Promotion, director of the Division of Adult and Community Health at the National Center for

Chronic Disease Prevention and Health Promotion (NCCDPHP), and chief of the national Behavioral Risk Factor Surveillance System. He also previously served as professor and director of the Research Unit in Health and Behavioral Change at the University of Edinburgh. He is the immediate past president of the International Union for Health Promotion and Education (IUHPE), as well as leader of the IUHPE Global Programme on Health Promotion Effectiveness. He holds an M.A. in history and philosophy of science and a Sc.D. in behavioral sciences, both from Johns Hopkins University.

Alberto Palloni is the Samuel H. Preston professor of sociology and population studies at the University of Wisconsin. His research explores the relation between early health and aggregate inequality in high-income countries. His work covers determinants of adult mortality and health disparities, adult health and mortality in low-income countries, statistical applications in population analyses, mathematical models for population dynamics, models for the analysis of self-reported health data, aging in developing countries, the effects of HIV/AIDS on families and households in sub-Saharan Africa, and the relationships between early health status and adult socioeconomic achievement and health status. His recent research has assessed the impact of selection mechanisms arising from early childhood experience on adult socioeconomic differentials in health and mortality in high-income countries. He is a member of the American Academy of Arts and Sciences. He holds a B.S. from the Catholic University of Chile and a Ph.D. from the University of Washington.

Samuel H. Preston is a professor of demography and sociology at the University of Pennsylvania. His major research interest is the health of populations, with primary attention to mortality trends and patterns in large aggregates, including 20th century mortality transitions and black/white differentials in the United States. His recent research has focused on the mortality effects of cigarette smoking and obesity in developed countries. He is a recipient of the Taeuber and Sheps Awards from the Population Association of America, and he was a laureate of the International Union for the Scientific Study of Population. He is a member of the National Academy of Sciences, the Institute of Medicine, and the American Philosophical Society, and a fellow of the American Statistical Association, the American Academy of Arts and Sciences, and the American Association for the Advancement of Science. He holds an M.A. and a Ph.D. in economics, both from Princeton University.

Index